AF831375

Fluorine Magnetic Resonance Imaging

Fluorine Magnetic Resonance Imaging

Methods and Applications in Biomedicine

edited by

Eric T. Ahrens

Ulrich Flögel

JENNY STANFORD
PUBLISHING

Published by

Jenny Stanford Publishing Pte. Ltd.
101 Thomson Road
#06-01, United Square
Singapore 307591

Email: editorial@jennystanford.com
Web: www.jennystanford.com

British Library Cataloguing-in-Publication Data
A catalogue record for this book is available from the British Library.

Fluorine Magnetic Resonance Imaging: Methods and Applications in Biomedicine

Copyright © 2025 by Jenny Stanford Publishing Pte. Ltd.
All rights reserved. This book, or parts thereof, may not be reproduced in any form or by any means, electronic or mechanical, including photocopying, recording or any information storage and retrieval system now known or to be invented, without written permission from the publisher.

For photocopying of material in this volume, please pay a copying fee through the Copyright Clearance Center, Inc., 222 Rosewood Drive, Danvers, MA 01923, USA. In this case permission to photocopy is not required from the publisher.

ISBN 978-981-5129-37-3 (Hardcover)
ISBN 978-1-003-53004-6 (eBook)

Contents

Preface

Research in fluorine-19 MRI is advancing at an impressive pace, producing tangible results at the forefront of biomedical research. Physically, the ^{19}F nucleus is an excellent MRI signal generator because it has 100% natural abundance, a nucleus with spin-1/2, and an intrinsic sensitivity per spin near that of a proton (^{1}H). Following fluorinated tracer administration, ^{19}F MRI produces pure 'hot spot' images with no background signals from the host. This image specificity enables robust ^{19}F tracer image quantification that can form the basis for novel *in vivo* biomarker development.

Molecular-cellular imaging applications have emerged as a potent use of ^{19}F MRI. Key applications of ^{19}F MRI include cell tracking, inflammation detection, and biosensing. ^{19}F MRI is also known to be effective for observing fluoro-pharmaceutical biodistribution and lung airways using fluorinated gases. Due to the absence of background signals from the host, ^{19}F MRI does not require a pre-scan prior to tracer administration to characterize background contrast, thus offering several advantages over metal–ion-based ^{1}H MRI contrast agents. ^{19}F tracers also do not require radioactive constituents. The challenges associated with fluorine MRI are well known, particularly image-sensitivity limitations due to the dilute concentration of tracer in the body. New developments in chemistry, MRI spin physics, and image reconstruction algorithms have diminished these sensitivity limitations, however, and innovations in ^{19}F tracer design have made it possible to image specific cell types and report physiologically important parameters (*e.g.*, local oxygenation) *in situ* with greater precision. Furthermore, fluorination is often used in medicinal chemistry to enhance a desired pharmacological activity. ^{19}F MRI detectability can be a useful by-product of drug fluorination to enable biodistribution, bioavailability, and drug metabolism assessment.

This compendium covers cutting-edge science and biomedical research in the field of ^{19}F MRI. Technical issues, such as pulse sequence considerations, limits of detection, clinical implementation, synthesis of novel ^{19}F MRI tracer agents, biosensing, and 'theranostic' agent design are discussed. Specific biomedical applications of ^{19}F MRI are presented in the context of cell therapy, acute inflammation, cancer, and neurological disease.

It has been a decade since the release of our first book on this topic, *Fluorine Magnetic Resonance Imaging* (eds. Flögel and Ahrens, 2016), and the field has seen remarkable progress in tracer designs, detection methods, and clinical translation. We hope this second book conveys the dynamic nature of the field and inspires next-generation ^{19}F MRI technologies that will push the boundaries of our ability to answer challenging and important biological questions.

Importantly, we offer our sincerest appreciation to the chapter authors for their contributions and for sharing their vision of the future of the field. We are honored to have had the opportunity to consolidate their various perspectives.

Eric T. Ahrens
Ulrich Flögel
May 2024

Chapter 1

Polymeric ^{19}F MRI Agents: Designing Responsive Imaging Agents

Hui Peng,[a,b] Cheng Zhang,[a,b] Changkui Fu[a,b] and Andrew K. Whittaker[a,b,]

[a]*Australian Institute for Bioengineering and Nanotechnology, The University of Queensland, Brisbane, QLD 4072, Australia*

[b]*ARC Centre of Excellence in Convergent Bio-Nano Science and Technology, The University of Queensland, Brisbane, QLD 4072, Australia*

a.whittaker@uq.edu.au

1.1 Introduction

The field of magnetic resonance imaging (MRI) was very much in its infancy when the potential for observation of the fluorine-19 nucleus was first discussed. The NMR-active isotope of fluorine has 100% natural abundance and is a spin-½ nucleus with high receptivity, having a gyromagnetic ratio which is 94% of that of ^{1}H. In their initial report, Holland, Bottomley and Hinshaw [1] demonstrated ^{19}F NMR images of tubes containing aqueous solutions of NaF, and in an indicator of future trends, reported images of the fluorocarbon, perfluorotributylamine. This remarkable short article pointed out

Fluorine Magnetic Resonance Imaging: Methods and Applications in Biomedicine
Edited by Eric T. Ahrens and Ulrich Flögel
Copyright © 2025 Jenny Stanford Publishing Pte. Ltd.
ISBN 978-981-5129-37-3 (Hardcover), 978-1-003-53004-6 (eBook)
www.jennystanford.com

the principal advantage of ^{19}F MRI of appropriate tracer molecules, namely the absence of confounding endogenous signals in biological tissue, and indicated several potential applications that have been realized since that time.

The first *in vivo* ^{19}F MRI study was reported by McFarland and co-workers in 1985 [2]. Their study of rats after exposure to fluorinated anesthetic molecules, or after chronic exposure to fluorocarbons, highlighted the importance of achieving a high concentration of fluorine spins within the imaging volume. Subsequent studies aimed to enhance the fluorine content of tracer molecules and maximize their localization within specific tissues in the animal bodies. Broadly speaking two approaches to achieving high fluorine content have been adopted by groups working in this field, namely either the self-assembly of small-molecule fluorinated molecules into stable particles suitable for *in vivo* injection, or the incorporation of highly-fluorinated segments into macromolecular (polymeric) structures which either behave as unimers or undergo self-assembly into nano-scale objects. The former approach has been dealt with by others in this text, and the design and use of fluorinated polymeric particles will be discussed below.

The final and most critical consideration in the design of a successful ^{19}F MRI agent is the maintenance of significant molecular mobility of the fluorinated segments within the imaging agent [3, 4]. The fluorine-19 nuclei with high gyromagnetic ratio participate, in the absence of averaging motion, in strong homo- and hetero-nuclear dipolar interactions with near-neighbor nuclei. Furthermore, the fluorine spins generally possess a large chemical shift anisotropy which further broadens the NMR signal. Both of these interactions will broaden NMR resonances and shorten the transverse or spin-spin relaxation times (T_2) of the fluorine nuclei. A long T_2 relaxation time is essential to allow for generation of spin echoes by the pulse trains used in imaging sequences. However, these interactions will be partially averaged by rapid molecular motion of the fluorinated segments. In the limit of rapid isotropic motion, the dipolar interactions are averaged to zero, and the anisotropic chemical shift evolves to the trace of the second rank tensor, *i.e.*, the isotropic chemical shift; the net effect is a lengthening of T_2. In real systems, only partial averaging is achieved, depending on the frequency and geometry of the molecular motion. For example, for a fluoromethyl

($-CF_3$) group, rapid rotation around the C_3-axis results in averaging of the dipolar interactions and chemical anisotropy by 50%. Additional librational motion of the rotational axis leads to further averaging and hence longer spin-spin relaxation times. With this in mind, researchers have concentrated on the development of imaging agents with highly mobile structures (for example fluoromethyl as opposed to fluorinated aromatic structures) and highly flexible structures, for example materials based on fluorinated ethers. These design trends will become apparent through the discussion presented within the next sections, and have also been highlighted in recent review articles [5–8].

1.2 Polymeric ^{19}F MRI Agents

Modern polymer science has provided chemists with an impressive array of approaches to building complex molecular architectures. Here we are referring primarily to the methods of controlled radical polymerization, reversible-addition fragmentation chain transfer (RAFT), atom-transfer radical polymerization (ATRP), nitroxide-mediated polymerization (NMP) employing so-called "click" reactions, the Huisgen cycloaddition, and thiol-ene(yne) reactions. These methods allow preparation of polymers with pre-defined molecular weight and dispersity, and with topologies ranging from linear, multi-block, branched, comb, star-shape, and combinations thereof. Structural control is particularly important in materials for bio-applications, as it has been repeatedly demonstrated that properties such as particle size and shape, surface chemistry including charge and hydrophilicity, and of course the presence of specific targeting moieties all play important roles in determining the biological fate of the nanoparticle [9].

The recognition of the potential for extended pharmacokinetics by enhancing molar mass prompted Mehta *et al.* in 1992 [10] to prepare a series of fluorinated dextrans and lysine by trifluoroacetylation of the amino groups on these polymers. In this paper the authors describe the challenges in maintaining solubility after substitution of amines and propose routes to fully soluble materials, balanced by the need for high fluorine content. They report *in vivo* MR images of the fluorinated monomer glucose (TFAG), but not of the polymeric species, although relative detections limits are reported and compared with TFAG. This group followed up with a study [11] of the potential of trifluoroacetylated (TFA) bovine serum albumin, γ-globulin, and immunoglobulin. Extended blood serum lifetime was reported for TFA-albumin, comparable to the gadolinium-based blood pool agents. *In vivo* imaging was not reported.

The importance of polymer architecture for imaging performance has been highlighted by a number of studies. The Wooley laboratory reported in 2008 the preparation of partly-fluorinated hyperbranched polymers and demonstrated *in vitro* ^{19}F MRI [12]. The hyperbranched molecules were prepared by atom transfer radical self-condensing vinyl copolymerization of a hydrophobic core of 4-chloromethyl styrene (CMS), lauryl acrylate (LA), and 1,1,1-tris(4′-(2″-bromoisobutyryloxy)phenyl)ethane. The core served as the scaffold for chain extension with a copolymer of 2,2,2-trifluoroethyl methacrylate (TFEMA) and tert-butyl acrylate; the tert-butyl groups of the latter were removed by acid hydrolysis to yield an overall hydrophilic corona. The ^{19}F MRI T_1 and T_2 at 11.7 T were approximately 550 and 55 ms, respectively, enabling the demonstration of *in vitro* imaging. In a later study [13] the same group reported that self-assembly of diblock copolymers of acrylic acid with styrene (PS) and 2,3,4,5,6-pentafluorostyrene (PPFS) resulted in rigid fluorinated cores with insufficient molecular mobility to allow ^{19}F MRI observation. The cores were also loaded with the small molecule perfluoro-15-crown-5-ether but only on swelling with an organic solvent could an NMR signal be observed. Similar observations were reported at the same time by Peng *et al.* [3, 4] who examined diblock copolymers of acrylic acid and n-butyl acrylate copolymerized with a range of fluoro(meth)acrylates. Peng's work is notable for articulating a clear set of design principles for effective polymeric ^{19}F MRI agents.

The high levels of functionality available with highly branched macromolecules has been exploited by a large number of groups [14–30]. For example, Criscione *et al.* [16] reported the self-assembly of fluorinated poly(amidoamine) (PAMAM) dendrimer and *in vivo* imaging of the PEGylated particles. Thurecht *et al.* [31] demonstrated around the same time that well-controlled hyperbranched structures could be prepared using RAFT polymerization of N,N-dimethylaminoethyl acrylate (DMAEA) with 2,2,2-trifluoroacrylate (TFEA) in the presence of ethylene glycol dimethacrylate crosslinking monomer. Functionalization of the chain ends with mannose units provided the single-molecule nanoparticles with targeting ability. High-quality *in vivo* images were obtained at a field strength of 16.4 T (T_1 = 480 ms, T_2 = 88 ms). In their subsequent study [32], these authors replaced the DMAEA units with oligoethylene glycol methyl ether methacrylate (OEGMA) segments and attached at the periphery of the molecules both folic acid for tumor targeting and Rhodamine B to enable fluorescence imaging. Upon intravenous injection of these multi-modal imaging agents into a mouse bearing a subcutaneous B16 melanoma, successful *in vivo* ^{19}F MRI and fluorescence imaging were demonstrated (Fig. 1.1) [33, 34].

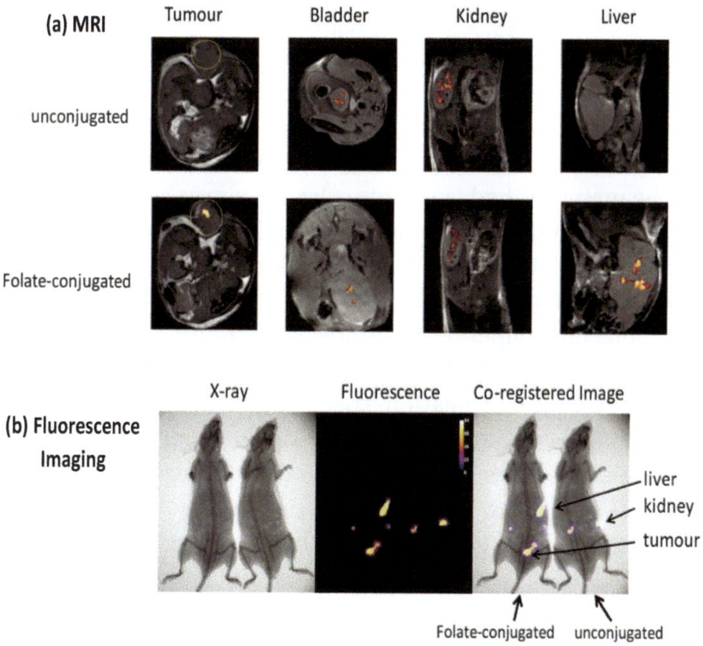

Figure 1.1 Molecular imaging of a TFEA-OEGMA hyperbranched polymer (HBP) in a subcutaneous B16 melanoma mouse model. (a) MRI of bladder, kidney, liver, or tumor (circled in image) in the tumor-bearing mice 1 h following intravenous injection of 100 μL of folate-conjugated or unconjugated (control) HBP (20 mg/mL in PBS). The high-resolution ^{1}H MR image is overlaid with the ^{19}F image. (b) Fluorescence images of mice following injection of the same two compounds at the same concentration. The fluorescence images are co-registered with X-ray images of the mice 1 h following subcutaneous injection. Reproduced from Ref. [32] with permission.

Despite numerous studies demonstrating the potential for *in vivo* ^{19}F MRI, researchers continued to push for higher imaging intensity. Other molecular approaches included placing highly polar groups, such as PEG segments [33, 35], close to the fluorinated segments so as to ensure high segmental mobility in aqueous environments, and the use of inherently flexible fluorinated moieties. The work by Ahrens and colleagues (for example [36]), Metrangolo and Resnati [37] and others has highlighted the potential of perfluorinated polyethers as a key element of ^{19}F MRI agents. Following this lead, Zhang *et al.* introduced several linear/block and comb copolymers incorporating perfluorinated poly(propylene ether) (PFPE) moieties [38–44]. In their first report, Zhang *et al.* [38] prepared a series of diblock copolymers of PFPE of degree of polymerization (DP) ~ 10, and oligoethylene glycol methyl ether acrylate (OEGA) with DP = 4, 10 and 20, and examined their aggregation behavior experimentally and through molecular dynamics simulations. All three materials had outstanding ^{19}F MRI performance and provided intense MR signals *in vivo*. The materials had favorable relaxation properties ($T_1 \sim 400$ ms and $T_2 \sim 80$–100 ms at 9.4 T) for MRI. The *in vivo* biodistribution determined by MRI in a murine model and by *ex vivo* fluorescence studies of excised organs shows that the biological fate of the macromolecules depend on the polymer composition (Fig. 1.2). In their subsequent study [44] the authors were able to demonstrate that details of the aggregation, specifically the extent of exposure of the fluorinated segments on the exterior of the particles, had a strong influence on the rates of cell uptake and penetration of MCF-7 breast cancer 3D spheroids. The chemotherapeutic molecule doxorubicin (DOX) was attached to the ends of the chains *via* an acid-cleavable hydrazone linker. The extent of delivery and cell killing by the DOX-loaded polymeric particles was directly related to the extent of uptake into the tumor spheroids. Finally, Zhang and coworkers [41] provided a means for enhancing tumor uptake *in vivo* by attaching at the terminal position the peptide aptamer specific for the heat shock protein 70, which is overexpressed in breast cancer cells. Greatly enhanced tumor uptake was observed by ^{19}F MRI of a sub-cutaneous MCF-7 murine model. *Ex vivo* fluorescence measurements confirmed enhanced tumor penetration for the aptamer-labelled polymers.

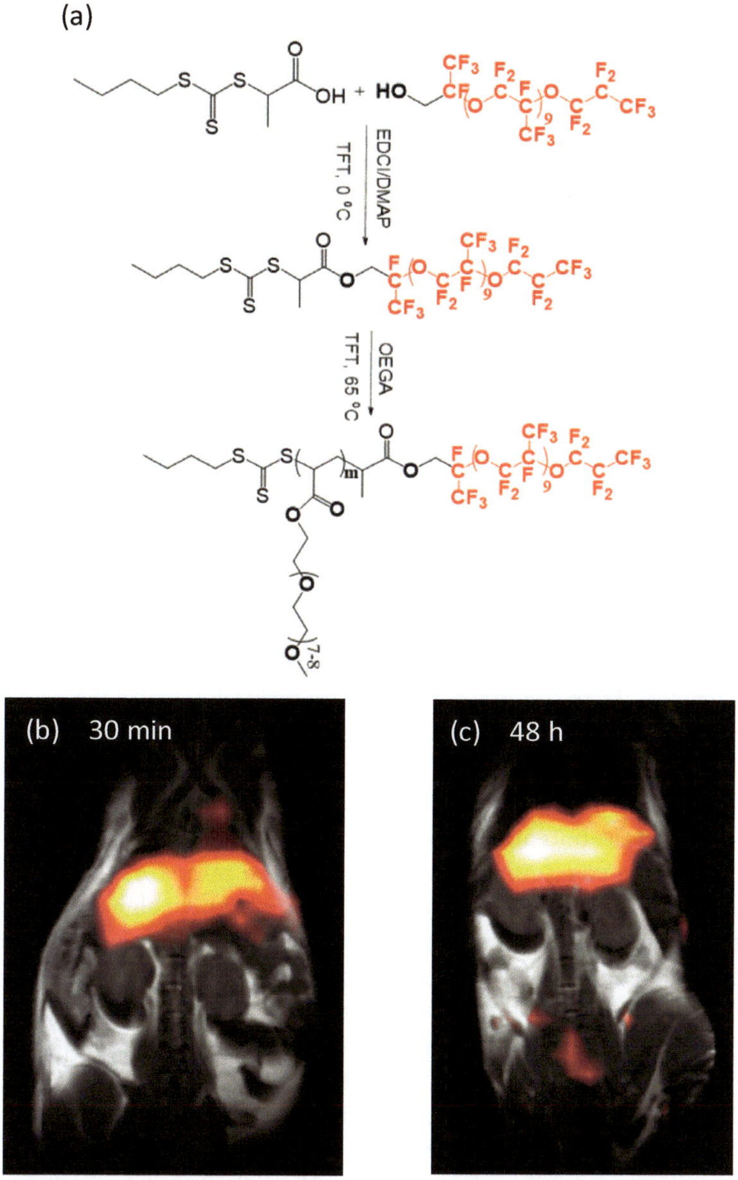

Figure 1.2 (a) Synthesis of PFPE RAFT agent and polymerization of OEGA. *In vivo* $^{1}H/^{19}F$ MRI at 9.4 T of a mouse following injection with a PFPE-*b*-POEGA diblock copolymer (b) 30 min and (c) 48 h after tail vein injection. High-resolution ^{1}H MR images are overlaid with the ^{19}F images. Reproduced from Ref. [38] with permission.

Comb copolymers have been used extensively in nanomedicine due to the potential for achieving high levels of functionality within a single molecule. This functionality can include ^{19}F imaging as demonstrated by Moonshi *et al.* [39] who prepared statistical comb polymers having both pendant PFPE segments and oligoethylene glycol units. Moonshi and her colleagues applied these polymers to cell tracking. Copolymers with greater than 20 wt.% fluorine were readily taken up upon incubation with mesenchymal stem cells (MSCs) in a dose-dependent manner as confirmed by *in vitro* ^{19}F NMR and fluorescence microscopy. The values of ^{19}F NMR T_1 and T_2 of the polymers within the labelled cells were 420 ms and 62 ms, respectively, at 9.4 T, similar to values of the polymers in saline solution. As a demonstration of the potential for cell tracking, 1×10^6 MSCs were injected into the subcutaneous flank of a mouse, and excellent images were obtained up to one week post-injection.

The self-assembly of polymeric nanoparticles during the so-called polymerization-induced self-assembly (PISA) of partly fluorinated block copolymers was demonstrated by Zhao *et al.* [45]. In this study, the authors prepared a poly(OEGA-*co*-TFEA) macro-chain transfer agent, and during chain extension with styrene monomer a range of particles of varying morphology were formed. Despite the particles having essentially identical surface chemistries, large differences in uptake by Chinese hamster ovarian (CHO) cells due to differences in shape were observed. The *in vitro* observation of strong MRI signals indicated the potential for specific cell imaging.

Sulfinyl-containing polymers have been attracting considerable attention over the past several years due to their high solubility in aqueous media. Originally proposed by Ringsdorf [46, 47] to assist drug delivery through cell membranes, these polymers have recently been explored for delivery and imaging applications [48–51]. Fu and colleagues [51] recognized the potential of the sulfinyl group for imparting high solubility to highly fluorinated macromolecules, and accordingly prepared a series of statistical copolymers of TFEA and 2-(methylsulfinyl)ethyl acrylate (MSEA) and found that they remained fully soluble up to ~15 wt.% fluorine, above which concentration aggregation of the chains and diminished imaging performance was noted. The relaxation times $T_1 \sim 450$ ms and $T_2 \sim 100$ ms were measured at 9.4 T. Furthermore, proof-of-principle imaging on a healthy murine model was demonstrated.

An alternative approach to reducing the tendency of fluorinated segments to associate and hence to maintain high segmental mobility is to position the F-moiety close to a highly polar group such as a hydroxyl, amide or sulfinyl. Guo *et al.* [52] prepared functionalized poly(ethylene imine)s (PEIs) having pendant trifluoroacetamide side chains (Fig. 1.3). In addition to ^{19}F MRI, the modified PEI was labelled with technetium-99m for single-photon emission computed

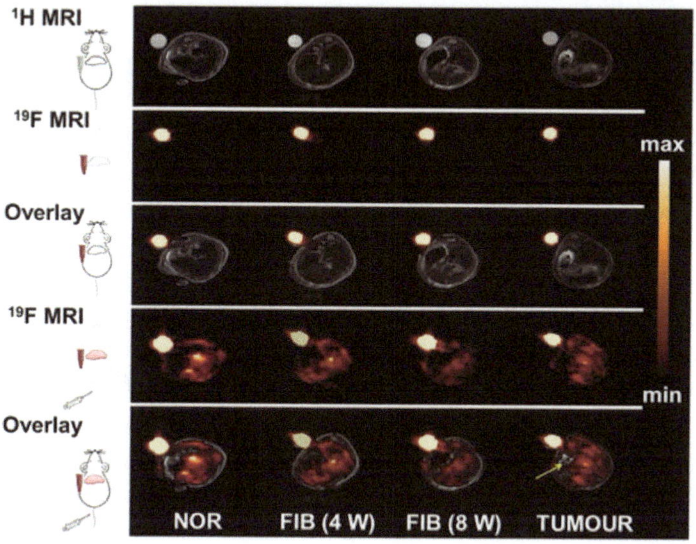

Figure 1.3 ^{19}F MRI of mice with liver fibrosis and bearing a xenografted HepG2 tumor injected with a fluorinated poly(ethylene imine) labelled with lactobionic acid. The upper three rows show the proton density image, the ^{19}F MRI, and the overlay image of the mouse prior to polymer injection. The area of intense MRI intensity is due to a reference sample of the polyimine in solution. The lower two rows show the ^{19}F MRI and the overlay image of the mouse after injection. The location of the tumor is indicated by the yellow arrow. Fibrosis was induced in the animals for either 4 or 8 weeks. Reproduced from Ref. [52] with permission.

tomography (SPECT) *via* association of the metal ions with the imine functionality. Additionally, the cyanine dye indocyanine green, used for photoacoustic imaging, was electrostatically attached to the PEI. More recently Kolouchova *et al.* [53] reported diblock copolymers of a hydrophilic monomer, either N-(2-hydroxypropyl)-methacrylamide] (PHPMA) or poly(2-methyl-2-oxazoline), and a second block of poly(N-(2,2-difluoroethyl)acrylamide). The inclusion of the fluorinated block led to the polymers exhibiting a lower critical solution temperature (LCST); above the LCST particles of hydrodynamic diameter 30–700 nm were observed depending on the polymer composition. The authors reported *in vitro* NMR and MRI. Several other groups have reported the self-assembly and MRI properties of oxazoline-containing macromolecules [43, 54–56]. Recently Jirak and colleagues [57] prepared homopolymers of ethyl acrylamide with a single fluorine substitution, *i.e.*, (N-(2-fluoroethyl) acrylamide)), and showed that these polymers are highly hydrophilic. When attached to gold nanoparticles the polymers give a strong ^{19}F MRI signal. At low field (1.5 T) the polymers had T_1 and T_2 relaxation times of 578 and 206 ms, respectively.

Fu *et al.* [58] later recognized that the polymers of N-(2,2,2-trifluoroethyl)acrylamide are insoluble in aqueous media and therefore proposed combining the polar sulfinyl and amide linkages in a single monomeric unit. Polymers of N-(2-((2,2,2-trifluoroethyl) sulfinyl) ethyl) acrylamide (FSAM) of low molecular weight were fully soluble at physiological temperatures, however, higher molecular weight homopolymers demonstrated aggregation behavior at 31°C. Notably, excellent ^{19}F imaging results were reported across the LCST [58], which contrasts with the behavior of other thermally-sensitive polymers. To demonstrate the capacity of these low-fouling polymers to extend the blood serum lifetime of biological macromolecules, serum albumin as a model protein was grafted with PFSAM using a RAFT "grafting from" approach. The pharmacokinetics of the protein conjugate was examined by ^{19}F MRI and *ex vivo* imaging of the major organs and was dependent on the size of the grafted chains. Significant ^{19}F MRI signal intensity was observed 12 weeks after tail-vein injection (Fig. 1.4).

Following the concept of close association of the fluorinated group with highly polar and hydrophilic moieties, Feng and coworkers [59] introduced zwitterionic polymers containing carboxybetaine units substituted at the quaternary ammonium position with fluorinated groups. The authors described excellent imaging performance in a mouse model and reduced retention in the liver and spleen compared with injected perfluoro-15-crown-5-ether nanoemulsions.

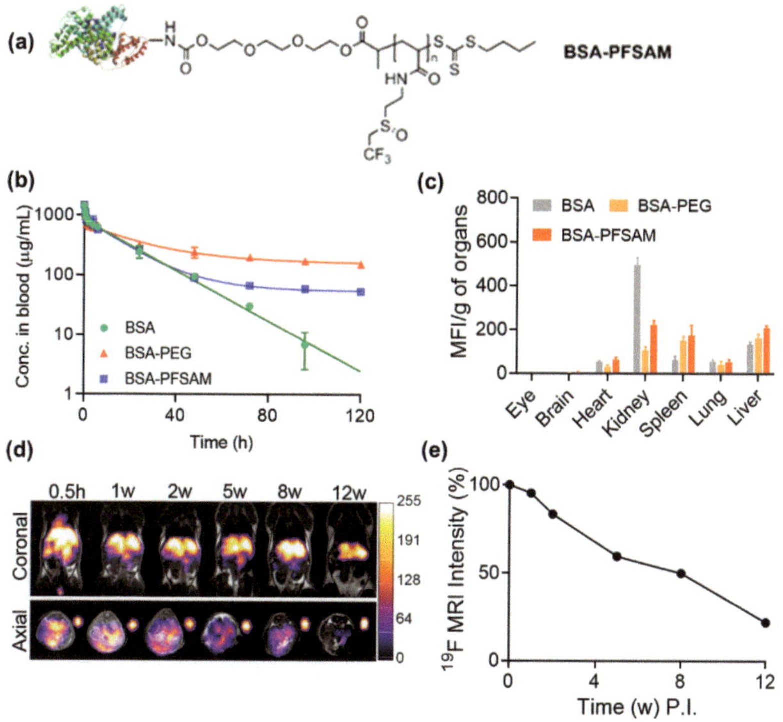

Figure 1.4 ^{19}F MRI study of pharmacokinetics of PFSAM-conjugated bovine serum albumin (BSA). (a) Structure of BSA-PFSAM conjugate. (b) Pharmacokinetics of BSA, BSA-PEG and BSA-PFSAM. (c) Biodistribution of BSA, BSA-PEG and BSA-PFSAM at 5 days post-injection. (d) Coronal and axial ^{19}F/^{1}H MR images of mice collected at various times after injection of BSA-PFSAM. (e) ^{19}F MRI signal intensity of the mice relative to a solution of FSAM (2 mg/mL in water). Reproduced from Ref. [58] with permission.

In summary, the development of polymeric ^{19}F MRI agents has consistently involved a tradeoff between higher fluorine content and favorable NMR relaxation parameters. A number of studies have reported that imaging performance is degraded when the fluorine content exceeds a certain limit, beyond which significant aggregation of the fluorinated segments limits segmental motion. Approaches for pushing towards higher fluorine content include judicious placement of the fluorine nuclei to reduce the extent of dipolar coupling, the use of inherently flexible fluorinated groups, notably the perfluorinated polyethers, and placement of the fluorine close to polar groups which impart high aqueous solubility. As will be seen in the next section the lessons learned have been extended into the design of responsive materials.

1.3 Responsive Polymeric ^{19}F MRI Agents

1.3.1 Designing Responsive Polymeric ^{19}F MRI Agents

The field of responsive ^{19}F NMR and MRI agents has a considerable history, and recent work on polymeric ^{19}F MRI agents has built on extensive studies of small molecules. A number of groups have reported changes in chemical shift and transverse and longitudinal relaxation times in response to exposure to an external stimulus. pH-responsive small molecule ^{19}F agents are often fluorinated acids. For example Boiadjiev and Lightner [60], building on earlier work by Deutsch and Taylor [61], have reported the changes in chemical shift of 26 fluorinated organic acids across the pKa. The applications of these and similar agents have included confirmation of intracellular and extracellular pH in red blood cells [62] and in perfused rat hearts [63].

The high affinity of oxygen for organofluorine compounds has prompted their application as oxygen carriers and sensors of local blood oxygen levels. Every NMR spectroscopist appreciates that molecular oxygen has paramagnetic properties and is familiar with the effects of dissolved oxygen on T_1 relaxation times. A significant body of research has been conducted on the effect of oxygen on relaxivity of fluorine spins in fluoroemulsions, including, as an example, the study by Fishman *et al.* of oxygen tension in a xenograft murine model of breast cancer [64]. The effect of oxygen on ^{19}F NMR chemical shifts can also be significant and has been used to investigate the penetration of the gas into membranes [65] and the extent of solvent exposure of proteins in oxygenated solution [66]. Mason and coworkers exploited this phenomenon to map the distribution and pO_2 of oxypherol (perfluorotributylamine) in a murine model of prostate cancer [67].

Finally, the responsiveness of a polymer to an external stimulus often leads to a change in hydrodynamics properties, and subsequently to assembly or disassembly of multi-polymer particles. The resultant changes in chain dynamics, and inter-nuclear separation of NMR-active spins (^{19}F and ^{1}H), result in enhancement or diminution of dipolar interactions involving the fluorine nuclei. As discussed above, high molecular mobility (*i.e.*, long T_2) is a necessary condition for strong MRI signals and so changes in mobility and internuclear interactions are routinely exploited in responsive MRI agents.

1.3.2 pH Responsive Polymeric ^{19}F MRI Agents

The field of responsive ^{19}F MR agents was pioneered by Ralph Mason. Mason and his colleagues have published extensively on small molecule fluorinated compounds in which the ^{19}F chemical shift and relaxivity is sensitive to external parameters such as pH and blood oxygenation [10, 62, 67, 68]. In 1996, Mason and others [69] reported the conjugation of a fluorinated form of vitamin-B6, 6-fluoropyridoxal, to three natural polymers and demonstrated a strong change in ^{19}F NMR chemical shift at the pKa of the conjugate. Despite this demonstration, the low fluorine content (a single aromatic fluorine on each pyridoxal moiety) will preclude use of this molecule for *in vivo* studies.

A series of studies have reported materials comprised in part by tertiary amines possessing a pKa close to physiological pH. The most-commonly used tertiary-amine monomer is 2-(dimethylamino) ethyl methacrylate (DMAEMA); homopolymers of DMAEMA have a pKa of ~7 [70]. However, substitution with other alkyl groups leads to small but significant changes in pKa, providing scope for control of the responsive behavior. For example, the diethyl amine polymer (poly(2-(diethylamino)ethyl methacrylate) has pKa = 7.4, the diisopropyl polymer has pKa = 6.3, the dibutyl polymer has pKa = 5.1, poly(2-(piperidine)ethyl methacrylate) has pKa = 6.8, and poly(N-(morpholino)ethyl methacrylate) has pKa = 4.9 [70, 71].

In 2007, Oishi *et al.* [72] reported the preparation of core-shell nanoparticles with a crosslinked pH responsive core containing the pH responsive 2-(diethylamino)ethyl methacrylate (DEAMA) and the partly-fluorinated methacrylate TFEMA. The corona of the particles comprised PEG chains terminated with an acetal moiety for further functionalization. The intensity of the NMR signal depended on TFEMA content in the core and on the pH, in an expected manner, however imaging was not demonstrated due to the relatively low signal-to-noise ratios in the ^{19}F NMR spectra, and presumably the short T_2 relaxation times. In a later study, Nurmi *et al.* [73] examined the effect of polymer architecture on ^{19}F NMR properties in block and statistical copolymers of DMAEMA (including the quaternized form of this monomer) with TFEMA. Again, this work highlighted the importance of maintaining high segmental mobility of the fluorinated segments.

Wang and co-workers adopted a similar approach in their design of pH-sensitive star polymers [23–27]. The stars were prepared by RAFT polymerization; initially a macro-RAFT agent of OEGMA was chain extended by polymerization of three monomers, 2-(dimethylamino)ethyl methacrylate (DMAEMA), TFEMA and EGDMA [23]. The inclusion of the difunctional crosslinking monomer EGDMA resulted in the formation of a core-shell structure. The pH sensitivity of the particles was demonstrated by cryo-TEM, ^{19}F NMR relaxometry and MRI. In solutions at pH below the pKa of the tertiary amine (pH ~ 6.5), the core became swollen, the transverse relaxation times of the fluorine spins within the core increased in magnitude, and the MRI signal was "switched on". In a follow up study [24] the authors replaced the EGDMA with a disulfide-containing difunctional monomer, to allow degradation of the star polymers under reductive environments and providing the potential for renal excretion.

The effect of structure of the tertiary amine, and hence the pKa as mentioned above, on pH-responsive ^{19}F MRI was reported by Huang *et al.* [74]. These researchers prepared a range of ionizable diblock copolymers of 2-(dibutylamino)ethyl, 2-(diisopropylamino) ethyl and 2-(pentamethylene imino)ethyl methacrylate with three fluorinated methacrylates and demonstrated multiple transitions in MRI intensity on changing the pH of solutions of mixtures of three polymers with different pKa.

An alternative approach to pH responsive behavior is the incorporation of acid-cleavable groups. For example, Munkhbat and colleagues [75] prepared disulfide-crosslinked nanogels, containing within the core the acid-cleavable tetrahydropyranyl groups in combination with TFEMA (Fig. 1.5). Prior to cleavage of the hydrophobic tetrahydropyranyl moieties the nanogels were MRI-silent, but cleavage to form carboxylic acid groups apparently enhanced local segmental mobility and consequently the ^{19}F MRI signal. The authors were able to demonstrate effective *in vitro* and *in vivo* MRI measurements in a mouse model of inflammation.

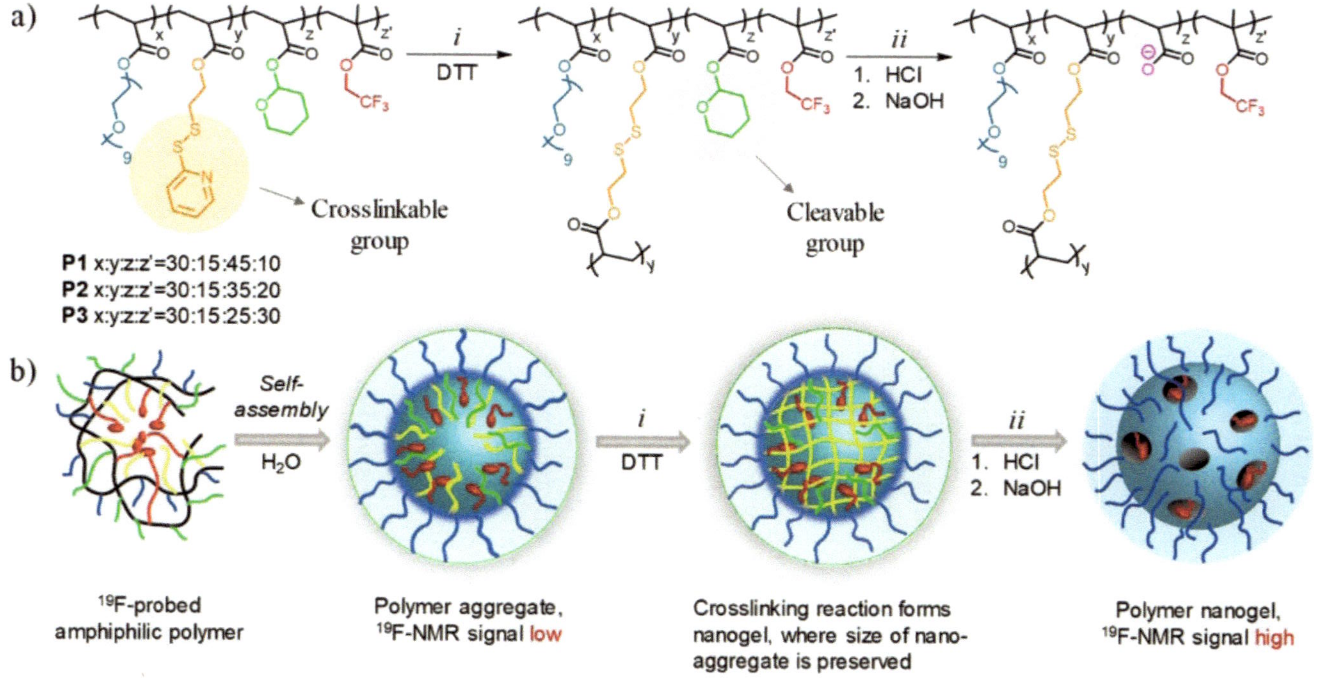

Figure 1.5 Design of fluorinated core shell particles. (a) Structure of the polymer and the nanogel. (*i*) Nanogel formation *via* crosslinking of pyridyldisulfide groups with dithiothreitol (DTT); (*ii*) cleavage of the tetrahydropyranyl group. (b) Schematic illustration of the preparation of the fluorinated nanogel with decreased interior density. Reproduced from Ref. [75] with permission.

1.3.3 Polymeric ^{19}F MRI Agents Measuring Reactive Oxygen Species

An important target for responsive imaging agents, or indeed for responsive delivery molecules, is the elevated levels of reactive oxygen species (ROS) within solid tumors [76]. Several groups have exploited changes in hydrophilicity of chain segments on oxidation to trigger disassembly of particles and activation of ^{19}F MRI signals. For example, Fu and colleagues [77] developed amphiphilic diblock copolymers of PEG and TFEA copolymerized with a thioether methacrylate monomer, 2-((2-((2-(ethylthio)ethyl)thio)ethyl)thio) ethyl methacrylate. The thioether moieties could be readily oxidized to trigger disassembly of the polymeric particles and "switching on" of the ^{19}F MRI signal. The disassembly process was examined using high-resolution ^{1}H and ^{19}F NMR, dynamic light scattering (DLS) and measurements of optical turbidity.

More recently Kolouchova *et al.* [78] introduced partly-fluorinated polyacrylamide copolymers comprising a hydrophilic block of poly(N-(2-hydroxypropyl) methacrylamide) and a hydrophobic block of N-(2,2-difluoroethyl)acrylamide copolymerized with a ferrocene-containing acrylamide monomer. The amphiphilic block copolymer could be self-assembled through a temperature switch and loaded with drug molecules. Of note, the particles could disassemble in the presence of ROS which reacted with the pendant ferrocene to form polar ferrocinium cationic species. The polymers could be imaged both before and after disassembly.

1.3.4 Redox-Responsive Polymeric ^{19}F MRI Agents

It is widely recognized that diseased tissues often possess an abnormal biochemical makeup that presents a target for both imaging and therapy. For example, as summarized by Gamcsik *et al.* [79], the levels of the strong reducing agent glutathione is often elevated several fold in tumorous tissue. The most-commonly adopted approach to the design of imaging agents sensitive to a reducing environment is the introduction of disulfide linkages capable of being reduced to sulfhydryl (–SH) groups. For example, Nakamura *et al.* [80] developed nanoparticles comprised of perfluoro[15] crown-5-ether encapsulated within a silica shell and decorated

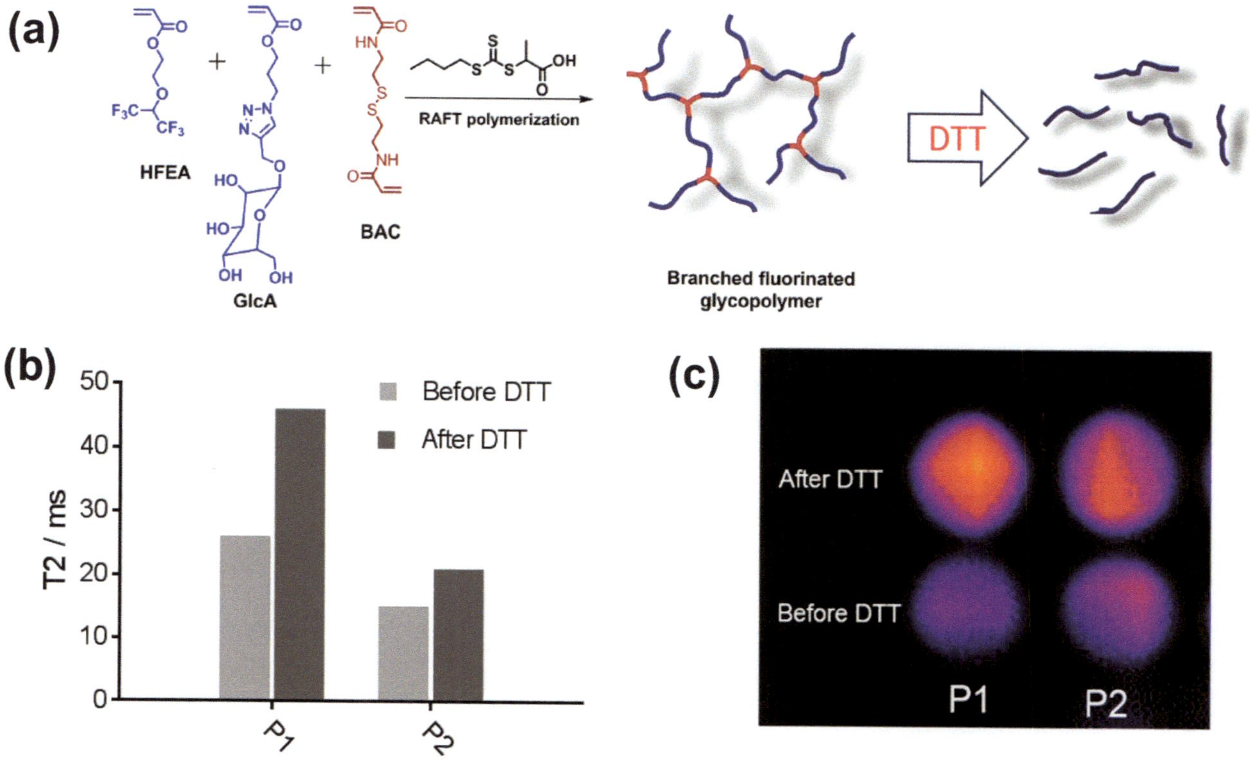

Figure 1.6 Hyperbranched glycopolymers. (a) Synthetic route for formation of branched fluorinated glycopolymers in a one-pot RAFT polymerization, and disassembly on addition of DTT. (b) ^{19}F NMR T_2 relaxation times of polymers with fluorine contents of 6.0 wt.% (labelled P1) and 10.5 wt.% (labelled P2) before and after treatment with 10 mM DTT. (c) ^{19}F MRI images of solutions of polymer P1 and P2.

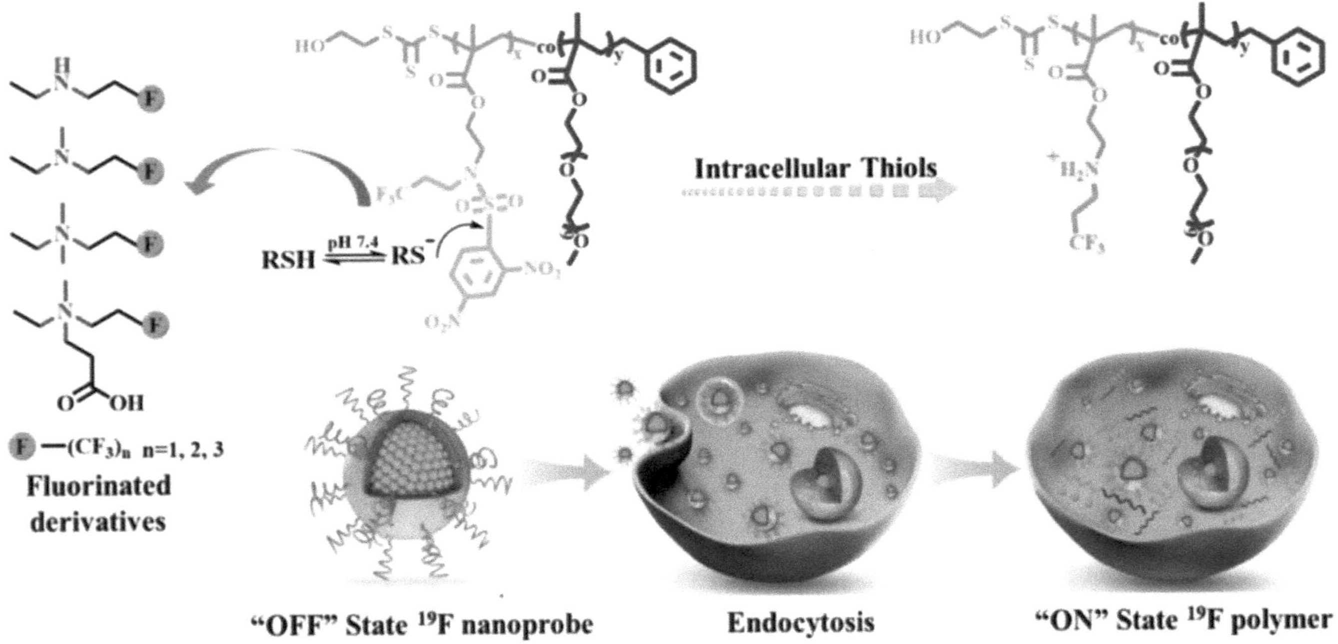

Figure 1.7 Design of amino-activatable nano-assemblies capable of ^{19}F MRI detection of an intracellular reducing microenvironment for imaging of biothiols.

with disulfide-linked paramagnetic Gd^{3+} chelates. On exposure to a model reducing agent, cleavage of the gadolinium chelate resulted in recovery of the ^{19}F NMR and MRI signals. Fu *et al.* [81] later prepared hyperbranched glycopolymers of a fluorinated acrylate monomer and a D-glucose glycoacrylate. The branched structure was formed by RAFT polymerization in the presence of a disulfide-crosslinking acrylamide monomer (Fig. 1.6). Increases in T_2 relaxation times and ^{19}F MR image intensities were observed on reduction of the disulfide and disassembly of the hyperbranched polymer. The inclusion of the glycopolymer monomer enhanced uptake by breast cancer cells.

The concept of disassembly on reduction was also exploited by Huang *et al.* [82] who prepared copolymers of OEGMA and an amino methacrylate substituted with a short partly-fluorinated alkane and 2,4-dinitrobenzenesulfonyl as a cleavable activating group (Fig. 1.7). Reaction with reducing agents resulted in loss of the aromatic group and protonation of the amine which triggered disassembly of the particles and "turning on" of the ^{19}F MR signal. Reaction of the self-assembled particles with a range of reducing and oxidizing agents was tested, and *in vivo* MRI in a liver hepatocellular carcinoma xenograph model was presented.

1.3.5 Polymeric ^{19}F MRI Agents Measuring Oxygen Tension

As discussed above, there exists a significant body of work on the measurement of oxygen tension *in vivo* using ^{19}F NMR and MRI. The vast majority of these studies have relied on measurements of either ^{19}F longitudinal relaxivity ($R_1 = 1/T_1$) or less commonly the ^{19}F chemical shift of small molecule fluorocarbons. Recently, Taylor *et al.* [83] have extended this approach to macromolecular ^{19}F MRI agents to take advantage of the control of structure afforded by controlled radical polymerization methods. Taylor prepared several series of statistical copolymers of fluorinated acrylates bearing 3, 6, or 9 equivalent fluorine atoms with the hydrophilic monomer OEGA. The ^{19}F MR intensity was measured and was found to increase with increasing content of fluorinated monomer up to a maximum level beyond which aggregation of the fluorinated segments degraded the MR signal. Impressively the authors found a linear relationship between relaxivity and dissolved oxygen content, pointing to the potential of polymeric agents for this application.

1.3.6 Enzyme-Responsive Polymeric ^{19}F MRI Agents

An attractive target for responsive imaging and delivery agents is the presence of specific endogenous enzymes. The group of Kikuchi has been especially active in this field over the past decade or so. In their early work, they introduced molecules containing a paramagnetic gadolinium chelate attached *via* a cleavable peptide linker to a partly-fluorinated aromatic moiety [84, 85]. The peptide linker was specifically cleaved by caspase-3, a marker of cell apoptosis. In their second paper [85], the fluorinated "reporter group", 7-amino-4-trifluoromethylcoumarin, is fluorescent, and the molecule can act as a dual-mode (MRI/fluorescence) imaging agent. A similar approach was adopted by Yue *et al.* [86] for the detection of MMP-2 secreted from the squamous carcinoma cell line SCC7.

Matsuo and coworkers have reported the development of self-assembled nanoparticles which are capable of disassembly on exposure to a specific substrate, for example a protein [87, 88]. In one version of this design, they included a short peptide sequence which could be selectively cleaved by the matrix metalloproteinase MMP-2 [87]. Upon cleavage a low molecular weight partly-fluorinated molecule was released, providing a strong ^{19}F NMR signal. The nanoparticles were able to detect MMP-2 secreted from HT1080 fibrosarcoma cells at a concentration less than 0.5 nM. Self-assembly and disassembly of peptide-containing molecules was also reported by Yuan *et al.* [89, 90] for detection of caspase-3/7, including in a zebra fish model.

Finally, the susceptibility of the ester linkage to enzymatic cleavage was exploited by Wang *et al.* [22] who prepared amphiphilic dendritic macromolecules with arms terminated with 3,5-bis(trifluoromethyl)benzoate groups. On exposure to porcine liver esterase, the fluorinated benzoic acid was released from the polymer structure, and the NMR intensity increased significantly.

1.3.7 Ion-Responsive Polymeric ^{19}F MRI Agents

Many pathologies are characterized by abnormal intra- or extra-cellular ionic concentrations, and consequently ionic strength and makeup have become a target for disease detection. For example, over 30 years ago Smith and co-workers reported X-ray microanalysis analysis of intracellular ionic concentrations of the major ionic species and found differences between normal and cancerous cells [91, 92]. Altered ionic makeup is also a consequence of ion channel disorders, indicated for a number of neurological diseases [93], and in bone diseases such as osteoporosis [94, 95]. As a consequence, non-invasive indicators of ionic content are actively being developed. For example, Smith and colleagues [96] developed a fluorinated chelator of sodium ions and demonstrated that the ^{19}F chemical shift of the compound was a function of Na^+ concentration. Preslar *et al.* [97, 98] have developed peptide amphiphiles in which the hydrophobe is a perfluorinated alkyl chain and showed that ^{19}F MRI signal intensity was dependent on dissolved calcium ion concentration across the range 2–6 mM.

The salting out effect of uncharged hydrophilic molecules is well known and is widely exploited, for example in protein purification. The properties of poly(ethylene glycol) in salt solutions have been extensively studied [99, 100] and form the basis of the design of a polymeric ^{19}F indicator of salt content in diseased cancer cells by Zhang and colleagues [101]. They showed that the rate of transverse ^{19}F relaxation is related to the hydrodynamic volume of statistical copolymers of OEGA and TFEA, and demonstrated differences in T_2 relaxation time in healthy and diseased MCF-7 breast cancer cells.

1.4 Summary and Perspective

The field of polymeric ^{19}F MRI agents has relied heavily on the introduction over the past several decades of new methods of free radical polymer synthesis to control the placement of fluorine spins throughout the polymer or nanoparticle. In addition, the use of highly flexible or strongly polar groups has gained prominence. The maximum atomic weight percentage of fluorine achievable to date lies between 20 and 30 percent, inferior to low molecular weight assemblies. However, fluorinated polymeric have significant advantages related to the relative ease of structural control. For example, reactive chain ends or side chains allow for attachment of a second imaging agent, and this is particularly useful for pre-clinical small animal imaging where fluorescence techniques have some advantages. Similarly, conjugation with targeting ligands has been repeatedly demonstrated in *in vivo* models of disease. As highlighted in Section 1.3, polymeric structures are especially capable of carrying moieties which react with or respond to changes in the environment conditions. As a consequence, responsive polymeric ^{19}F MRI agents are increasingly gaining prominence, and it is anticipated that over the next decade additional examples of *in vivo* imaging of realistic animal models of disease will be presented to our community.

References

1. Holland, G. N., Bottomley, P. A., and Hinshaw, W. S. (1977). 19F magnetic resonance imaging. *J. Magn. Reson. (1969)*, **28**, 133–136.
2. McFarland, E. W., Koutcher, J. A., Rosen, B. R., Teicher, B. A., and Brady, T. J. (1985). *In Vivo* 19F NMR Imaging. *J. Comput Assist. Tomogr.*, **9**, X 15.
3. Peng, H., Blakey, I., Dargaville, B., Rasoul, F., Rose, S., and Whittaker Andrew, K. (2009). Synthesis and evaluation of partly fluorinated block copolymers as MRI imaging agents. *Biomacromolecules*, **10**, 374–381.
4. Peng, H., Thurecht, K. J., Blakey, I., Taran, E., and Whittaker, A. K. (2012). Effect of solvent quality on the solution properties of assemblies of partially fluorinated amphiphilic diblock copolymers. *Macromolecules*, **45**, 8681–8690.
5. Jirak, D., Galisova, A., Kolouchova, K., Babuka, D., and Hruby, M. (2019). Fluorine polymer probes for magnetic resonance imaging: quo vadis? *Magn. Reson. Mater. Phys., Biol. Med.*, **32**, 173–185.
6. Tirotta, I., Dichiarante, V., Pigliacelli, C., Cavallo, G., Terraneo, G., Bombelli, F. B., Metrangolo, P., and Resnati, G. (2015). 19F magnetic resonance imaging (MRI): from design of materials to clinical applications. *Chem. Rev.*, **115**, 1106–1129.
7. Fu, C., Yu, Y., Xu, X., Wang, Q., Chang, Y., Zhang, C., Zhao, J., Peng, H., and Whittaker, A. K. (2020). Functional polymers as metal-free magnetic resonance imaging contrast agents. *Prog. Polym. Sci.*, **108**, 101286.
8. Zhang, C., Yan, K., Fu, C., Peng, H., Hawker, C. J., and Whittaker, A. K. (2021). Biological utility of fluorinated compounds: from materials design to molecular imaging, therapeutics and environmental remediation. *Chem. Rev.*, Published 5 Oct 2021.
9. Kamaly, N., Xiao, Z., Valencia, P. M., Radovic-Moreno, A. F., and Farokhzad, O. C. (2012). Targeted polymeric therapeutic nanoparticles: design, development and clinical translation. *Chem. Soc. Rev.*, **41**, 2971–3010.

10. Mehta, V. D., Kulkarni, P. V., Mason, R. P., Babcock, E. E., Constaninescu, A., and Antich, P. P. (1992). Novel molecular probes for fluorine-19 magnetic resonance imaging: synthesis and characterization of fluorinated polymers. *Bioorg. Med. Chem. Lett.,* **2**, 527–532.

11. Mehta, V. D., Kulkarni, P. V., Mason, R. P., Constantinescu, A., and Antich, P. P. (1994). Fluorinated proteins as potential 19F magnetic resonance imaging and spectroscopy agents. *Bioconjugate Chem.,* **5**, 257–261.

12. Du, W., Nystrom, A. M., Zhang, L., Powell, K. T., Li, Y., Cheng, C., Wickline, S. A., and Wooley, K. L. (2008). Amphiphilic hyperbranched fluoropolymers as nanoscopic 19F magnetic resonance imaging agent assemblies. *Biomacromolecules,* **9**, 2826–2833.

13. Nystrom, A. M., Bartels, J. W., Du, W., and Wooley, K. L. (2009). Perfluorocarbon-loaded shell crosslinked knedel-like nanoparticles: lessons regarding polymer mobility and self-assembly. *J. Polym. Sci., Part A Polym. Chem.,* **47**, 1023–1037.

14. Ardana, A., Whittaker, A. K., and Thurecht, K. J. (2014). PEG-based hyperbranched polymer theranostics: optimizing chemistries for improved bioconjugation. *Macromolecules,* **47**, 5211–5219.

15. Celentano, W., Neri, G., Distante, F., Li, M., Messa, P., Chirizzi, C., Chaabane, L., De Campo, F., Metrangolo, P., Baldelli Bombelli, F., and Cellesi, F. (2020). Design of fluorinated hyperbranched polyether copolymers for 19F MRI nanotheranostics. *Polym. Chem.,* **11**, 3951–3963.

16. Criscione, J. M., Le, B. L., Stern, E., Brennan, M., Rahner, C., Papademetris, X., and Fahmy, T. M. (2009). Self-assembly of pH-responsive fluorinated dendrimer-based particulates for drug delivery and noninvasive imaging. *Biomaterials,* **30**, 3946–3955.

17. Liu, X., Yuan, Y., Bo, S., Li, Y., Yang, Z., Zhou, X., Chen, S., and Jiang, Z.-X. (2017). Monitoring fluorinated dendrimer-based self-assembled drug-delivery systems with 19F magnetic resonance. *Eur. J. Org. Chem.,* **2017**, 4461–4468.

18. Ogawa, M., Kataoka, H., Nitahara, S., Fujimoto, H., Aoki, H., Ito, S., Narazaki, M., and Matsuda, T. (2012). Water-soluble fluorinated polymer nanoparticle as 19F MRI contrast agent prepared by living random copolymerizaiton from dendrimer initiator. *Bull. Chem. Soc. Jpn.,* **85**, 79–86.

19. Ogawa, M., Nitahara, S., Aoki, H., Ito, S., Narazaki, M., and Matsuda, T. (2010). Fluorinated polymer nanoparticles as a novel 19F MRI contrast agent prepared by dendrimer-initiated living radical polymerization. *Macromol. Chem. Phys.,* **211**, 1369–1376.

20. Ogawa, M., Nitahara, S., Aoki, H., Ito, S., Narazaki, M., and Matsuda, T. (2010). Synthesis and evaluation of water-soluble fluorinated dendritic block-copolymer nanoparticles as a 19F-MRI contrast agent. *Macromol. Chem. Phys.*, **211**, 1602–1609.
21. Porsch, C., Zhang, Y., Oestlund, A., Damberg, P., Ducani, C., Malmstroem, E., and Nystroem, A. M. (2013). *In vitro* evaluation of non-protein adsorbing breast cancer theranostics based on 19F-polymer containing nanoparticles. *Part. Part. Syst. Charact.*, **30**, 381–390.
22. Wang, H., Raghupathi, K. R., Zhuang, J., and Thayumanavan, S. (2015). Activatable dendritic 19F probes for enzyme detection. *ACS Macro Lett.*, **4**, 422–425.
23. Wang, K., Peng, H., Thurecht, K. J., Puttick, S., and Whittaker, A. K. (2013). pH-responsive star polymer nanoparticles: potential 19F MRI contrast agents for tumour-selective imaging. *Polym. Chem.*, **4**, 4480–4489.
24. Wang, K., Peng, H., Thurecht, K. J., Puttick, S., and Whittaker, A. K. (2014). Biodegradable core crosslinked star polymer nanoparticles as 19F MRI contrast agents for selective imaging. *Polym. Chem.*, **5**, 1760–1771.
25. Wang, K., Peng, H., Thurecht, K. J., Puttick, S., and Whittaker, A. K. (2016). Multifunctional hyperbranched polymers for CT/19F MRI bimodal molecular imaging. *Polym. Chem.*, **7**, 1059–1069.
26. Wang, K., Peng, H., Thurecht, K. J., and Whittaker, A. K. (2016). Fluorinated POSS-star polymers for 19F MRI. *Macromol. Chem. Phys.*, **217**, 2262–2274.
27. Wang, K., Peng, H., Thurecht Kristofer, J., Puttick, S., and Whittaker Andrew, K. (2015). Segmented highly branched copolymers: rationally designed macromolecules for improved and tunable (19)F MRI. *Biomacromolecules*, **16**, 2827–2839.
28. Wang, Z., Yue, X., Wang, Y., Qian, C., Huang, P., Lizak, M., Niu, G., Wang, F., Rong, P., Kiesewetter, D. O., Ma, Y., and Chen, X. (2014). A symmetrical fluorous dendron-cyanine dye-conjugated bimodal nanoprobe for quantitative 19F MRI and NIR fluorescence bioimaging. *Adv. Healthcare Mater.*, **3**, 1326–1333.
29. Yu, W., Yang, Y., Bo, S., Li, Y., Chen, S., Yang, Z., Zheng, X., Jiang, Z.-X., and Zhou, X. (2015). Design and synthesis of fluorinated dendrimers for sensitive 19F MRI. *J. Org. Chem.*, **80**, 4443–4449.
30. Yu, Y. B. (2013). Fluorinated dendrimers as imaging agents for 19F MRI. *Wiley Interdiscip. Rev. Nanomed. Nanobiotechnol.*, **5**, 646–661.

31. Thurecht, K. J., Blakey, I., Peng, H., Squires, O., Hsu, S., Alexander, C., and Whittaker, A. K. (2010). Functional hyperbranched polymers: toward targeted *in vivo* 19F magnetic resonance imaging using designed macromolecules. *J. Am. Chem. Soc.,* **132**, 5336–5337.

32. Rolfe, B. E., Blakey, I., Squires, O., Peng, H., Boase, N. R. B., Alexander, C., Parsons, P. G., Boyle, G. M., Whittaker, A. K., and Thurecht, K. J. (2014). Multimodal polymer nanoparticles with combined 19F magnetic resonance and optical detection for tunable, targeted, multimodal imaging *in vivo*. *J. Am. Chem. Soc.,* **136**, 2413–2419.

33. Biaggi, C., Benaglia, M., Ortenzi, M., Micotti, E., Perego, C., and De Simoni, M.-G. (2013). Easily available, low cost 19F MRI agents: poly(ethylene-glycol)-functionalized fluorinated ethers. *J. Fluorine Chem.,* **153**, 172–177.

34. Rossi, S., Benaglia, M., Ortenzi, M., Micotti, E., Perego, C., and De Simoni, M. G. (2011). Poly(ethylene-glycol)-based fluorinated esters: a readily available entry for novel 19F-MRI agents. *Tetrahedron Lett.,* **52**, 6581–6583.

35. Rossini, A. J., Hildebrand, M. P., Hazendonk, P. A., and Schurko, R. W. (2014). Multinuclear solid-state NMR studies of polymer-supported scandium triflate catalysts. *J. Phys. Chem. C,* **118**, 22649–22662.

36. Ahrens, E. T., Flores, R., Xu, H., and Morel, P. A. (2005). *In vivo* imaging platform for tracking immunotherapeutic cells. *Nat. Biotechnol.,* **23**, 983–987.

37. Tirotta, I., Mastropietro, A., Cordiglieri, C., Gazzera, L., Baggi, F., Baselli, G., Grazia Bruzzone, M., Zucca, I., Cavallo, G., Terraneo, G., Baldelli Bombelli, F., Metrangolo, P., and Resnati, G. (2014). A superfluorinated molecular probe for highly sensitive *in vivo* 19F-MRI. *J. Am. Chem. Soc.,* **136**, 8524–8527.

38. Zhang, C., Moonshi, S. S., Han, Y., Puttick, S., Peng, H., Magoling, B. J. A., Reid, J. C., Bernardi, S., Searles, D. J., Kral, P., and Whittaker, A. K. (2017). PFPE-based polymeric 19F MRI agents: a new class of contrast agents with outstanding sensitivity. *Macromolecules,* **50**, 5953–5963.

39. Moonshi, S. S., Zhang, C., Peng, H., Puttick, S., Rose, S., Fisk, N. M., Bhakoo, K., Stringer, B. W., Qiao, G. G., Gurr, P. A., and Whittaker, A. K. (2018). A unique 19F MRI agent for the tracking of non phagocytic cells *in vivo*. *Nanoscale,* **10**, 8226–8239.

40. Zhang, C., Kim, D. S., Lawrence, J., Hawker, C. J., and Whittaker, A. K. (2018). Elucidating the impact of molecular structure on the 19F NMR dynamics and MRI performance of fluorinated oligomers. *ACS Macro Lett.,* **7**, 921–926.

41. Zhang, C., Moonshi, S. S., Wang, W., Ta, H. T., Han, Y., Han, F. Y., Peng, H., Kral, P., Rolfe, B. E., Gooding, J. J., Gaus, K., and Whittaker, A. K. (2018). High F-content perfluoropolyether-based nanoparticles for targeted detection of breast cancer by 19F magnetic resonance and optical imaging. *ACS Nano,* **12**, 9162–9176.

42. Zhang, C., Li, L., Han, F. Y., Yu, X., Tan, X., Fu, C., Xu, Z. P., and Whittaker, A. K. (2019). Integrating fluorinated polymer and manganese-layered double hydroxide nanoparticles as pH-activated 19F MRI agents for specific and sensitive detection of breast cancer. *Small,* **15**.

43. Zhang, C., Sanchez, R. J. P., Fu, C., Clayden-Zabik, R., Peng, H., Kempe, K., and Whittaker, A. K. (2019). Importance of thermally induced aggregation on 19F magnetic resonance imaging of perfluoropolyether-based comb-shaped poly(2-oxazoline)s. *Biomacromolecules,* **20**, 365–374.

44. Zhang, C., Liu, T., Wang, W., Bell, C. A., Han, Y., Fu, C., Peng, H., Tan, X., Kral, P., Gaus, K., Gooding, J. J., and Whittaker, A. K. (2020). Tuning of the aggregation behavior of fluorinated polymeric nanoparticles for improved therapeutic efficacy. *ACS Nano,* **14**, 7425–7434.

45. Zhao, W., Ta, H. T., Zhang, C., and Whittaker, A. K. (2017). Polymerization-induced self-assembly (PISA) – control over the morphology of 19F-containing polymeric nano-objects for cell uptake and tracking. *Biomacromolecules,* **18**, 1145–1156.

46. Batz, H. G., Hofmann, V., and Ringsdorf, H. (1973). Pharmakologisch aktive polymere, 4. Monomere und polymere alkylsulfinyl-alkylacrylate und -methacrylate als mögliche resorptionsvermittler. *Makromol. Chem.,* **169**, 323–325.

47. Hofmann, V., Ringsdorf, H., and Muacevic, G. (1975). Pharmakologisch aktive polymere, 8. Poly[2-(methylsulfinyl)äthylacrylat]e und ihre vermittelnde wirkung auf die transkutane resorption von pharmaka. *Makromol. Chem.,* **176**, 1929–1943.

48. Li, S., Chung, H. S., Simakova, A., Wang, Z., Park, S., Fu, L., Cohen-Karni, D., Averick, S., and Matyjaszewski, K. (2017). Biocompatible polymeric analogues of DMSO prepared by atom transfer radical polymerization. *Biomacromolecules,* **18**, 475–482.

49. Yan, J., Li, S., Cartieri, F., Wang, Z., Hitchens, T. K., Leonardo, J., Averick, S. E., and Matyjaszewski, K. (2018). Iron oxide nanoparticles with grafted polymeric analogue of dimethyl sulfoxide as potential magnetic resonance imaging contrast agents. *ACS Appl. Mater. Interfaces,* **10**, 21901–21908.

50. Qiao, R., Fu, C., Li, Y., Qi, X., Ni, D., Nandakumar, A., Siddiqui, G., Wang, H., Zhang, Z., Wu, T., Zhong, J., Tang, S.-Y., Pan, S., Zhang, C., Whittaker, M. R., Engle, J. W., Creek, D. J., Caruso, F., Ke, P. C., Cai, W., Whittaker, A. K., and Davis, T. P. (2020). Sulfoxide-containing polymer-coated nanoparticles demonstrate minimal protein fouling and improved blood circulation. *Adv. Sci.,* **7**, 2000406.

51. Fu, C., Zhang, C., Peng, H., Han, F., Baker, C., Wu, Y., Ta, H., and Whittaker, A. K. (2018). Enhanced performance of polymeric 19F MRI contrast agents through incorporation of highly water-soluble monomer MSEA. *Macromolecules,* **51**, 5875–5882.

52. Guo, Z., Gao, M., Song, M., Li, Y., Zhang, D., Xu, D., You, L., Wang, L., Zhuang, R., Su, X., Liu, T., Du, J., and Zhang, X. (2016). Superfluorinated PEI derivative coupled with 99mTc for ASGPR targeted 19F MRI/SPECT/PA tri-modality imaging. *Adv. Mater.,* **28**, 5898–5906.

53. Kolouchova, K., Sedlacek, O., Jirak, D., Babuka, D., Blahut, J., Kotek, J., Vit, M., Trousil, J., Konefal, R., Janouskova, O., Podhorska, B., Slouf, M., and Hruby, M. (2018). Self-assembled thermoresponsive polymeric nanogels for 19F MR imaging. *Biomacromolecules,* **19**, 3515–3524.

54. Kaberov, L. I., Verbraeken, B., Hruby, M., Riabtseva, A., Kovacik, L., Kereiche, S., Brus, J., Stepanek, P., Hoogenboom, R., and Filippov, S. K. (2017). Novel triphilic block copolymers based on poly(2-methyl-2-oxazoline)-block-poly(2-octyl-2-oxazoline) with different terminal perfluoroalkyl fragments: synthesis and self-assembly behaviour. *Eur. Polym. J.,* **88**, 645–655.

55. Sedlacek, O., Jirak, D., Vit, M., Ziolkowska, N., Janouskova, O., and Hoogenboom, R. (2020). Fluorinated water-soluble poly(2-oxazoline)s as highly sensitive 19F MRI contrast agents. *Macromolecules,* **53**, 6387–6395.

56. Kaberov, L. I., Kaberova, Z., Murmiliuk, A., Trousil, J., Sedlacek, O., Konefal, R., Zhigunov, A., Pavlova, E., Vit, M., Jirak, D., Hoogenboom, R., and Filippov, S. K. (2021). Fluorine-containing block and gradient copoly(2-oxazoline)s based on 2-(3,3,3-trifluoropropyl)-2-oxazoline: a quest for the optimal self-assembled structure for 19F imaging. *Biomacromolecules,* **22**, 2963–2975.

57. Jirak, D., Svoboda, J., Filipova, M., Pop-Georgievski, O., and Sedlacek, O. (2021). Antifouling fluoropolymer-coated nanomaterials for 19F MRI. *Chem. Commun.,* **57**, 4718–4721.

58. Fu, C., Tan, X., Yu, Y., Zhao, J., Zhang, C., Peng, H., Whittaker Andrew, K., Demir, B., Searles Debra, J., Alcantara, S., Kelly Hannah, G., Kent

Stephen, J., Kumar, V., Han, F., Xu, W., Woodruff Trent, M., and Boyer, C. (2020). Low-fouling fluoropolymers for bioconjugation and *in vivo* tracking. *Angew. Chem., Int. Ed.,* **59**, 4729–4735.

59. Feng, Z., Li, Q., Wang, W., Ni, Q., Wang, Y., Song, H., Zhang, C., Kong, D., Liang, X.-J., and Huang, P. (2020). Superhydrophilic fluorinated polymer and nanogel for high-performance 19F magnetic resonance imaging. *Biomaterials,* **256**, 120184.
60. Boiadjiev, S. E., and Lightner, D. A. (1999). Carboxylic acid ionization constants by 19F NMR spectroscopy. *J. Phys. Org. Chem.,* **12**, 751–757.
61. Deutsch, C. J., and Taylor, J. S. (1987). Intracellular pH as measured by 19F NMRa. *Ann. N. Y. Acad. Sci.,* **508**, 33–47.
62. Mehta, V. D., Kulkarni, P. V., Mason, R. P., Constantinescu, A., Aravind, S., Goomer, N., and Antich, P. P. (1994). 6-fluoropyridoxol: a novel probe of cellular pH using 19F NMR spectroscopy. *FEBS Lett.,* **349**, 234–238.
63. Hunjan, S., Mason, R. P., Mehta, V. D., Kulkarni, P. V., Aravind, S., Arora, V., and Antich, P. P. (1998). Simultaneous intracellular and extracelular pH measurement in the heart by 19F NMR of 6-fluoropyridoxol. *Magn. Reson. Med.,* **39**, 551–556.
64. Fishman, J. E., Joseph, P. M., Carvlin, M. J., Saadi-Elmandjra, M., Mukherji, B., and Sloviter, H. A. (1989). *In vivo* measurements of vascular oxygen tension in tumors using MRI of a fluorinated blood substitute. *Invest. Radiol.,* **24**, 65–71.
65. Luchette, P. A., Prosser, R. S., and Sanders, C. R. (2002). Oxygen as a paramagnetic probe of membrane protein structure by cysteine mutagenesis and 19F NMR spectroscopy. *J. Am. Chem. Soc.,* **124**, 1778–1781.
66. Bezsonova, I., Forman-Kay, J., and Prosser, R. S. (2008). Molecular oxygen as a paramagnetic NMR probe of protein solvent exposure and topology. *Concepts Magn. Reson. Part A,* **32A**, 239–253.
67. Mason, R. P., Antich, P. P., Babcock, E. E., Constantinescu, A., Peschke, P., and Hahn, E. W. (1994). Non-invasive determination of tumor oxygen tension and local variation with growth. *Int. J. Radiat. Oncol. Biol. Phys.,* **29**, 95–103.
68. Mason, R. P., Rodbumrung, W., and Antich, P. P. (1996). Hexafluorobenzene: a sensitive 19F NMR indicator of tumor oxygenation. *NMR Biomed.,* **9**, 125–134.
69. Mehta, V. D., Sivasubramanian, A., Kulkarni, P. V., Mason, R. P., and Antich, P. P. (1996). 6-Fluoropyridoxal polymer conjugates: novel 19F pH indicators for magnetic resonance spectroscopy. *Bioconjugate Chem.,* **7**, 536–540.

70. Hu, J., Zhang, G., Ge, Z., and Liu, S. (2014). Stimuli-responsive tertiary amine methacrylate-based block copolymers: synthesis, supramolecular self-assembly and functional applications. *Prog. Polym. Sci.,* **39**, 1096–1143.

71. Zhu, L., Powell, S., and Boyes, S. G. (2015). Synthesis of tertiary amine-based pH-responsive polymers by RAFT polymerization. *J. Polym. Sci., Part A Polym. Chem.,* **53**, 1010–1022.

72. Oishi, M., Sumitani, S., and Nagasaki, Y. (2007). On-off regulation of 19F magnetic resonance signals based on pH-sensitive PEGylated nanogels for potential tumor-specific smart 19F MRI probes. *Bioconjugate Chem.,* **18**, 1379–1382.

73. Nurmi, L., Peng, H., Seppala, J., Haddleton, D. M., Blakey, I., and Whittaker, A. K. (2010). Synthesis and evaluation of partly fluorinated polyelectrolytes as components in 19F MRI-detectable nanoparticles. *Polym. Chem.,* **1**, 1039–1047.

74. Huang, X., Huang, G., Zhang, S., Sagiyama, K., Togao, O., Ma, X., Wang, Y., Li, Y., Soesbe Todd, C., Sumer Baran, D., Takahashi, M., Sherry, A. D., and Gao, J. (2013). Multi-chromatic pH-activatable 19F-MRI nanoprobes with binary ON/OFF pH transitions and chemical-shift barcodes, *Angew. Chem., Int. Ed.,* **52**, 8074–8078.

75. Munkhbat, O., Canakci, M., Zheng, S., Hu, W., Osborne, B., Bogdanov, A. A., and Thayumanavan, S. (2019). 19F MRI of polymer nanogels aided by improved segmental mobility of embedded fluorine moieties. *Biomacromolecules,* **20**, 790–800.

76. Brieger, K., Schiavone, S., Miller, F. J., Jr., and Krause, K. H. (2012). Reactive oxygen species: from health to disease. *Swiss Med. Wkly.,* **142**, w13659.

77. Fu, C., Herbst, S., Zhang, C., and Whittaker, A. K. (2017). Polymeric 19F MRI agents responsive to reactive oxygen species. *Polym. Chem.,* **8**, 4585–4595.

78. Kolouchova, K., Groborz, O., Cernochova, Z., Skarkova, A., Brabek, J., Rosel, D., Svec, P., Starcuk, Z., Slouf, M., and Hruby, M. (2021). Thermo- and ROS-responsive self-assembled polymer nanoparticle tracers for 19F MRI theranostics. *Biomacromolecules,* **22**, 2325–2337.

79. Gamcsik, M. P., Kasibhatla, M. S., Teeter, S. D., and Colvin, O. M. (2012). Glutathione levels in human tumors. *Biomarkers,* **17**, 671–691.

80. Nakamura, T., Matsushita, H., Sugihara, F., Yoshioka, Y., Mizukami, S., and Kikuchi, K. (2015). Activatable 19F MRI nanoparticle probes for the detection of reducing environments. *Angew. Chem., Int. Ed.,* **54**, 1007–1010.

81. Fu, C., Tang, J., Pye, A., Liu, T., Zhang, C., Tan, X., Han, F., Peng, H., and Whittaker, A. K. (2019). Fluorinated glycopolymers as reduction-responsive 19F MRI agents for targeted imaging of cancer. *Biomacromolecules*, **20**, 2043–2050.

82. Huang, P., Song, H., Wang, C., Kong, D., Wang, W., Guo, W., Wang, Y., Yang, G., and Wang, C. (2018). Fluorine meets amine: reducing microenvironment-induced amino-activatable nanoprobes for (19)F-magnetic resonance imaging of biothiols. *ACS Appl. Mater. Interfaces*, **10**, 18532–18542.

83. Taylor Nicholas, G., Johnson Rob, R., 3rd, Teator Aaron, J., Leibfarth Frank, A., Chung Sang, H., Lee Yueh, Z., Kwansa Albert, L., Milliken Nina, J. B., Yingling Yaroslava, G., and Koshlap Karl, M. (2020). Partially fluorinated copolymers as oxygen sensitive (19) F MRI agents. *Chemistry*, **26**, 9982–9990.

84. Mizukami, S., Takikawa, R., Sugihara, F., Hori, Y., Tochio, H., Waelchli, M., Shirakawa, M., and Kikuchi, K. (2008). Paramagnetic relaxation-based 19F MRI probe to detect protease activity. *J. Am. Chem. Soc.*, **130**, 794–795.

85. Mizukami, S., Takikawa, R., Sugihara, F., Shirakawa, M., and Kikuchi, K. (2009). Dual-function probe to detect protease activity for fluorescence measurement and 19F MRI. *Angew. Chem., Int. Ed.*, **48**, 3641–3643, S3641/1-S3641/6.

86. Yue, X., Wang, Z., Zhu, L., Wang, Y., Qian, C., Ma, Y., Kiesewetter, D. O., Niu, G., and Chen, X. (2014). Novel 19F activatable probe for the detection of matrix metalloprotease-2 activity by MRI/MRS. *Mol. Pharm.*, **11**, 4208–4217.

87. Matsuo, K., Kamada, R., Mizusawa, K., Imai, H., Takayama, Y., Narazaki, M., Matsuda, T., Takaoka, Y., and Hamachi, I. (2013). Specific detection and imaging of enzyme activity by signal-amplifiable self-assembling 19F MRI probes. *Chem. - Eur. J.*, **19**, 12875–12883.

88. Takaoka, Y., Kiminami, K., Mizusawa, K., Matsuo, K., Narazaki, M., Matsuda, T., and Hamachi, I. (2011). Systematic study of protein detection mechanism of self-assembling 19F NMR/MRI nanoprobes toward rational design and improved sensitivity. *J. Am. Chem. Soc.*, **133**, 11725–11731.

89. Yuan, Y., Ge, S., Sun, H., Dong, X., Zhao, H., An, L., Zhang, J., Wang, J., Hu, B., and Liang, G. (2015). Intracellular self-assembly and disassembly of 19F nanoparticles confer respective "Off" and "On" 19F NMR/MRI

signals for legumain activity detection in zebrafish. *ACS Nano,* **9**, 5117–5124.

90. Yuan, Y., Sun, H., Ge, S., Wang, M., Zhao, H., Wang, L., An, L., Zhang, J., Zhang, H., Hu, B., Wang, J., and Liang, G. (2015). Controlled intracellular self-assembly and disassembly of 19F nanoparticles for MR imaging of caspase 3/7 in zebrafish. *ACS Nano,* **9**, 761–768.

91. Smith, N. K. R., Stabler, S. B., Cameron, I. L., and Medina, D. (1981). X-ray microanalysis of electrolyte content of normal, preneoplastic, and neoplastic mouse mammary tissue. *Cancer Res.,* **41**, 3877–3880.

92. Smith, N. R., Sparks, R. L., Pool, T. B., and Cameron, I. L. (1978). Differences in the intracellular concentration of elements in normal and cancerous liver cells as determined by X-ray microanalysis. *Cancer Res.,* **38**, 1952–1959.

93. Cooper, E. C., and Jan, L. Y. (1999). Ion channel genes and human neurological disease: recent progress, prospects, and challenges. *Proc. Natl. Acad. Sci. USA,* **96**, 4759–4766.

94. Cremers, S., and Garnero, P. (2006). Biochemical markers of bone turnover in the clinical development of drugs for osteoporosis and metastatic bone disease: potential uses and pitfalls. *Drugs,* **66**, 2031–2058.

95. Lauffenburger, T., Olah, A. J., Dambacher, M. A., Guncaga, J., Lentner, C., and Haas, H. G. (1977). Bone remodeling and calcium metabolism: a correlated histomorphometric, calcium kinetic, and biochemical study in patients with osteoporosis and Paget's Disease. *Metabolism,* **26**, 589–606.

96. Smith, G. A., Morris, P. G., Hesketh, T. R., and Metcalfe, J. C. (1986). Design of an indicator of intracellular free Na+ concentration using 19F-NMR. *Biochim. Biophys. Acta.,* **889**, 72–83.

97. Preslar, A. T., Lilley, L. M., Sato, K., Zhang, S., Chia, Z. K., Stupp, S. I., and Meade, T. J. (2017). Calcium-induced morphological transitions in peptide amphiphiles detected by 19F-magnetic resonance imaging. *ACS Appl. Mater. Interfaces,* **9**, 39890–39894.

98. Preslar, A. T., Tantakitti, F., Park, K., Zhang, S., Stupp, S. I., and Meade, T. J. (2016). 19F magnetic resonance imaging signals from peptide amphiphile nanostructures are strongly affected by their shape. *ACS Nano,* **10**, 7376–7384.

99. Ataman, M. (1987). Properties of aqueous salt solutions of poly(ethylene oxide). Cloud points, θ temperatures. *Colloid Polym. Sci.,* **265**, 19–25.

100. Florin, E., Kjellander, R., and Eriksson, J. C. (1984). Salt effects on the cloud point of the poly(ethylene oxide) + water system. *J. Chem. Soc., Faraday Trans. 1,* **80**, 2889–2910.

101. Zhang, C., Moonshi, S. S., Peng, H., Puttick, S., Reid, J., Bernardi, S., Searles, D. J., and Whittaker, A. K. (2016). Ion-responsive 19F MRI contrast agents for the detection of cancer cells. *ACS Sens.,* **1**, 757–765.

Chapter 2

Nanofluorides: Inorganic Fluoride Nanocrystals for ^{19}F MRI

Dana Cohen, Reut Mashiach, Hyla Allouche-Arnon and Amnon Bar-Shir
Department of Molecular Chemistry and Materials Science, Weizmann Institute of Science, Rehovot, Israel
amnon.barshir@weizmann.ac.il

Recent advances in ^{19}F-MRI tracers have focused on a novel type of ^{19}F-nanoformulation utilizing inorganic fluoride-based nanocrystals (NCs). Called *nanofluorides*, these nanoformulations combine the advantages of inorganic NCs (*i.e.*, small size, dense fluoride content, monodispersity, colloidal stability, and surface modifiability) with the merits of ^{19}F MRI (*i.e.*, quantitative, 'hot-spot' images). This chapter highlights important principles for the design, synthesis and use of nanofluorides as tracer agent for ^{19}F-MRI studies. Among the issues we will discuss are the crystallographic requirements of the designed M_xF_y (M = cation and F = fluoride anion) NCs, their

Fluorine Magnetic Resonance Imaging: Methods and Applications in Biomedicine
Edited by Eric T. Ahrens and Ulrich Flögel
Copyright © 2025 Jenny Stanford Publishing Pte. Ltd.
ISBN 978-981-5129-37-3 (Hardcover), 978-1-003-53004-6 (eBook)
www.jennystanford.com

relaxation characteristics, as well as approaches for surface coating, targeting and "multicolor" MRI capabilities. We will also examine strategies to shorten the T_1 relaxation time of nanofluorides, such as paramagnetic relaxation enhancement (PRE) and the introduction of nanocrystalline-defects for relaxation enhancement (NDRE). The design of glyconanofluoride immunotracers for *in vivo* ^{19}F-MRI mapping of inflammatory processes is shown as a specific application of these novel ^{19}F agents. Given the relatively fast tumbling of small nanofluorides in aqueous solution and size stability *in vivo*, we foresee nanofluorides enabling new uses of ^{19}F MRI that have remained unapproachable due to the inherent limitations of the nano-tracers currently in use, namely, organofluorine-based formulations.

2.1 Introduction

The nanotechnology revolution has accelerated the development of medicine through the development of numerous types of nanoformulations for medical applications. Of these formulations, inorganic nanocrystals (NCs) stand out due to the ability to control both their chemical and physical properties for a desired application. In contrast to phospholipid-based or polymeric-based nanoformulations, inorganic NCs offer a great variety of design possibilities, including precise control of size, shape and content. The ability to control the size and shape of gold (Au) NCs, for example, has enabled utility in profiling biological environments in terms of both biodistribution and cell-uptake [1, 2]. Additionally, optimized nanoformulations for computational tomography (CT) exhibit improved *in vivo* imaging performance [3, 4] beyond what is achievable using clinically-approved organic agents. This development has led to the design and synthesis of other types of non-Au-based inorganic NC imaging agents for CT [5].

In MRI, metal-oxide-based NCs have been developed and proposed as contrast agents, due to their paramagnetic or superparamagnetic characteristics that affect the surrounding water's relaxation properties. In the same manner as shown for Au NCs, both the chemical and physical properties of metal oxide NCs (*e.g.*, iron oxide, manganese oxide [6], gadolinium oxide [7] and others) can be precisely controlled, making them unique imaging agents that exhibit supreme sensitivity [8] with controllable size [9] and shape [10] and a tunable contrast-enhancement mechanism (*i.e.*, T_1- or T_2-based relaxation enhancement) [6]. However, similar to other contrast agents used for MRI which are designed to manipulate the water-relaxation properties, metal oxide NCs lack specificity and involve high background signals, which often generate either false-positive or false-negative signal interpretations. Another limitation is the inability to qualify the *in vivo* content of metal oxides based on their MRI signal modulation of surrounding water. These impediments have stimulated efforts to develop alternative detection approches.

Inorganic fluoride-based NCs, namely nanofluorides, serve as functional NCs [11] in a variety of fields, including photonics [12], bioimaging [13], biomedicine [14], sensing [15] and catalysis [16]. The wide range of applications of nanofluorides often relies on their unique ability to display up-conversion luminescence following core doping with lanthanide ions [17, 18]. Although rich in ^{19}F content, nanofluorides (M_xF_y, M = cation, F = F^-) have not been studied in solution with high-resolution ^{19}F-NMR (nuclear magnetic resonance) or used for ^{19}F MRI until recently. This is because in nanocrystal-based formulations, the restricted mobility of the elements within the crystal frequently results in NMR line-broadening due to high dipole-dipole interactions of the crystal's immobilized elements, which prevents detectability in liquid-state NMR experiments.

This limitation has recently been overcome by using ultra-small-sized (<10 nm) inorganic nanofluorides [19–23] that experience sufficient averaging of the homonuclear dipolar interactions of their fluoride content. The nanoformulations have been successfully used for high-resolution ^{19}F-NMR experiments in solutions. These formulations merge the advantages of inorganic NCs (*e.g.*, small and controllable size, dense fluoride content, monodispersity, colloidal stability, surface modifiability) with the merits of ^{19}F MRI, as described in this book. In this chapter, we describe recent developments in adapting nanofluorides for ^{19}F MRI and highlight important principles for obtaining detectable, robust ^{19}F-MRI signals and synthetic strategies. We discuss the importance of crystallographic properties for MRI applications and summarize strategies to shorten the T_1-relaxation properties of the crystalline formulations. We also describe the use of the NC formulations for mapping inflammation, as well as the potential use for multiplexed imaging.

2.2 ^{19}F-NMR Signals of Dispersed Nanofluorides

Water-soluble CaF_2 NCs, coated with citrate as the capping ligand, were the first type of nanofluorides studied with high-resolution ^{19}F-NMR [19]. The small size of the nanofluoride, with an average hydrodynamic diameter of 4.4 ± 1.4 nm, allows fast tumbling in water thereby averaging-out homonuclear dipolar interactions of ^{19}F nuclear spins within their core obtain narrow high-resolution ^{19}F-NMR spectra (Fig. 2.1a). The elements in the CaF_2 NCs obey a typical face-centered cubic (fcc) arrangement in a fluorite-type structure result in magnetically equivalent fluorides within the crystal and a singlet peak at −109.6 ppm, similar to the frequency obtained for CaF_2 powder *via* solid-state NMR [24]. This is an advantageous property for ^{19}F-MRI agents, as the high magnetic equivalence of the ^{19}F source maximizes signal intensity [25, 26].

The crucial need for the fcc arrangement in order to obtain a singlet ^{19}F-NMR peak while dispersed in water is evident by comparing the liquid-state ^{19}F-NMR characteristics of CaF_2 NCs and other nanofluorides with different crystal properties. For example, LaF_3 NC, which has a tysonite-type crystal structure, exhibits a more complex ^{19}F-NMR spectrum compared to CaF_2 NCs. LaF_3 NC displays peaks at the expected chemical shifts obtained from solid-state ^{19}F-NMR studies of LaF_3 crystal (Fig. 2.1b), where three types of fluorides are observed from nonequivalent crystallographic sites [27]. Generally, this more complex ^{19}F-NMR spectrum leads to chemical shift artifacts in ^{19}F MRI, which can be alleviated using advanced pulse sequences [28, 29], but can lead to a reduction in the overall ^{19}F-MRI signal and tracer sensitivity.

In contrast to LaF_3 NCs, SrF_2 NCs has the same fcc crystal structure as CaF_2 and thus also displays a singlet ^{19}F-NMR peak, as expected from magnetically equivalent fluorides in the SrF_2 fluorite-like crystal structure (Fig. 2.1c). These findings demonstrate that small nanofluorides in aqueous solution can be visualized and studied with ^{19}F-NMR and may be useful for ^{19}F MRI with appropriate formulation.

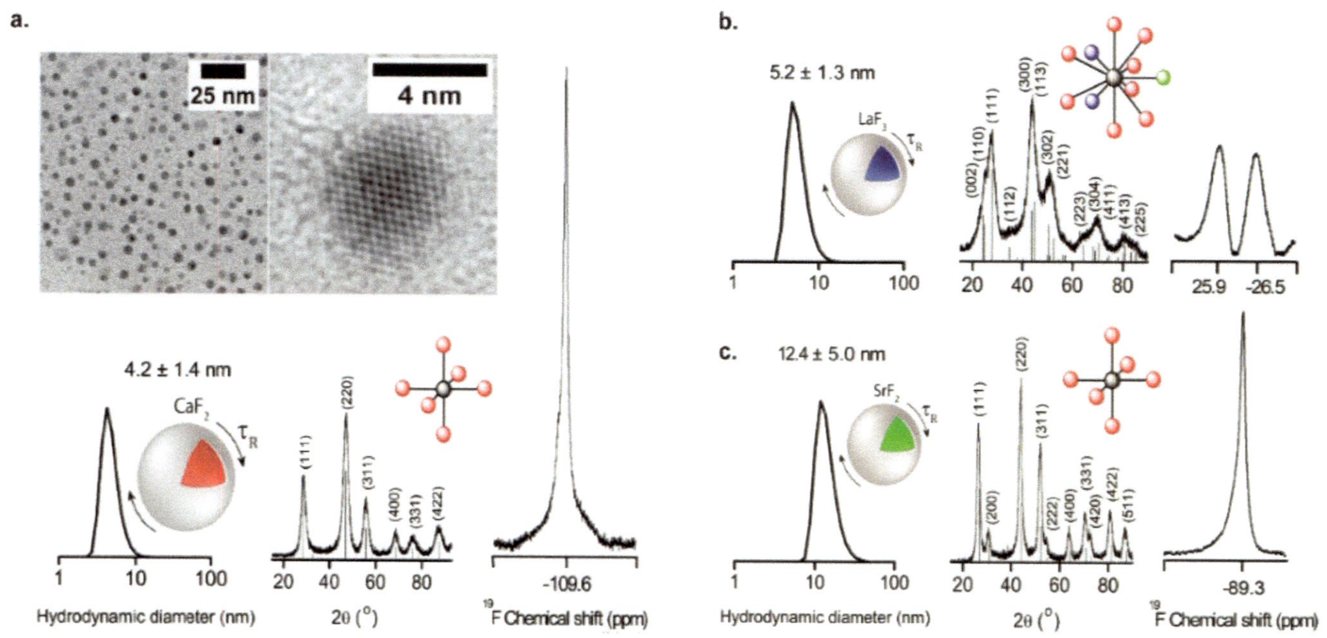

Figure 2.1 Water-soluble, citrate-coated nanofluorides. (a) TEM images, DLS histogram, XRD pattern and ^{19}F-NMR of CaF_2 NCs. (b) DLS histogram, XRD pattern and ^{19}F-NMR of LaF_3 NCs. (c) DLS histogram, XRD pattern and ^{19}F-NMR of SrF_2 NCs. Shown are schematic representations of the positions of the fluorides (red, green or blue) at the cation M (Ca^{2+}, Sr^{2+} or La^{3+}, gray) first coordination sphere in fluorite-type (CaF_2 and SrF_2) and tysonite-type (LaF_3) crystal structures. Figure was modified with permission from Ref. [19].

2.3 ^{19}F-NMR Relaxation Properties of Nanofluorides

Due to their solid, crystalline nature, nanofluorides have restricted elemental (*i.e.*, ^{19}F) mobility and thus are expected to have long longitudinal (T_1) and short transverse (T_2) relaxation times. These relaxation times determine suitability as imaging agents and guide the MRI acquisition parameters and pulse sequence type required for data acquisition. Indeed, the T_2 values of nanofluorides are very low, of the order of 2–3 ms, and is much shorter than the T_2 values of organofluorine-based materials which have a faster ^{19}F-correlation time. Nevertheless, the ability to monitor extremely short T_2 values using modern MRI systems and methods opens new opportunities for the design and development of novel inorganic NCs as "X"-nuclei MRI-tracers. Pulse sequences, such as zero echo time (ZTE) [30] and ultrashort TE (UTE) [31], have been used for ^{19}F MRI in cases where the fluorinated material had extremely short T_2 values [32, 33]; signals were undetectable using conventional spin echo (SE)- or gradient echo (GE)-based protocols.

As implied from solid-state NMR measurements of CaF_2 powders, whose longitudinal relaxation times is of the order of tens of seconds to minutes [34, 35], the ^{19}F T_1 of CaF_2 NCs is indeed relatively long (~10 s) [19]. This large T_1 value is obtained even for the smallest CaF_2 NCs synthesized (~4 nm), which are well dispersed in water. This limitation requires the use of low flip angle excitation pulses (Ernst angle) to shorten the repetition time (TR) in order to increase signal averaging and the ^{19}F-MRI signal-to-noise ratio (SNR) for a given acquisition time. Indeed, it was found that when using the UTE sequence with a TE = 20 μs (TR = 150 ms with a flip angle of 10° to achieve Ernst angle conditions), a clear ^{19}F-MRI signal is observed from water-dispersed CaF_2 NCs (Fig. 2.2a). This was also the case when polyethylene glycol-coated CaF_2 NCs were injected into the footpad of inflamed mice. In that particular study, the accumulation of the nanofluorides in the lymph node of the injected leg could be clearly visualized with ^{19}F MRI, representing the first-ever use of small-sized inorganic nanofluorides as imaging agents for *in vivo* ^{19}F MRI (Fig. 2.2b) [19].

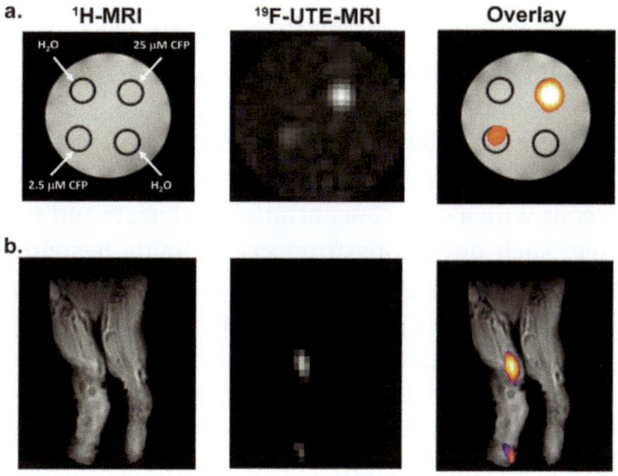

Figure 2.2 ^{19}F MRI of water-dispersed CaF_2 NCs. From left to right: ^{1}H-MRI; ^{19}F-MRI (UTE, acquired with TE = 20 µs) and ^{19}F-MR signal of (a) phantoms and (b) a representative mouse, displayed as a pseudo-color map on ^{1}H image. CFP: CaF_2 with PEG coating (used for both phantom and *in vivo* studies). Figure was modified with permission from Ref. [19].

2.4 Introducing Defects in CaF_2 NCs for T_1-Relaxation Enhancement

Despite nanofluorides' demonstrated suitability as imaging agents for ^{19}F MRI, their relatively long T_1 relaxation time (*i.e.*, several seconds) requires a long TR which limits data acquisition speed. This restricts the number of signal averages for a given acquisition time and, hence, limits the achievable ^{19}F-MRI SNR. Thus, significant shortening of the nanofluoride T_1 value is key to realizing sensitive NC detection. While the obvious solution is to induce paramagnetic relaxation enhancement (PRE) using paramagnetic elements (see section below), other options may hold promise, especially because of concerns regarding lanthanide toxicity and the need to simplify the ^{19}F-MRI signal quantification, which can be affected by dopants with PRE characteristics [36].

The shortening of the T_1 observed in solid-state NMR studies of solid crystalline materials without inducing PRE was attributed to several factors, including reduced crystal size and the induction of crystal defects. Specifically, in the case of CaF_2 powders, it was demonstrated, again using solid-state NMR, that crystal defects facilitate element mobility at the defect site, serving as "relaxation sinks" that essentially lead to T_1 shortening [37]. Subjecting raw CaF_2 powders to elongated mechanical stress *via* ball milling, not only resulted in the restructuring of the large grains but also yielded small crystals with a high concentration of defects and strong lattice distortions. Such deviations from crystallinity, beyond increasing the surface-to-volume ratio of disordered interfacial regions, has the potential to improve the performance of CaF_2 as solid F^- electrolytes [38, 39] and thus to shorten T_1 relaxation times [22, 38, 39]. Indeed, a significant spin-lattice relaxation enhancement (*ca.* one order of magnitude) was found to highly correlate with the presence of lattice defects, as confirmed by TEM and XRD measurements, which were attributed to facilitate element mobility at the defect site and enhanced dipolar interactions [40–42]. The realization that lattice defects are reflected in the NMR properties of nanofluorides open a unique opportunity to overcome the relatively long T_1 of the originally proposed highly-crystalline CaF_2 NCs [19] by introducing intentional defects.

The surfactant capping ligands of colloidal nanofluorides are vital for their production and stabilization as mono-dispersed fabrications, which is a key feature of their development for ^{19}F-MRI applications. Organic capping agents are also essential components in the synthesis of NCs, as they may bind to both the formed NCs and to the available precursors during the formation of NCs. Thus, the affinity of the ligands to the NCs' surface or to the precursors in the solution has the potential to determine the mechanism of formation and final morphologies of the NCs, such as their crystallinity, size and shape. Synthesizing colloidal CaF_2 NCs with oleyl phosphate (OP, with a phosphate head group) or oleic acid (OA, with a carboxylate head group) as their capping (surfactant) agent clearly demonstrates this versatiliy [20]. The stronger binding affinity of the phosphate group (as compared to carboxylate group) to Ca^{2+} [43] contributed to the formation of two clearly different crystal architectures (Fig. 2.3). While OA-CaF_2 NCs featured a highly-ordered, crystalline lattice (Fig. 2.3a), distinct crystal-defects, *i.e.*, grain boundaries, were observed in OP-CaF_2 NCs (Fig. 2.3b). Importantly, this synthetic induction of crystal defects (specifically, grain boundaries) in colloidal CaF_2 NCs resulted in a much shorter T_1-relaxation time of the fluoride within the NC—by about 10-fold, from 11 sec to 1 sec (Fig. 2.3c). This enhanced relaxation improved the SNR of ^{19}F-MRI images by a factor of 4 at a given scan time.

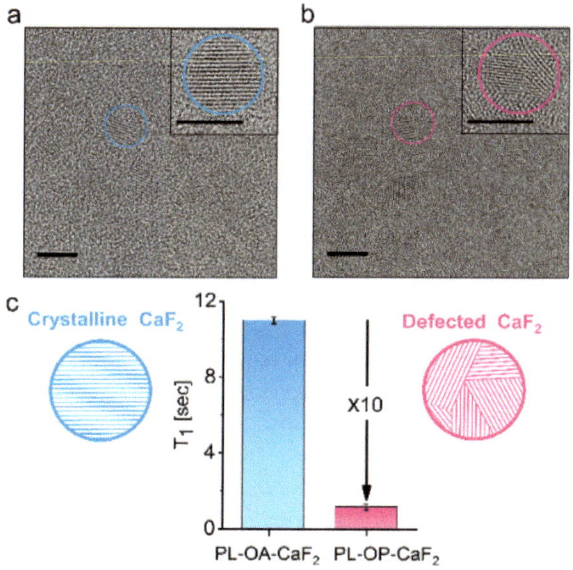

Figure 2.3 Synthetically induced defects in CaF_2 NCs using a phosphate-based ligand. (a, b) HR-TEM image (scale bar 10 nm, in inset 5 nm) of (a) OA-CaF_2 and (b) OP-CaF_2 NCs. (c) T_1-relaxation times for water-dispersed PL-OA-CaF_2 and PL-OP-CaF_2 NCs. PL: phospholipid; OA: oleic acid; OP: oleyl phosphate. Reprinted with permission from Ref. [20].

The relationship between CaF_2 NCs' lattice parameters, measured with TEM, and the T_1 of the fluoride content, measured with ^{19}F-NMR, was demonstrated by precisely controlling the size and polydispersity of CaF_2 NCs using OA and oleylamine (OM) as ligands and 1-octadecen (ODE) as the solvent (Fig. 2.4). Here too, the T_1 values were shorten by one-order-of magnitude without the aid of paramagnetic elements [44]. The ability to shorten the interplanar distances within the NCs using different synthetic routes toward relaxation enhancement could be attributed to enhanced dipolar integrations, which are proportional to the distance between the nuclei [45].

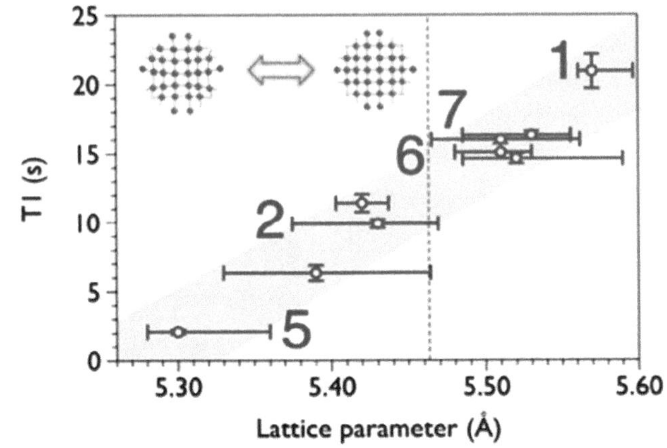

Figure 2.4 The correlation between T_1-relaxation times and the lattice parameters of CaF_2 NCs. Numbering represents different ratios of the ligands and solvent, OA:OM:ODE, used for the synthesis of CaF_2 NCs. (#1 0.5:0.125:0.375, #2 0.5:0.25:0.25, #5 0.5:0.5:0, #6 0.375:0.625:0, #7 0.25:0.75:0). In inset, schematic illustration of lattice distortion. Reprinted with permission from Ref. [44].

The two above demonstrations (Figs. 2.3 and 2.4) highlight that by inducing defects into small-sized nanofluorides, one can shorten the T_1 of their fluoride content by as much as 10-fold. However, the 1 s T_1 obtained in defected nanofluorides should ideally be shorter to gain more sensitivity. Thus, nanocrystalline-defects relaxation enhancement (NDRE) should be further explored in pursuit of novel crystal-engineering approaches to rationalize architecture-relaxation relationships in nanofluorides.

2.5 Paramagnetic Doping of CaF_2 NCs for T_1-Relaxation Enhancement

Although introducing crystal defects into CaF_2 NCs was found to be applicable for shortening T_1-relaxation times, much shorter nanofluoride T_1 values are needed for their implementation as robust ^{19}F-MRI tracers. In this regard, PRE is still the mechanism of choice for shortening T_1-relaxation times of nuclear spins with Gd^{3+}- and, very recently, Mn^{2+}-complexes [46, 47] for clinical ^1H MRI. PRE was also used to improve ^{19}F-MRI performance, with several studies demonstrating the ability to reduce T_1 values of perfluorocarbon (PFC) nanoemulsions using paramagnetic metals [32, 48–52]. Some designs demonstrated effective relaxation enhancements, attributed to a modified phospholipid coating with paramagnetic elements. Other fabrications showed that embedding high-spin paramagnetic metal ions within the PFC matrix could lead to effective PRE and improved ^{19}F MRI.

Interestingly, many of the available NCs with up conversion fluorescence properties are nanofluorides doped with paramagnetic lanthanides [17, 18]. An inorganic up conversion NC composed of a crystalline host (commonly used nanofluorides are CaF_2 and $NaYF_4$) and a dopant (typically a combination of lanthanide ions), added at a low concentration, endows the nanoformulation with its unique and versatile up conversion fluorescence property. While the host lattice with its crystal structure serves as the matrix that brings these centers into optimal positions, the lanthanide dopant provides the luminescent centers. To successfully incorporate impurity ions into the host lattice, it is common to use ions with the same valence as, and similar ionic radii to, those of the host (*i.e.*, nanofluoride crystal core) [53]. In this regard, the versatility of the nanofluorides used in up conversion fluorescent studies provides an established synthetic approach for introducing paramagnetic elements to generate paramagnetic nanofluorides (Fig. 2.5a). The use of such dopants, mostly lanthanides, was also reported to alter both the size and shape of nanofluorides [54].

Obviously, paramagnetic lanthanides (Ln^{3+}, except for La^{3+} and Lu^{3+}) can affect the relaxation properties of neighboring nuclei in crystalline materials, mainly through PRE induction [55, 56]. Therefore, they could be used to fabricate paramagnetic nanofluorides [22]. Nevertheless, it should be emphasized that lanthanide ions that induce a very strong PRE effect (Gd produces the highest effect) can result in a "blind-sphere" radius in crystals [36], reflected by the NMR signal nullification of neighboring atoms, which may complicate MRI signal quantification. Therefore, the design and synthesis of nanofluorides doped with Ln^{3+} ions should be approached with great care to achieve a strong PRE effect to shorten T_1-relaxation, but with negligible blind-sphere radius induction to preserve ^{19}F-MRI detectability of all fluorines in the NC.

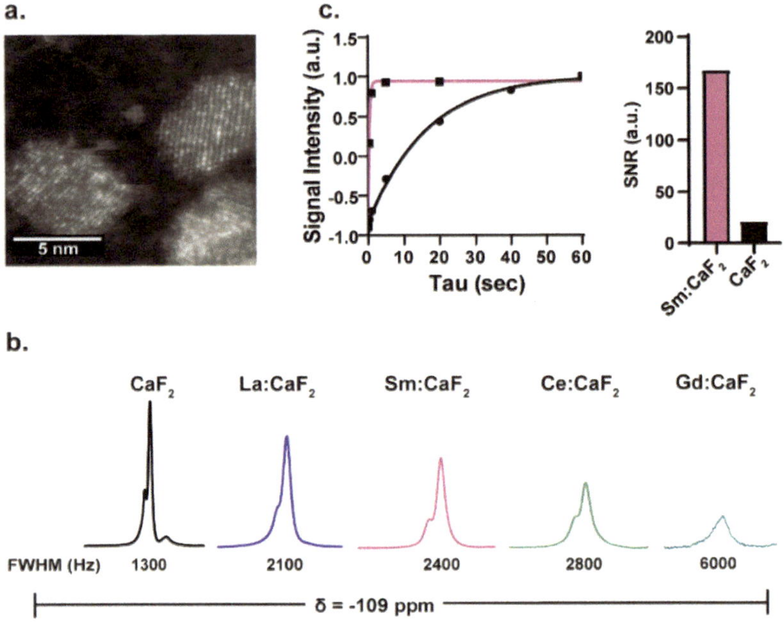

Figure 2.5 Paramagnetic CaF_2 NCs. (a) High-resolution STEM/HAADF image of Sm^{3+}-doped CaF_2 NCs with bright spots associate with the heavier (Sm^{3+}) atoms that are incorporated into the crystals' lattice sites. (b) ^{19}F-NMR spectra of non-doped and lanthanide-doped CaF_2 NCs, $\delta = -109$ ppm. (c) Inversion recovery plots (left) of Sm:CaF_2 (pink line) and CaF_2 (black line) NCs and their corresponding SNR obtained from ^{19}F MRI of a representative phantom, shown as a bar graph (right). Adapted with permission from Ref. [22].

Using CaF_2 NCs, it was demonstrated that doping the crystal core with different Ln^{3+} ions results in a different ^{19}F-NMR spectra (Fig. 2.5b). As expected, diamagnetic dopant La^{3+} had no noted impact on the T_1-relaxation properties of the fluoride content and only a mild effect on ^{19}F-NMR line-broadening. In contrast, doping CaF_2 NCs with Ce^{3+} and Gd^{3+} resulted in significant line-broadening and a reduction in SNR of ^{19}F-NMR spectra due to the strong PRE and blind-sphere inductions by these dopants. Therefore, the use of these and other paramagnetic elements with strong PRE induction capabilities is less preferable for ^{19}F-MRI applications. However, Ln^{3+}, with mild PRE, is expected to induce a negligible blind-sphere radius in the nanofluoride lattice. While producing a minor effect on the ^{19}F-NMR signal line width (Fig. 2.5b), Sm^{3+}-doped CaF_2 (Sm:CaF_2) NCs exhibited significantly shorter T_1 values compared to non-doped nanofluorides (Fig. 2.5c). Increasing the amount of the Sm^{3+} dopant in Sm:CaF_2 NCs further shortened the T_1-relaxation times, but also results in a ^{19}F-NMR SNR reduction, which cancels out the sensitivity improvement. For a low Sm^{3+} dopant concentration of 3.5%, effective T_1 shortening was achieved, generating 140-fold shorter T_1 values compared to non-doped CaF_2 NCs ($T_{1,non\text{-}doped}$ = 10 s) with extremely short T_1 values ($T_{1,doped}$ = 0.07 s). The short T_1 obtained for Sm:CaF_2 NCs resulted in an 8-fold enhancement in ^{19}F-MRI sensitivity for a given acquisition time (Fig. 2.5c) [22]. This observation is crucial for implementation of these materials as effective ^{19}F tracers.

2.6 Glyconanofluorides for ^{19}F MRI of Inflammatory Activity

The ability to construct a small-sized inorganic core scaffold on which biomolecules can be assembled to form nanobiomimetics holds great promise for both medical diagnostics and therapeutics. One such type of nanobiomimetic is glyconanoparticles (GNPs), which are composed of an inorganic crystal core [57–60], to which multiple carbohydrate copies are attached. The formed carbohydrate surface allows GNPs to specifically bind lectin receptors, which are highly expressed on the membrane of immune cells, making them efficient immunomodulation and immunolabeling agents [61]. Given their biocompatibility, GNPs hold great potential as an imaging platform for mapping inflammation-associated events. While fluorescent GNPs (based on quantum dot cores) enable the efficient labeling of activated immune cells [62], they cannot be used for non-invasive *in vivo* imaging. In contrast, magnetic GNPs based on an iron oxide core [59] have been proposed for *in vivo* MRI of inflammation [63], but their detection relies on the interpretation of ^1H-MRI signal voids, which can lead to false-positive results, poor specificity and lack of quantifiability.

The ability to tune nanofluorides' surface throughout synthesis, along with their ^{19}F-MRI detectability, make nanofluorides promising candidates for targeted imaging agents. GNPs composed of Sm:CaF_2 decorated with multiple copies of lactose at their surface have been studied [22]. This formulation can be obtained by first synthesizing Sm:CaF_2 NCs with OA as their capping ligand. Then, hydrophobic interactions between OA tails and the organic tail of lactose-modified phospholipids leads to the formation of lactose-modified nanofluorides (Fig. 2.6a).

In good agreement with previous reports showing increased cellular uptake of lactose-modified NCs by macrophages [64, 65], lactose-coated nanofluorides (*i.e.*, lactose-phospholipid (LPL) CaF_2) were shown to accumulate in activated immune cells, demonstrating their utility as immunotracers for mapping inflammation with ^{19}F MRI. As shown in Fig. 2.6b, a six-fold higher amount of glyconanofluorides (*i.e.*, LPL-CaF_2 NCs) accumulated in lymphatic regions of mice induced with focal inflammation following injection into the footpad compared to mice injected with non-glycosylated nanofluorides (PL-CaF_2). This observation, which was based on the fluorescence characteristics of the nanofluoride coating, was obtained *ex vivo via* a FACS analysis of harvested lymphatic cells. In agreement with other reports showing that several types of immune cells, including CD11b leukocytes, B cells and T cells, can recognize and bind synthetic carbohydrate formulations [62, 66, 67], a FACS analysis showed significant uptake of the glyconanofluorides (compared to non-glycosylated nanofluorides) by CD11b leukocytes (7-fold higher), CD19 B-cells (6-fold higher) and CD8 T-cells (3-fold higher).

In vivo ^{19}F-MRI experiments performed ten days after the injection of the immunogenic emulsion to induce the inflammation further validated the higher accumulation of the glyconanofluorides in inflamed lymph nodes. Here (Fig. 2.6c), a much higher ^{19}F-MRI signal was obtained in the lymph nodes of inflamed mice one hour after the footpad injection of glyconanofluorides (LPL-CaF_2) compared to the leg injected with non-glycosylated nanofluorides (PL-CaF_2). Quantifying this difference for a group of mice injected with both formulations confirmed the significantly higher accumulation of LPL-CaF_2 within the inflamed lymph node, translating into a 2-fold higher ^{19}F-MRI signal (P-value = 0.033). This observation demonstrates that imaging inflammatory activity is feasible with ^{19}F MRI following the systemic administration of glyconanofluorides.

Similar promising results were obtained in the context of acute cerebral ischemia. Here, glyconanofluorides were used for mapping

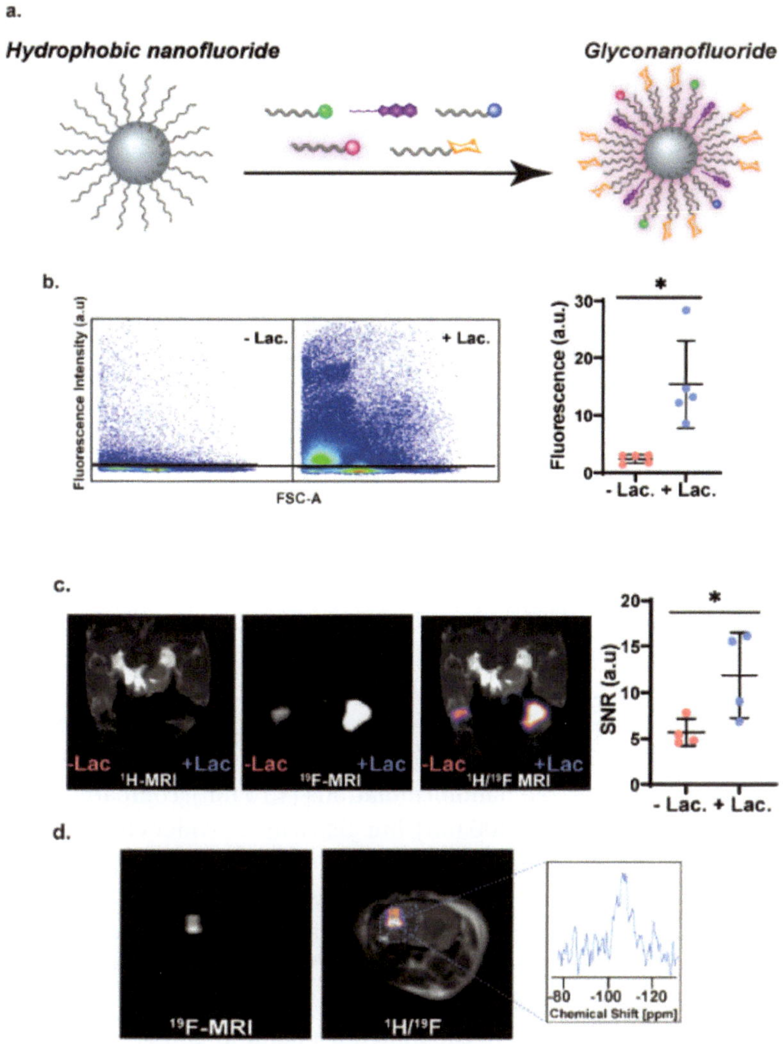

Figure 2.6 Paramagnetic glyconanofluorides for mapping inflammation with ^{19}F MRI. (a) Schematic representation of the preparation of glyconanofluorides. (b) FACS analysis of cells excised from the lymph nodes of inflamed mice post injection with PL- and LPL-CaF_2. (c) *In vivo* ^{19}F-MRI study of inflamed mice. From left to right: ^{1}H-MRI, ^{19}F-MRI overlay of ^{1}H/^{19}F and SNR dot-graph of ^{19}F signal from the relevant ROI. (d) ^{19}F-MRI of LPL-CaF_2 NCs in stroked mice. PL: phospholipid; LPL: lactose-modified phospholipid; -Lac: PL-CaF_2; +Lac: LPL-CaF_2. Adjusted with permission from Ref. [22].

neuroinflammation following an ischemic event, which is expected one [68, 69]-to-two weeks [70–72] after insult. Indeed, the systemic administration of LPL-CaF_2 NCs (*i.e.*, glyconanofluorides) two weeks post stroke induction was clearly visualized using ^{19}F MRI (Fig. 2.6d). Furthermore, a characteristic ^{19}F-NMR spectrum of CaF_2 NCs ($\delta = -109$ ppm, Fig. 2.6d, right panel) was obtained from a voxel located at the stroke area, confirming the accumulation of intact nanofluorides in that region. Fluorescent microscopy of a sectioned brain of the scanned mouse confirmed the accumulation of the glyconanofluorides in Iba-1-positive cells recruited to the vicinity of the stroke region.

2.7 Nanofluorides Enable "Multicolor" ^{19}F MRI

Taking advantage of their chemical shift span of over ~30 ppm, researchers have already attempted to use PFC-based formulations for "multicolor" ^{19}F-MRI studies [73–79]. However, the multiple, non-equivalent ^{19}F atoms in several PFC-based formulations generate multiple ^{19}F-peaks in the ^{19}F-NMR spectrum, which reduce their sensitivity (often not all ^{19}F spins count toward the ^{19}F-MR signal), raising the need for advanced acquisition schemes to reduce chemical shift artifacts [29]. Inorganic nanofluorides offer not only relatively small-sized nanoformulations (<10 nm), compared to PFC nanoemulsions (100–200 nm), but also a much wider chemical-shift range in their ^{19}F-NMR spectrum, spanning from *ca.* −10 ppm (BaF_2) to *ca.* −200 ppm (MgF_2), as revealed by solid-state NMR of powders [24]. Such a feature provides a platform for the development of a series of nanofluorides with different ^{19}F-NMR chemical shifts which can serve as "multicolor" tracer agents for MRI.

To demonstrate this capability, a phantom composed of three tubes, consisting of Sm:CaF_2 NC, Sm:SrF_2 NCs, and a third with a mixture of both (Fig. 2.7a). The ^{19}F-MRI signals from SrF_2 ($\delta = -88$ ppm) and CaF_2 ($\delta = -109$ ppm) were assigned yellow and blue colors, respectively. A mixture of equal concentrations of glyconanofluorides (*i.e.*, LPL-CaF_2) and non-glycosylated nanofluorides (*i.e.*, PL-SrF_2) were injected into the footpad of inflamed mice. Multicolor ^{19}F MRI generated two distinct signals from immune-targeted nanofluorides (LPL-CaF_2) and non-targeted nanofluorides (PL-SrF_2), with a much higher accumulation of the former (Fig. 2.7b).

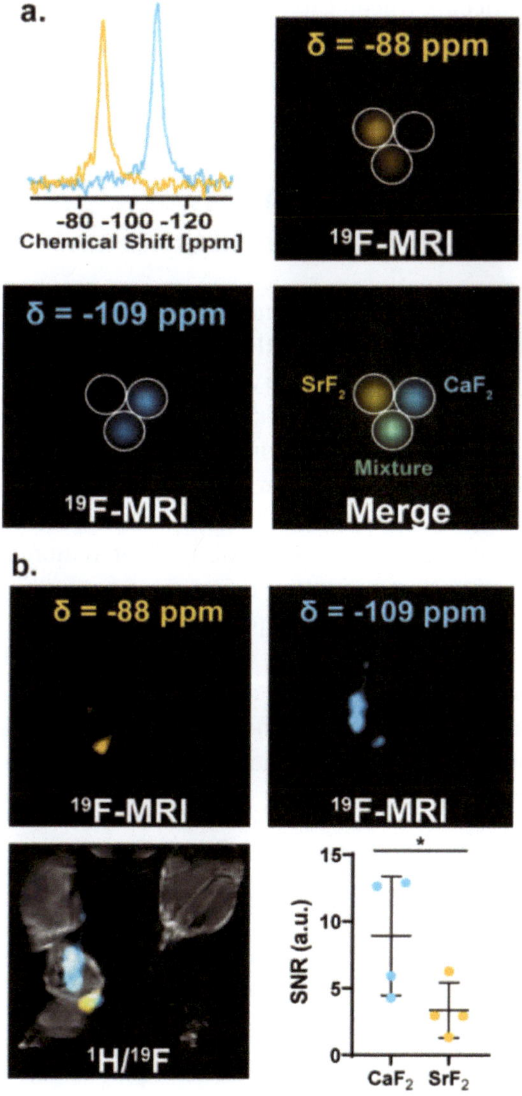

Figure 2.7 Multicolor ^{19}F MRI with nanofluorides. (a) ^{19}F-NMR spectra of PL-Sm:SrF_2 (yellow) and LPL-Sm:CaF_2 (light-blue) NCs along with multicolor ^{19}F MRI of a phantom containing PL-Sm:SrF_2, LPL-Sm:CaF_2 and a mixture of both NCs. (b) *In vivo* ^{19}F MRI of an inflamed mouse injected in the footpad with a PL-Sm:SrF_2 and LPL-Sm:CaF_2 mixture. ^{19}F-MRI data sets were acquired with the center of frequency offset at either –88 ppm (left, yellow) or –109 ppm (right, light-blue). Quantitative data from four individual mice is shown as a dot-plot graph. Adjusted with permission from Ref. [22].

The multicolor tracer capability could be further extended by designing and adding other nanofluorides, such as BaF_2 and MgF_2, with each one potentially marking a different target—together yielding a multicolor map of biological activity.

2.8 *In Situ* ^{19}F-NMR Studies of Rationally Designed Nanofluorides

One of the advantages of using inorganic NCs over organofluorine-based formulations for ^{19}F MRI is the ability to control the NC architecture (size, shape and crystal properties) by adjusting the synthetic protocol. Thus, understanding nanofluoride formation pathways is crucial for further optimization. While gaining control over NC formation pathways in solutions is usually achieved by experimentation to elucidate the key controllable parameters (Figs. 2.3 and 2.4), the NMR "visibility" of nanofluorides offers unique opportunity to monitor their formation in real time with ^{19}F-NMR (Fig. 2.8) [21]. By acquiring a series of high-resolution ^{19}F-NMR spectra over the entire course of nanofluoride formation, quantitative information can be obtained *in situ* regarding the size and concentration of the as-synthesized nanofluorides as well as the F^- consumption rate. This information can assist in revealing growth mechanisms.

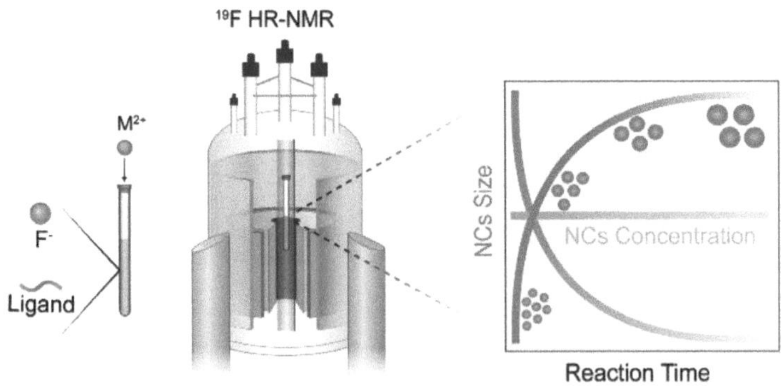

Figure 2.8 *In situ* ^{19}F-NMR of nanofluoride formation. Schematic illustration of the setup used in a real-time NMR study of the formation of nanofluorides in solution. Reprinted with permission from Ref. [21].

In one such attempt, it was shown that *in situ* ^{19}F-NMR can differentiate between two growth mechanisms by which CaF_2 NCs are formed. Interestingly, in this example it was demonstrated that phosphate-based ligands induce coalescence grow dynamics that result in crystallographic twining defects, which shorten the T_1 of the fluorides in the NCs. Thus, this *in situ* ^{19}F-NMR approach could potentially be used to shed light on the formation of nanofluorides under various conditions within a single experimental setup, without the need to interfere with the synthesis, as a means to formulate novel materials with interesting ^{19}F-MRI properties.

2.9 Outlook

Although colloidal inorganic NCs are widely used as imaging agents, ranging from gold nanoparticles for CT, to quantum dots for fluorescent imaging, and iron oxide nanoparticles for MRI, until very recently, fluoride-based NCs have not been studied. The initial ^{19}F-NMR experiments involving nanofluorides dispersed in solutions achieved sufficient averaging of homonuclear dipolar interactions within the nanofluorides' cores to show potential for ^{19}F MRI. With the recent advances in fluoride-containing materials and their extensive development and use in other fields, further development of nanofluorides as ^{19}F-MRI tracers looks promising.

References

1. Blanco, E.; Shen, H.; Ferrari, M. Principles of nanoparticle design for overcoming biological barriers to drug delivery. *Nat Biotechnol* **2015,** *33* (9), 941–951.
2. Mitchell, M. J.; Billingsley, M. M.; Haley, R. M.; Wechsler, M. E.; Peppas, N. A.; Langer, R. Engineering precision nanoparticles for drug delivery. *Nat Rev Drug Discov* **2021,** *20* (2), 101–124.
3. Cormode, D. P.; Roessl, E.; Thran, A.; Skajaa, T.; Gordon, R. E.; Schlomka, J. P.; Fuster, V.; Fisher, E. A.; Mulder, W. J.; Proksa, R.; Fayad, Z. A. Atherosclerotic plaque composition: analysis with multicolor CT and targeted gold nanoparticles. *Radiology* **2010,** *256* (3), 774–782.
4. Betzer, O.; Shwartz, A.; Motiei, M.; Kazimirsky, G.; Gispan, I.; Damti, E.; Brodie, C.; Yadid, G.; Popovtzer, R. Nanoparticle-based CT imaging

technique for longitudinal and quantitative stem cell tracking within the brain: application in neuropsychiatric disorders. *ACS Nano* **2014,** *8* (9), 9274–9285.

5. Hsu, J. C.; Nieves, L. M.; Betzer, O.; Sadan, T.; Noel, P. B.; Popovtzer, R.; Cormode, D. P. Nanoparticle contrast agents for X-ray imaging applications. *Wiley Interdiscip Rev Nanomed Nanobiotechnol* **2020,** *12* (6), e1642.
6. Kim, T.; Momin, E.; Choi, J.; Yuan, K.; Zaidi, H.; Kim, J.; Park, M.; Lee, N.; McMahon, M. T.; Quinones-Hinojosa, A.; Bulte, J. W.; Hyeon, T.; Gilad, A. A. Mesoporous silica-coated hollow manganese oxide nanoparticles as positive T_1 contrast agents for labeling and MRI tracking of adipose-derived mesenchymal stem cells. *J Am Chem Soc* **2011,** *133* (9), 2955–2961.
7. Park, J. Y.; Baek, M. J.; Choi, E. S.; Woo, S.; Kim, J. H.; Kim, T. J.; Jung, J. C.; Chae, K. S.; Chang, Y.; Lee, G. H. Paramagnetic ultrasmall gadolinium oxide nanoparticles as advanced T_1 MRI contrast agent: account for large longitudinal relaxivity, optimal particle diameter, and in vivo T_1 MR images. *ACS Nano* **2009,** *3* (11), 3663–3669.
8. Shapiro, E. M.; Skrtic, S.; Sharer, K.; Hill, J. M.; Dunbar, C. E.; Koretsky, A. P. MRI detection of single particles for cellular imaging. *Proc Natl Acad Sci U S A* **2004,** *101* (30), 10901–10906.
9. Park, J.; Lee, E.; Hwang, N. M.; Kang, M.; Kim, S. C.; Hwang, Y.; Park, J. G.; Noh, H. J.; Kim, J. Y.; Park, J. H.; Hyeon, T. One-nanometer-scale size-controlled synthesis of monodisperse magnetic iron oxide nanoparticles. *Angew Chem Int Ed Engl* **2005,** *44* (19), 2873–2877.
10. Zhao, Z.; Zhou, Z.; Bao, J.; Wang, Z.; Hu, J.; Chi, X.; Ni, K.; Wang, R.; Chen, X.; Chen, Z.; Gao, J. Octapod iron oxide nanoparticles as high-performance T(2) contrast agents for magnetic resonance imaging. *Nat Commun* **2013,** *4*, 2266.
11. Fedorov, P. P.; Luginina, A. A.; Kuznetsov, S. V.; Osiko, V. V. Nanofluorides. *J Fluor Chem* **2011,** *132* (12), 1012–1039.
12. Wu, S.; Han, G.; Milliron, D. J.; Aloni, S.; Altoe, V.; Talapin, D. V.; Cohen, B. E.; Schuck, P. J. Non-blinking and photostable upconverted luminescence from single lanthanide-doped nanocrystals. *Proc Natl Acad Sci U S A* **2009,** *106* (27), 10917–10921.
13. Maurizio, S. L.; Tessitore, G.; Krämer, K. W.; Capobianco, J. A. BaYF5:Yb3+,Tm3+ Upconverting nanoparticles with improved population of the visible and near-infrared emitting states: implications for bioimaging. *ACS Appl Nano Mater* **2021,** *4* (5), 5301–5308.

14. Zhang, Y.; Yu, Z.; Li, J.; Ao, Y.; Xue, J.; Zeng, Z.; Yang, X.; Tan, T. T. Ultrasmall-superbright neodymium-upconversion nanoparticles via energy migration manipulation and lattice modification: 808 nm-activated drug release. *ACS Nano* **2017,** *11* (3), 2846–2857.

15. Zhao, J.; Gao, J.; Xue, W.; Di, Z.; Xing, H.; Lu, Y.; Li, L. Upconversion luminescence-activated DNA nanodevice for ATP sensing in living cells. *J Am Chem Soc* **2018,** *140* (2), 578–581.

16. Prechtl, M. H.; Teltewskoi, M.; Dimitrov, A.; Kemnitz, E.; Braun, T. Catalytic C-H bond activation at nanoscale Lewis acidic aluminium fluorides: H/D exchange reactions at aromatic and aliphatic hydrocarbons. *Chem Eur J* **2011,** *17* (51), 14385–14388.

17. Li, X.; Zhang, F.; Zhao, D. Lab on upconversion nanoparticles: optical properties and applications engineering via designed nanostructure. *Chem Soc Rev* **2015,** *44* (6), 1346–1378.

18. Zhou, B.; Shi, B.; Jin, D.; Liu, X. Controlling upconversion nanocrystals for emerging applications. *Nat Nanotechnol* **2015,** *10* (11), 924–936.

19. Ashur, I.; Allouche-Arnon, H.; Bar-Shir, A. Calcium fluoride nanocrystals: tracers for *in vivo* ^{19}F magnetic resonance imaging. *Angew Chem Int Ed Engl* **2018,** *57* (25), 7478–7482.

20. Mashiach, R.; Cohen, D.; Avram, L.; Harris, T.; Pinkas, I.; Houben, L.; Allouche-Arnon, H.; Bar-Shir, A. Inducing defects in ^{19}F-nanocrystals provides paramagnetic-free relaxation enhancement for improved *in vivo* hotspot MRI. *Nano Lett* **2020,** *20* (10), 7207–7212.

21. Mashiach, R.; Weissman, H.; Avram, L.; Houben, L.; Brontvein, O.; Lavie, A.; Arunachalam, V.; Leskes, M.; Rybtchinski, B.; Bar-Shir, A. *In situ* NMR reveals real-time nanocrystal growth evolution via monomer-attachment or particle-coalescence. *Nat Commun* **2021,** *12* (1), 229.

22. Cohen, D.; Mashiach, R.; Houben, L.; Galisova, A.; Addadi, Y.; Kain, D.; Lubart, A.; Blinder, P.; Allouche-Arnon, H.; Bar-Shir, A. Glyconanofluorides as immunotracers with a tunable core composition for sensitive hotspot magnetic resonance imaging of inflammatory activity. *ACS Nano* **2021,** *15* (4), 7563–7574.

23. Mashiach, R.; Weissman, H.; Avram, L.; Houben, L.; Diskin-Posner, Y.; Arunachalam, V.; Leskes, M.; Rybtchinski, B.; Bar-Shir, A. Cation-ligand complexation mediates the temporal evolution of colloidal fluoride nanocrystals through transient aggregation. *Nano Lett* **2021,** *21* (23), 9916-9921.

24. Sadoc, A.; Body, M.; Legein, C.; Biswal, M.; Fayon, F.; Rocquefelte, X.; Boucher, F. NMR parameters in alkali, alkaline earth and rare earth

fluorides from first principle calculations. *Phys Chem Chem Phys* **2011,** *13* (41), 18539–18550.

25. Tirotta, I.; Mastropietro, A.; Cordiglieri, C.; Gazzera, L.; Baggi, F.; Baselli, G.; Bruzzone, M. G.; Zucca, I.; Cavallo, G.; Terraneo, G.; Baldelli Bombelli, F.; Metrangolo, P.; Resnati, G. A superfluorinated molecular probe for highly sensitive in vivo ^{19}F-MRI. *J Am Chem Soc* **2014,** *136* (24), 8524–8527.

26. Ahrens, E. T.; Flores, R.; Xu, H.; Morel, P. A. *In vivo* imaging platform for tracking immunotherapeutic cells. *Nat Biotechnol* **2005,** *23* (8), 983–987.

27. Lo, A. Y.; Sudarsan, V.; Sivakumar, S.; van Veggel, F.; Schurko, R. W. Multinuclear solid-state NMR spectroscopy of doped lanthanum fluoride nanoparticles. *J Am Chem Soc* **2007,** *129* (15), 4687–4700.

28. Goette, M. J.; Keupp, J.; Rahmer, J.; Lanza, G. M.; Wickline, S. A.; Caruthers, S. D. Balanced UTE-SSFP for ^{19}F MR imaging of complex spectra. *Magn Reson Med* **2015,** *74* (2), 537–543.

29. Schoormans, J.; Calcagno, C.; Daal, M. R. R.; Wust, R. C. I.; Faries, C.; Maier, A.; Teunissen, A. J. P.; Naidu, S.; Sanchez-Gaytan, B. L.; Nederveen, A. J.; Fayad, Z. A.; Mulder, W. J. M.; Coolen, B. F.; Strijkers, G. J. An iterative sparse deconvolution method for simultaneous multicolor (19) F-MRI of multiple contrast agents. *Magn Reson Med* **2020,** *83* (1), 228–239.

30. Weiger, M.; Pruessmann, K. P.; Hennel, F. MRI with zero echo time: hard versus sweep pulse excitation. *Magn Reson Med* **2011,** *66* (2), 379–389.

31. Robson, M. D.; Gatehouse, P. D.; Bydder, M.; Bydder, G. M. Magnetic resonance: an introduction to ultrashort TE (UTE) imaging. *J Comput Assist Tomogr* **2003,** *27* (6), 825–846.

32. Kislukhin, A. A.; Xu, H.; Adams, S. R.; Narsinh, K. H.; Tsien, R. Y.; Ahrens, E. T. Paramagnetic fluorinated nanoemulsions for sensitive cellular fluorine-19 magnetic resonance imaging. *Nat Mater* **2016,** *15* (6), 662–668.

33. Prinz, C.; Starke, L.; Ramspoth, T. F.; Kerkering, J.; Martos Riano, V.; Paul, J.; Neuenschwander, M.; Oder, A.; Radetzki, S.; Adelhoefer, S.; Ramos Delgado, P.; Aravina, M.; Millward, J. M.; Fillmer, A.; Paul, F.; Siffrin, V.; von Kries, J. P.; Niendorf, T.; Nazare, M.; Waiczies, S. Pentafluorosulfanyl (SF_5) as a superior ^{19}F magnetic resonance reporter group: signal detection and biological activity of teriflunomide derivatives. *ACS Sens* **2021,** *6* (11), 3948–3956.

34. Lysiak, R. J.; Mahendroo, P. P. Nuclear magnetic resonance studies of pure and samarium-doped CaF_2 crystals. *J Chem Phys* **1966,** *44* (10), 4025–4029.

35. Cho, H.; Cappellaro, P.; Cory, D. G.; Ramanathan, C. Decay of highly correlated spin states in a dipolar-coupled solid: NMR study of CaF_2. *Phys Rev B* **2006,** *74* (22), 224434.

36. Li, W.; Zhang, Q.; Joos, J. J.; Smet, P. F.; Schmedt Auf der Gunne, J. Blind spheres of paramagnetic dopants in solid state NMR. *Phys Chem Chem Phys* **2019,** *21* (19), 10185–10194.

37. Dempah, K. E.; Lubach, J. W.; Munson, E. J. Characterization of the particle size and polydispersity of dicumarol using solid-state NMR spectroscopy. *Mol Pharm* **2017,** *14* (3), 856–865.

38. Ruprecht, B.; Wilkening, M.; Steuernagel, S.; Heitjans, P. Anion diffusivity in highly conductive nanocrystalline BaF_2: CaF_2 composites prepared by high-energy ball milling. *J Mater Chem* **2008,** *18* (44), 5412–5416.

39. Scholz, G.; Dörfel, I.; Heidemann, D.; Feist, M.; Stösser, R. Nanocrystalline CaF_2 particles obtained by high-energy ball milling. *J Solid State Chem* **2006,** *179* (4), 1119–1128.

40. Abdellatief, M.; Abele, M.; Leoni, M.; Scardi, P. Combined X-ray diffraction and solid-state 19F magic angle spinning NMR analysis of lattice defects in nanocrystalline CaF_2. *J Appl Crystallogr* **2013,** *46* (4), 1049–1057.

41. Heitjans, P.; Indris, S. Fast diffusion in nanocrystalline ceramics prepared by ball milling. *J Mater Chem* **2004,** *39* (16–17), 5091–5096.

42. Abdellatief, M.; Abele, M.; Leoni, M.; Scardi, P. Solid state nuclear magnetic resonance and X-ray diffraction line profile analysis of heavily deformed fluorite. *Thin Solid Films* **2013,** *530*, 44–48.

43. Gebauer, D.; Cölfen, H.; Verch, A.; Antonietti, M. The multiple roles of additives in $CaCO_3$ crystallization: a quantitative case study. *Adv Mater* **2009,** *21* (4), 435–439.

44. Xu, R.; Meng, F.; Liu, Y.; Duosiken, D.; Sun, K.; Pan, S.; Tao, K. Lattice distortion of CaF_2 nanocrystals for shortening their ^{19}F longitude relaxation time. *Chem Commun* **2021,** *57* (72), 9148–9151.

45. Bloembergen, N.; Purcell, E. M.; Pound, R. V. Relaxation effects in nuclear magnetic resonance absorption. *Phys Rev* **1948,** *73* (7), 679–712.

46. Caravan, P.; Ellison, J. J.; McMurry, T. J.; Lauffer, R. B. Gadolinium(III) chelates as MRI contrast agents: structure, dynamics, and applications. *Chem Rev* **1999,** *99* (9), 2293–2352.

47. Wahsner, J.; Gale, E. M.; Rodriguez-Rodriguez, A.; Caravan, P. Chemistry of MRI contrast agents: current challenges and new frontiers. *Chem Rev* **2019,** *119* (2), 957–1057.

48. Neubauer, A. M.; Myerson, J.; Caruthers, S. D.; Hockett, F. D.; Winter, P. M.; Chen, J.; Gaffney, P. J.; Robertson, J. D.; Lanza, G. M.; Wickline, S. A. Gadolinium-modulated ^{19}F signals from perfluorocarbon nanoparticles as a new strategy for molecular imaging. *Magn Reson Med* **2008,** *60* (5), 1066–1072.

49. Schmid, F.; Holtke, C.; Parker, D.; Faber, C. Boosting ^{19}F-MRI-SNR efficient detection of paramagnetic contrast agents using ultrafast sequences. *Magn Reson Med* **2013,** *69* (4), 1056–1062.

50. de Vries, A.; Moonen, R.; Yildirim, M.; Langereis, S.; Lamerichs, R.; Pikkemaat, J. A.; Baroni, S.; Terreno, E.; Nicolay, K.; Strijkers, G. J.; Grull, H. Relaxometric studies of gadolinium-functionalized perfluorocarbon nanoparticles for MR imaging. *Contrast Media Mol Imaging* **2014,** *9* (1), 83–91.

51. Jahromi, A. H.; Wang, C.; Adams, S. R.; Zhu, W.; Narsinh, K.; Xu, H.; Gray, D. L.; Tsien, R. Y.; Ahrens, E. T. Fluorous-soluble metal chelate for sensitive fluorine-19 magnetic resonance imaging nanoemulsion probes. *ACS Nano* **2019,** *13* (1), 143–151.

52. Peng, Q.; Li, Y.; Bo, S.; Yuan, Y.; Yang, Z.; Chen, S.; Zhou, X.; Jiang, Z. X. Paramagnetic nanoemulsions with unified signals for sensitive ^{19}F-MRI cell tracking. *Chem Commun* **2018,** *54* (47), 6000–6003.

53. Chen, D.; Wang, Y. Impurity doping: a novel strategy for controllable synthesis of functional lanthanide nanomaterials. *Nanoscale* **2013,** *5* (11), 4621–4637.

54. Chen, D.; Yu, Y.; Huang, F.; Huang, P.; Yang, A.; Wang, Y. Modifying the size and shape of monodisperse bifunctional alkaline-earth fluoride nanocrystals through lanthanide doping. *J Am Chem Soc* **2010,** *132* (29), 9976–9978.

55. Viswanathan, S.; Kovacs, Z.; Green, K. N.; Ratnakar, S. J.; Sherry, A. D. Alternatives to gadolinium-based metal chelates for magnetic resonance imaging. *Chem Rev* **2010,** *110* (5), 2960–3018.

56. Pintacuda, G.; John, M.; Su, X. C.; Otting, G. NMR structure determination of protein-ligand complexes by lanthanide labeling. *ACC Chem Res* **2007,** *40* (3), 206–212.

57. Barrientos, Á. G.; de la Fuente, J. M.; Rojas, T. C.; Fernández, A.; Penadés, S. Gold glyconanoparticles: synthetic polyvalent ligands mimicking glycocalyx-like surfaces as tools for glycobiological studies. *Chem Eur J* **2003,** *9* (9), 1909–1921.

58. Babu, P.; Sinha, S.; Surolia, A. Sugar-quantum dot conjugates for a selective and sensitive detection of lectins. *Bioconjug Chem* **2007,** *18* (1), 146–151.

59. El-Boubbou, K.; Zhu, D. C.; Vasileiou, C.; Borhan, B.; Prosperi, D.; Li, W.; Huang, X. Magnetic glyco-nanoparticles: a tool to detect, differentiate, and unlock the glyco-codes of cancer via magnetic resonance imaging. *J Am Chem Soc* **2010,** *132* (12), 4490–4499.

60. Garcia, I.; Gallo, J.; Genicio, N.; Padro, D.; Penades, S. Magnetic glyconanoparticles as a versatile platform for selective immunolabeling and imaging of cells. *Bioconjug Chem* **2011,** *22* (2), 264–273.

61. Johannssen, T.; Lepenies, B. Glycan-based cell targeting to modulate immune responses. *Trends Biotechnol* **2017,** *35* (4), 334–346.

62. Yu, M.; Yang, Y.; Han, R.; Zheng, Q.; Wang, L.; Hong, Y.; Li, Z.; Sha, Y. Polyvalent lactose–quantum dot conjugate for fluorescent labeling of live leukocytes. *Langmuir* **2010,** *26* (11), 8534–8539.

63. van Kasteren, S. I.; Campbell, S. J.; Serres, S.; Anthony, D. C.; Sibson, N. R.; Davis, B. G. Glyconanoparticles allow pre-symptomatic in vivo imaging of brain disease. *Proc Natl Acad Sci U S A* **2009,** *106* (1), 18–23.

64. Sur, I.; Cam, D.; Kahraman, M.; Baysal, A.; Culha, M. Interaction of multi-functional silver nanoparticles with living cells. *Nanotechnol* **2010,** *21* (17), 175104.

65. García, I.; Sánchez-Iglesias, A.; Henriksen-Lacey, M.; Grzelczak, M.; Penadés, S.; Liz-Marzán, L. M. Glycans as biofunctional ligands for gold nanorods: stability and targeting in protein-rich media. *J Am Chem Soc* **2015,** *137* (10), 3686–3692.

66. Chabre, Y. M.; Roy, R. Chapter 6 - Design and creativity in synthesis of multivalent neoglycoconjugates, in *Advances in Carbohydrate Chemistry and Biochemistry* (Horton, D., ed.), Academic Press, 2010; Vol. 63, pp. 165–393.

67. Avci, F. Y.; Li, X.; Tsuji, M.; Kasper, D. L. Carbohydrates and T cells: a sweet twosome. *Semin Immunol* **2013,** *25* (2), 146–151.

68. Schilling, M.; Besselmann, M.; Leonhard, C.; Mueller, M.; Ringelstein, E. B.; Kiefer, R. Microglial activation precedes and predominates over macrophage infiltration in transient focal cerebral ischemia: a study in green fluorescent protein transgenic bone marrow chimeric mice. *Exp Neurol* **2003,** *183* (1), 25–33.
69. Hamzei Taj, S.; Kho, W.; Aswendt, M.; Collmann, F. M.; Green, C.; Adamczak, J.; Tennstaedt, A.; Hoehn, M. Dynamic modulation of microglia/macrophage polarization by miR-124 after focal cerebral ischemia. *J Neuroimmune Pharmacol* **2016,** *11* (4), 733–748.
70. Wang, M.; Iliff, J. J.; Liao, Y.; Chen, M. J.; Shinseki, M. S.; Venkataraman, A.; Cheung, J.; Wang, W.; Nedergaard, M. Cognitive deficits and delayed neuronal loss in a mouse model of multiple microinfarcts. *J Neurosci* **2012,** *32* (50), 17948–17960.
71. Jones, K. A.; Maltby, S.; Plank, M. W.; Kluge, M.; Nilsson, M.; Foster, P. S.; Walker, F. R. Peripheral immune cells infiltrate into sites of secondary neurodegeneration after ischemic stroke. *Brain Behav Immun* **2018,** *67*, 299–307.
72. Kanazawa, M.; Ninomiya, I.; Hatakeyama, M.; Takahashi, T.; Shimohata, T. Microglia and monocytes/macrophages polarization reveal novel therapeutic mechanism against stroke. *Int J Mol Sci* **2017,** *18* (10).
73. Akazawa, K.; Sugihara, F.; Nakamura, T.; Matsushita, H.; Mukai, H.; Akimoto, R.; Minoshima, M.; Mizukami, S.; Kikuchi, K. Perfluorocarbon-based ^{19}F-MRI nanoprobes for in vivo multicolor imaging. *Angew Chem Int Ed Engl* **2018,** *57* (51), 16742–16747.
74. Chirizzi, C.; De Battista, D.; Tirotta, I.; Metrangolo, P.; Comi, G.; Bombelli, F. B.; Chaabane, L. Multispectral MRI with dual fluorinated probes to track mononuclear cell activity in mice. *Radiology* **2019,** *291* (2), 351–357.
75. Lim, Y. T.; Noh, Y. W.; Cho, J. H.; Han, J. H.; Choi, B. S.; Kwon, J.; Hong, K. S.; Gokarna, A.; Cho, Y. H.; Chung, B. H. Multiplexed imaging of therapeutic cells with multispectrally encoded magnetofluorescent nanocomposite emulsions. *J Am Chem Soc* **2009,** *131* (47), 17145–17154.
76. Partlow, K. C.; Chen, J.; Brant, J. A.; Neubauer, A. M.; Meyerrose, T. E.; Creer, M. H.; Nolta, J. A.; Caruthers, S. D.; Lanza, G. M.; Wickline, S. A. ^{19}F magnetic resonance imaging for stem/progenitor cell tracking with multiple unique perfluorocarbon nanobeacons. *FASEB J* **2007,** *21* (8), 1647–1654.
77. Tanifum, E. A.; Patel, C.; Liaw, M. E.; Pautler, R. G.; Annapragada, A. V. Hydrophilic fluorinated molecules for spectral ^{19}F-MRI. *Sci Rep* **2018,** *8* (1), 2889.

78. Weise, G.; Basse-Lusebrink, T. C.; Kleinschnitz, C.; Kampf, T.; Jakob, P. M.; Stoll, G. In vivo imaging of stepwise vessel occlusion in cerebral photothrombosis of mice by ^{19}F-MRI. *PLoS One* **2011,** *6* (12), e28143.

79. Flogel, U.; Temme, S.; Jacoby, C.; Oerther, T.; Keul, P.; Flocke, V.; Wang, X.; Bonner, F.; Nienhaus, F.; Peter, K.; Schrader, J.; Grandoch, M.; Kelm, M.; Levkau, B. Multi-targeted ^{1}H/^{19}F-MRI unmasks specific danger patterns for emerging cardiovascular disorders. *Nat Commun* **2021,** *12* (1), 5847.

Chapter 3

Multibranched Superfluorinated Molecular Probes for ^{19}F MRI

Cristina Chirizzi, Valentina Dichiarante, Pierangelo Metrangolo and Francesca Baldelli Bombelli
Laboratory of Supramolecular and Bio-Nanomaterials (SupraBioNanoLab), Department of Chemistry, Materials, and Chemical Engineering "Giulio Natta", Politecnico di Milano, Via Luigi Mancinelli 7, 20131 Milan, Italy
francesca.baldelli@polimi.it

3.1 Introduction

^{19}F MRI has been emerging as promising whole-body imaging technique complementary to ^{1}H MRI. In fact, ^{19}F has 100% natural abundance, a gyromagnetic ratio very close to that of the proton and 83% sensitivity of ^{1}H. Moreover, in the human body there are only low amounts of inorganic fluorine in bones and teeth, which are not detectable by ^{19}F MRI due to their low T_2. Thus, the only *in vivo* detectable fluorine comes from exogenous tracers that can be clearly imaged and quantified by ^{19}F MRI. However, the intrinsically low sensitivity of MRI has prompted efforts to develop effective

Fluorine Magnetic Resonance Imaging: Methods and Applications in Biomedicine
Edited by Eric T. Ahrens and Ulrich Flögel
Copyright © 2025 Jenny Stanford Publishing Pte. Ltd.
ISBN 978-981-5129-37-3 (Hardcover), 978-1-003-53004-6 (eBook)
www.jennystanford.com

^{19}F MRI tracers. An ideal ^{19}F MRI tracer should have a high number of equivalent ^{19}F atoms to yield a single sharp resonance signal [1].

In this sense, the first molecules used as ^{19}F MRI tracers were perfluorocarbons (PFC) containing a high number of non-equivalent ^{19}F atoms, such as perfluoro-octyl-bromide (PFOB), perfluorodecaline (PFD) and perfluorotributylamine (PFTBA). Later, PFC bearing many equivalent ^{19}F atoms, such as perfluoro-15-crown ether (PFCE), and blends of perfluoropolyethers (PFPE) containing a huge number of ^{19}F atoms (pseudoequivalent), were also proposed as more sensitive ^{19}F MRI tracers [2, 3] (see Fig. 3.1). Despite PFC being widely used as ^{19}F MRI agents, they were not specifically designed for this function, and thus, it should be possible to produce tailored polyfluorinated molecules for sensitive^{19}F MRI.

As it has been recently shown in an interesting review from Jiang *et al.* [4], perfluoro-*tert*-butanol (PFTB) represents an excellent building block for developing customized ^{19}F MRI agents thanks to its nine equivalent ^{19}F atoms and enhanced reactivity of the –OH group. In this work the authors nicely summarized the development, characteristics and main applications of PFTB-based derivatives as ^{19}F MRI tracers. In this context, in 2014 our research group [5] proposed a biocompatible formulation of a PFTB4-dendron bearing 36 equivalent ^{19}F atoms called PERFECTA (from suPERFluorinatEdContrasT Agent, see Fig. 3.1). PERFECTA can be obtained in high yield *via* a Mitsunobu reaction between pentaerythritol and PFTB providing a crystalline white powder insoluble in water and common organic solvents. In contrast to PFOB, PFCE, and PFPE, PERFECTA is not perfluorinated, but possesses a hydrocarbon polar core and four ether bonds, which may promote molecular degradation in the environment through the cleavage of such bonds in physiological conditions, thus overcoming bioaccumulation issues. PERFECTA is also characterized by an intrinsic chemical versatility for further functionalization, differently from other fluorinated probes. In fact, its molecular structure allows the substitution of one of the four symmetrical arms with an alkyl chain ending with vary functional groups such as thiol, carboxylic acid, alcohol, azide, and amine, while maintaining 27 magnetically equivalent ^{19}F in the molecule. We believe that PERFECTA and its three-arm derivatives not only are high-performance ^{19}F MRI agents, but also represent a valuable alternative to PFC as they combine a huge ^{19}F payload with a short-chain multibranched structure that it is considered more environmentally sustainable with respect to compounds bearing perfluoroalkyl long chains (≥6 carbon atoms) [6].

In this contribution, we will first describe in detail the characteristics of all the biocompatible formulations of PERFECTA developed for ^{19}F MRI use since 2014 in terms of colloidal stability, magnetic properties and biological applications. We will also briefly report about the possibility of using polyfluorinated molecules as dual imaging multiscale agents thanks to their specific Raman C-F vibrational mode, falling in a silent region of the biological background, which renders them visible in cells and tissues by Raman microscopy [7]. In this regard, the superiority of PERFECTA with respect to poly(lactic-co-glycolic acid) (PLGA) formulated PFOB and PFCE, in terms of Raman and ^{19}F MRI sensitivities will be also shown. In a second paragraph, we will present the superfluorinated compounds, bearing the PERFECTA-derived multibranched synthon with 27 magnetically equivalent fluorine atoms, developed so far, reporting their synthesis, characterization, relaxometry features and biological applications.

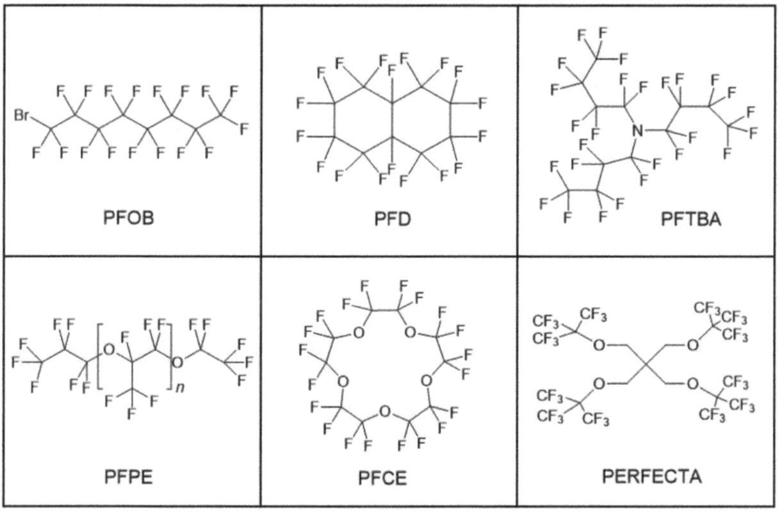

Figure 3.1 Chemical structure of perfluorooctyl bromide (PFOB), perfluorodecalin (PFD), perfluorotributylamine (PFTBA), perfluoropolyether (PFPE), perfluoro-15-crown-5-ether (PFCE) and PERFECTA.

3.2 Applications of Biocompatible Formulations of PERFECTA

As explained in the introduction, among all branched fluorinated probes based on PFTB derivatives, PERFECTA gained attention in last decades thanks to its high sustainability and excellent MRI sensitivity. Indeed, this superfluorinated solid compound can be considered an ideal ^{19}F-based contrast agent (CA) due to its easy and scalable synthesis procedure, high payload of chemically equivalent ^{19}F atoms (36/molecule, Fig. 3.1) that accounts for a single intense well-resolved ^{19}F NMR signal and enhanced sensitivity without chemical shift artifacts. Additionally, PERFECTA can be easily formulated in nanoparticle (NP) dispersions of biocompatible surfactants, which have been shown to be suitable for both *in vitro* and *in vivo* applications with optimal relaxometry properties, including a short T_1 and a long T_2 (690/202 T_1/T_2 measured at a magnetic field of 7 T) [8].

PERFECTA-based formulations typically consist of a fluorinated core that either can be coated by a lipid monolayer, further stabilized by different emulsifiers, or can be directly coated by biocompatible surfactants or encapsulated in polymer NPs. The first formulation of PERFECTA has been proposed by Tirotta *et al.* in which the fluorinated probe was emulsified into an aqueous solution containing lecithin and safflower oil [5]. Resulting emulsions showed high fluorine payload and optimal colloidal properties as shown by the NP size (of about 200 nm), which was invariant up to 4 weeks in several fluids (water, PBS and cell culture media). High PERFECTA payload (74.40 ± 1.98 mM) together with low cytoxicity enabled very efficient labelling of bone-marrow derived DCs (1.92×10^{13} ^{19}F atoms/cell) without affecting cellular viability (>90%) after only 2 h of incubation. Importantly, the PERFECTA-labelled DCs were implanted in naive Lewis rats to preliminarily assess feasibility for *in vivo* ^{19}F MRI. The results demonstrated that labelled and administered cells were detectable by acquisition times suitable for clinical applications (*i.e.*, 10 min).

Taking inspiration from the application described above, Chirizzi and colleagues proposed an alternative formulation of PERFECTA using common polymeric emulsifiers (*i.e.*, Pluronic) [8]. PERFECTA loaded-Pluronic NPs could readily be obtained with stable colloidal properties, suitable for *in vitro* applications. As for the first emulsion, this formulation showed a large ^{19}F payload which provided an excellent MR-sensitivity and thus a sensitive labelling of microglial cells. Importantly, when compared to Pluronic formulated PFCE NPs, PERFECTA demonstrated that a very a small number of cells (n = 8000) could be detected by ^{19}F MRI with a 3-fold higher label uptake. This optimized formulation was then exploited for simultaneous use with PFCE loaded-Pluronic NPs *in vivo*, taking advantage of the high chemical shift (19.3 ppm) related to the specific resonance frequencies of the two fluorinated probes. Indeed, this approach enabled monitor of the phagocytic activity of mononuclear cells by ^{19}F MRI, and in particular, the phagocytosis boost in the early phase of mononuclear cell regeneration after CSF1Ri cell depletion (Fig. 3.2).

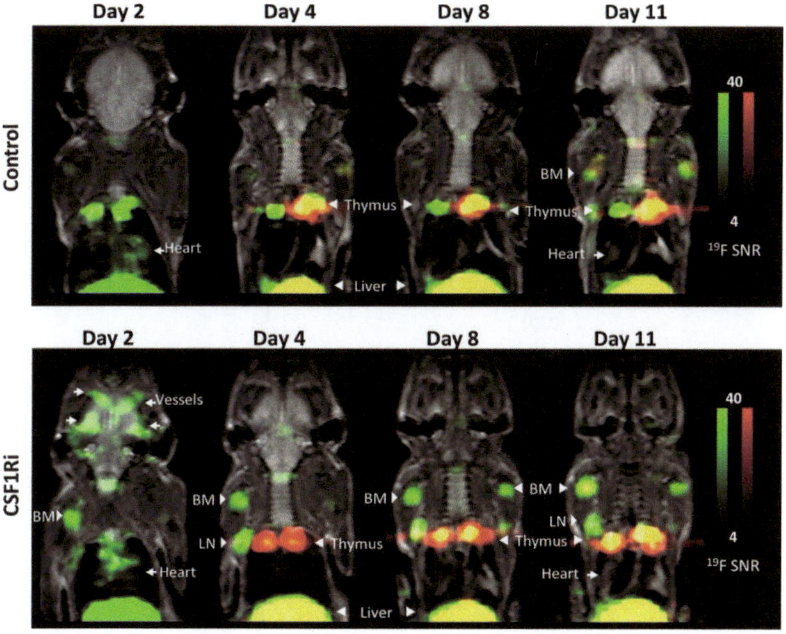

Figure 3.2 *In vivo* multicolour ^{19}F MRI acquisitions in control and mononuclear cell depleted mice at different days after interruption of the depletion diet and administration of PFCE (green) or PERFECTA (red). At the first MRI scan (day 2), 24 hours after PFCE administration, a higher uptake was detected in the bone marrow (BM) of depleted mice, while with PERFECTA administered at a later phase of mononuclear cells regeneration, the ^{19}F-signal was similar to control mice. Fluorine signal in the BM remained unchanged after the second administration of both tracers (day 9 after depletion). Reference [8] contains additional details.

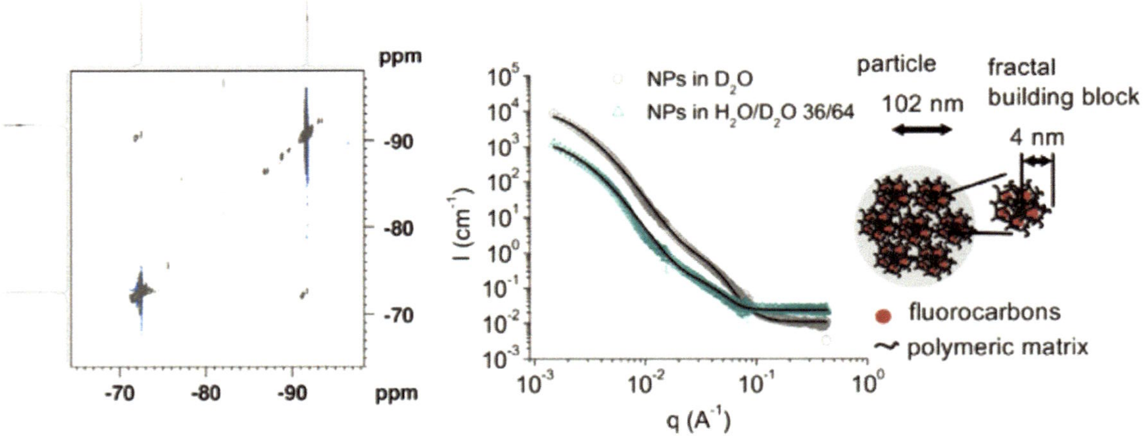

Figure 3.3 NOESY experiments in NPs. (Left) ^{19}F-^{19}F NOESY measurements of PFCE (−92 ppm) and PERFECTA (−73 ppm). The cross-peak between both probes demonstrated that PFCE and PERFECTA are close to each other. (Right) SANS scattering patterns of PERFECTA-PFCE PLGA-NPs in solution. SANS demonstrated a fractal-like structure of PERFECTA-PFCE PLGA-NPs with spherical building blocks that contain PFCE, PERFECTA, and polymer. Adapted from Ref. [9].

The simultaneous use of PFCE and PERFECTA was also studied by Koshkina and colleagues, which developed "two-colour" polymeric NPs [9]. In this case, co-encapsulation of the two fluorinated probes seemed to be essential for guaranteeing a sufficient PERFECTA payload into the NP core. The resulting PLGA-PERFECTA/PFCE NPs were characterized to obtain both colloidal and morphological information. Results showed that these NPs were colloidally stable with an average size of 145 nm. Small angle neutron scattering experiments of these dispersions elucidated a local structure suggestive of a fractal-like assembly of spherical building blocks, each of them containing both ^{19}F probes and PLGA, as also confirmed by ^{19}F-^{19}F Nuclear Overhauser Enhancement Spectroscopy (NOESY) experiments (Fig. 3.3). The NP uptake and ^{19}F MRI sensitivity after labelling DCs cells was also assessed, and displayed high cell viability and excellent ^{19}F MRI response compared to pure PFCE PLGA NPs.

In addition to the previously described applications, our group recently developed an innovative and sustainable system to efficiently disperse PERFECTA in aqueous solutions using the film-forming protein hydrophobin (HFBII) [10]. This method represents the first evidence of doping a solid fluorinated NP (FNP) through a self-assembled layer of a naturally amphiphilic protein (Fig. 3.4). The resulting biocompatible FNPs showed a strong colloidal stability, also in the biological environment. Indeed, the monodisperse population of 176 nm and the low PDI (0.15) were stable and constant over one week. Importantly, HFBII-FNPs dispersed in both human serum and plasma maintained their native protein coating with the formation of a limited protein corona, suggesting a stealth effect. Moreover, a high PERFECTA payload ($8.9 \pm 0.8 \times 10^{19}$ ^{19}F/ml) and optimal relaxometry properties ($T_1/T_2 = 510/236$) of the agent are well-suited for cellular labelling applications. HFBII-FNPs were successfully used for microglial cell labelling and showed 10% uptake efficiency after 1 h of incubation and, the maximum uptake was reached after 6 h (1.72×10^{12} atoms per cell).

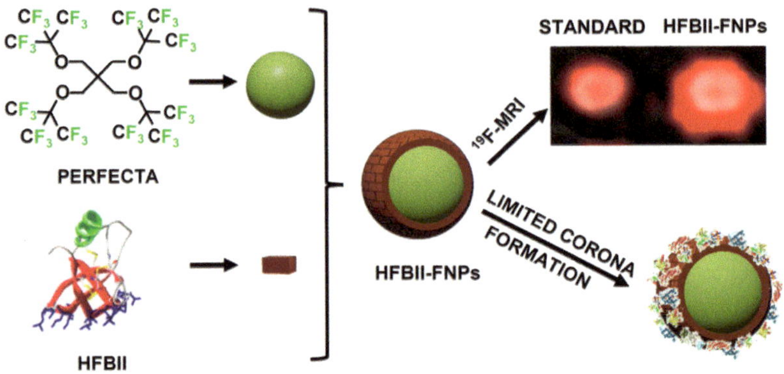

Figure 3.4 Representation of PERFECTA NPs stabilized by a HFBII layer (left, HFBII-FNPs) and (right) subsequent exploration of their ^{19}F MRI sensitivity and interfacial interactions with proteins in biological fluid. Source is Ref. [10].

Additionally, Jamgotchian and colleagues proposed biocompatible micelles loaded with PERFECTA for targeting solid tumors [11]. The micelle product, of approximately 20 nm in size, were composed of an amphiphilic di-block-co-polymer with a perfluorinated hydrophobic domain and a hydrophilic poly(ethylene glycol) polar headgroups. The fluorinated chain promotes high PERFECTA encapsulation efficiency (41 wt.%), while pegylation imparts stealth properties ensuring colloidal stability in the biological environment. *In vitro* validation of the obtained micelles in cancer HEK293 and MC38 cell lines showed significant uptake without affecting cell viability, even at a high concentration (1 mM). Importantly, PERFECTA-PFTD-PEG demonstrated excellent *in vivo* tumor homing ability, where 3.6% of the injected fluorinated probe accumulated in the cancerous tissues, as detected by ^{19}F MRI (Fig. 3.5).

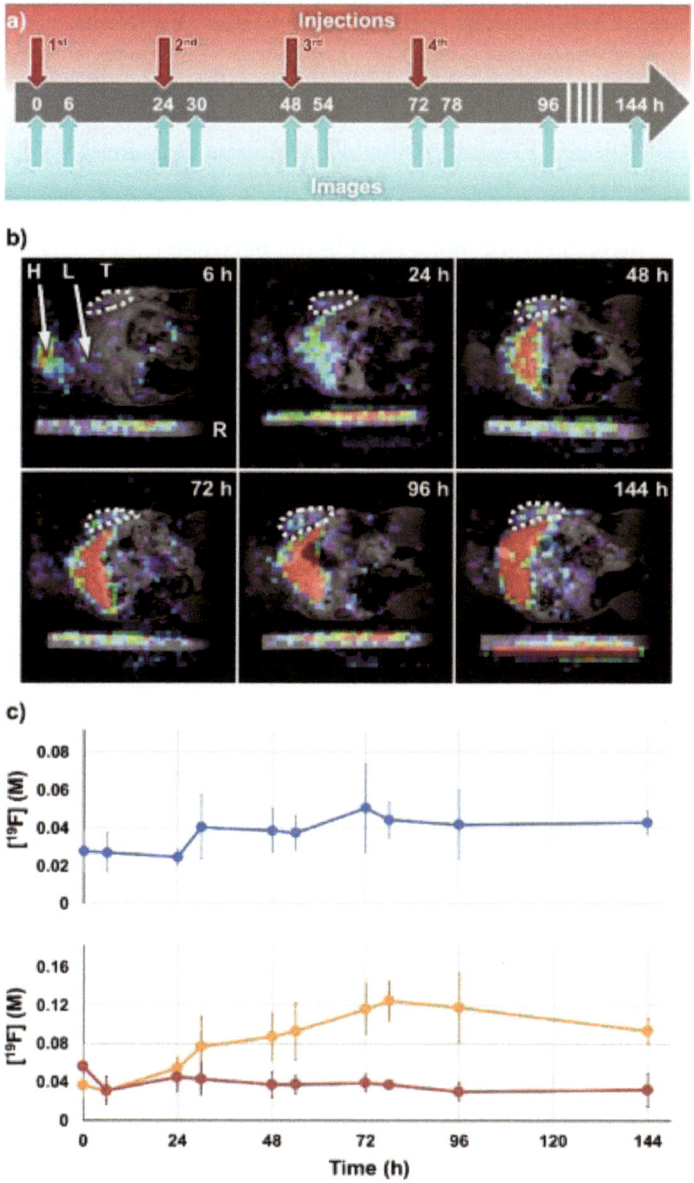

Figure 3.5 PERFECTA tumor studies in mice. (a) Description of the *in vivo* imaging protocol used in tumor-bearing mice after PERFECTA-PFTD-PEG micelles intravenous injection; (b) Combined ^{1}H and ^{19}F MRI images of diseased mice at specific times after micelles administration (T = tumor; L = liver; H = heart; R = reference tube. (c) Quantification of ^{19}F amount in the tumor site (blue) in comparison to the liver (orange), and to the heart (red). Source is Ref. [11].

3.2.1 PERFECTA as Multimodal Imaging Probe

More recently, PERFECTA proved to work as a biocompatible multimodal imaging probe able to provide significant information at different length scales with high specificity and spatial resolution. Indeed, its intrinsic molecular features rendered it active in both ^{19}F MRI (whole body scale) and Raman microscopy (tissue/cell scale) and allowed a direct signal correlation between the two techniques using the same tracer [7]. Until now, ^{19}F MRI has been often combined with fluorescence imaging, enabling analysis at cellular level. An alternative imaging approach suitable for cell/tissue detection is based on the probe's intrinsic C-F Raman vibrational modes. In this way, Raman microscopy allows the direct intracellular detection of the fluorinated molecule overcoming possible drawbacks of the use of fluorescent dyes co-encapsulated with the ^{19}F probe such as dissociation from the ^{19}F NPs, and fluorescence quenching. In this context, PERFECTA exhibits a specific Raman signature that is distinguishable from other standard ^{19}F probes, such as PFCE and PFOB. Additionally, the characteristic Raman signal of PERFECTA falls in a region that is well separated from resident biomolecules and it can be clearly detected in the cytoplasm of labelled cells. PERFECTA NPs administrated to well-established autoimmune allergic encephalitis (EAE) mouse models were detected by ^{19}F MRI mostly in the liver, but also along the spinal cord as small hot spots, where leukocytes end up in this model. After ^{19}F MRI analysis, the mice were sacrificed, and sections of spinal cord were directly analysed by Raman microscopy. PERFECTA Raman signals were observed in the analysed tissues, thus validating *in vivo* MRI observations. Additionally, Raman spectroscopy can be used to detect chemical fingerprints of biological molecules of the tissue where the fluorinated probe was localized. Overall, the combination of *in vivo* ^{19}F MRI and *ex vivo* Raman microscopy enables PERFECTA biodistribution detection in the whole body, direct detection at tissue level, as well as additional information on a tissue's biomolecular composition. It was observed that PERFECTA accumulated in regions with reduced lipid content and high protein-related signals (demyelinated white matter tracts) in agreement with this inflammation model (Fig. 3.6).

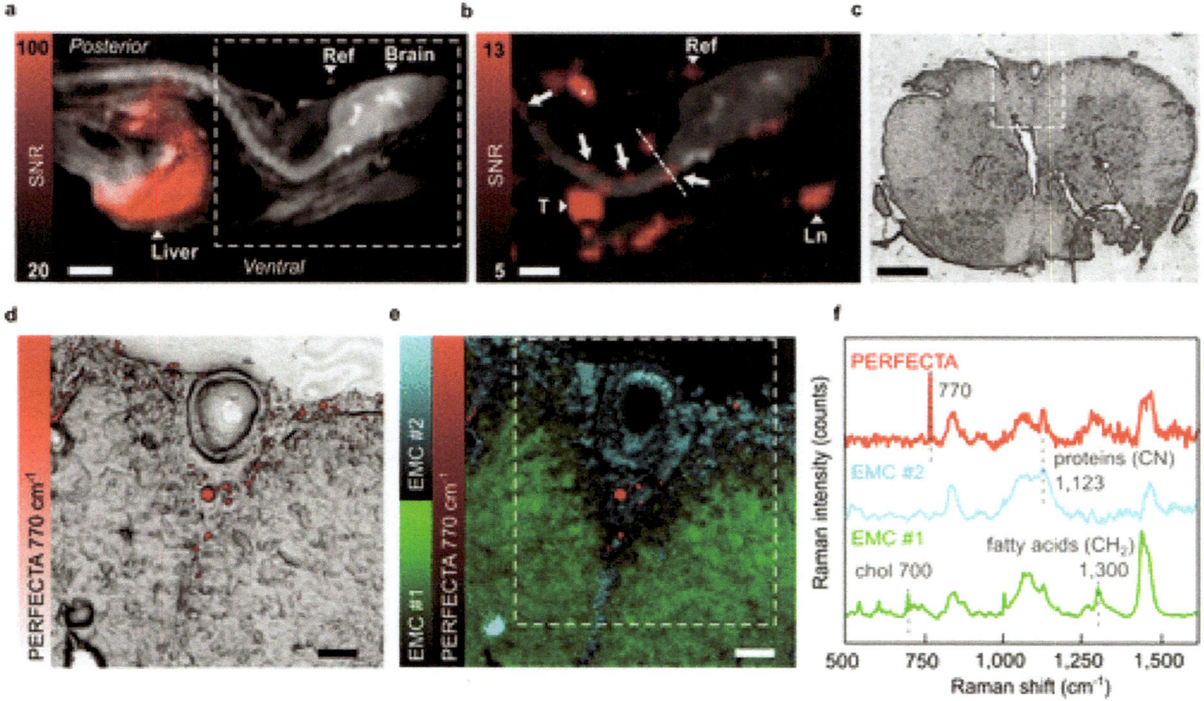

Figure 3.6 PERFECTA imaging in EAE model. (a) Acquisition of whole body *in vivo* ^{19}F MRI of PERFECTA (red) in mice and merged with the anatomical image. (b) PERFECTA signal (red spot) was detected along the spinal cord (white arrows) where Raman imaging was successively performed. (c) Bright field (BF) image of the unstained tissue section collected from mouse spinal cord and analysed by Raman imaging (white dashed square). (d) BF image of the selected biological regions merged with the Raman PERFECTA map. (e) Merging of Raman maps related to empty modeling component (EMC) #1 and #2 and to PERFECTA. (f) Representative Raman spectra obtained by *ex vivo* Raman imaging. EMC #1: white matter (green) (chol: cholesterol); EMC #2: epithelium and immune cell infiltration (cyan). Source is Ref. [7].

3.3 Superfluorinated Short-Chain Multibranched F_{27}-Probes

Further refinements of PERFECTA were explored *via* a series of studies aimed at the modification of its molecular structure, by selectively replacing one of the four fluorinated arms with a different reactive group. This synthetic approach resulted in several derivatives, each bearing 27 equivalent fluorine atoms (F_{27}-probes), that are useful building blocks for further functionalization and conjugation steps. The starting point in this direction was provided by Yu and collaborators, who were able to synthesize a family of branched fluorinated compounds and amphiphiles (F_{27}-FG, Fig. 3.7a), exploiting the Mitsunobu reaction as key step for introducing three perfluoro-*tert*-butoxyl groups on the pentaerythritol scaffold [12]. Thanks to the symmetrical arrangement of all 27 fluorine atoms, each of the reported substrates gave one sharp singlet signal in the ^{19}F NMR spectrum, and thus appeared highly promising as an MRI probe. The same authors applied this route to the synthesis of a bispherical F_{27}-substituted molecule, named ^{19}FIT (Fig. 3.7b), which was water-soluble (CMC ~ 7 mM in PBS), and interestingly, showed an *in vivo* residence half-life of about 0.5 day in mice, with no evidence of organ retention or *in vivo* degradation [13]. In a subsequent study to optimize the properties of ^{19}FIT, the authors developed a divergent fluorous mixture synthesis of asymmetric fluorinated dendrimers, with an increasing number of hydrophilic branches [14]. Four generations of dendrimers bearing the same F_{27} moiety were prepared in high yields and purities, with both longitudinal and transverse relaxation times decreasing from G0 to G3; in solutions of deuterated methanol at 25 °C and 25 mM, T_1 values ranging from 1.1 s to 741 ms were measured, and T_2 decreased from 878 ms to 500 ms, while increasing hydrophilicity. The exponential growth in the number of branches, without changing their length, eventually led to steric congestion and defects in the dendrimers; proportionate branching was proposed as alternative strategy. According to this method, both number and length of branches exponentially increase but in opposite directions. Although a monodisperse fluoro-dendron with 243 equivalent fluorine atoms was effectively obtained by this way, no results concerning its imaging performance were reported [15].

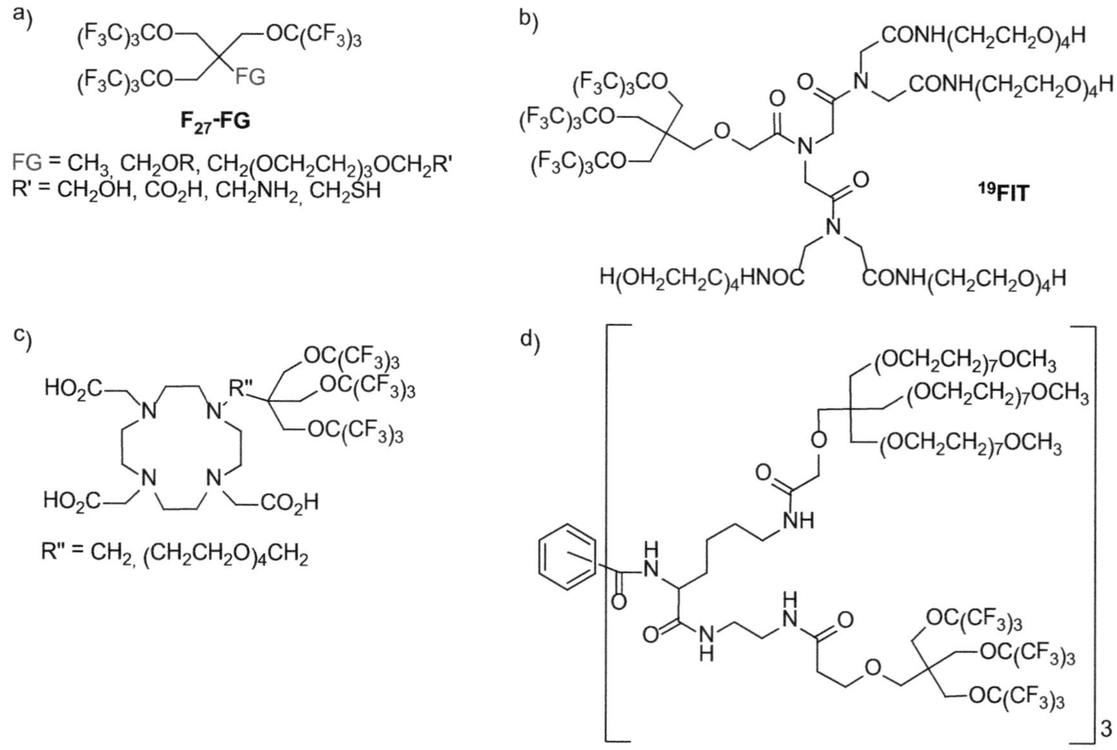

Figure 3.7 Molecular structures of selected branched fluorinated derivatives of pentaerythritol. (a) First series of F_{27}-functionalized compounds reported by Yu *et al.* [12]. (b) ^{19}FIT. (c) F_{27}-derivatives of DOTA. (d) $(F_{27})_3$-amphiphilic contrast agent for liposomal drug tracking.

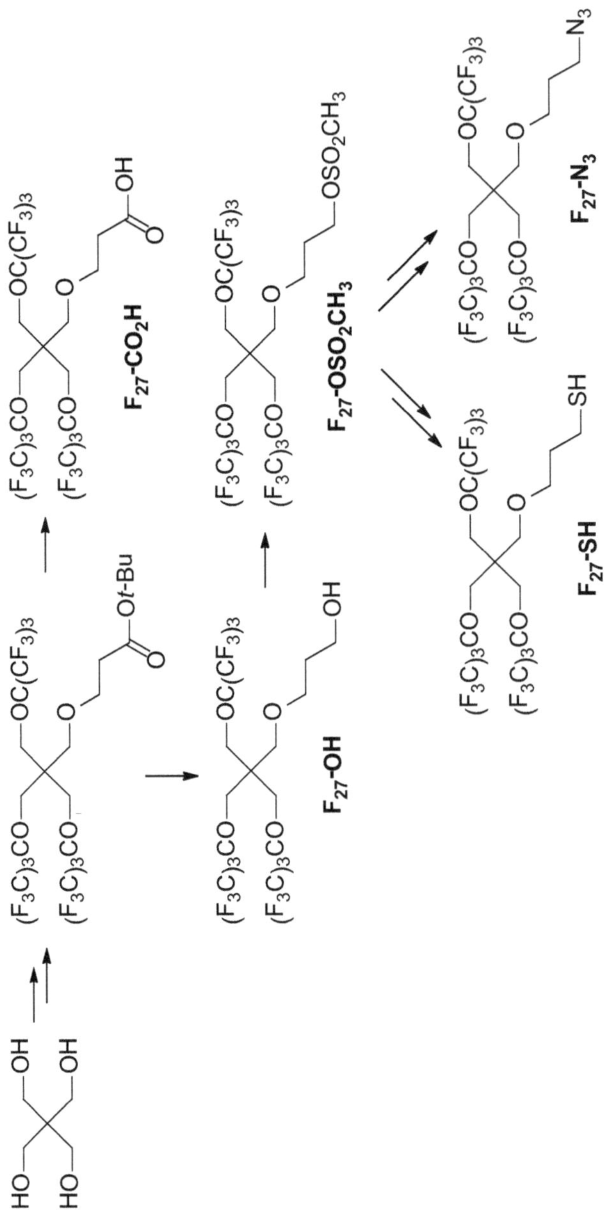

Scheme 3.1 Synthetic pathways for obtaining F27-probes with different functional groups.

Taking inspiration from Yu's achievements, our group recently focused on the synthesis of a series of F_{27}-derivatives of pentaerythritol with different functional groups, to broaden the set of possible derivatization paths. As detailed in Scheme 3.1, we first optimized the synthesis of thiol F_{27}-SH. Due to the high tendency of sulfhydryl group to bind metals, F_{27}-SH was able to self-assemble on gold and copper surfaces into stable (super) hydrophobic monolayers, with a better frictional behavior if compared to commonly used long-chain perfluoroalkylthiols [16]. At the same time, it was effectively employed as a stabilizing ligand in the synthesis of atomically precise gold nanoclusters coated by a fluorinated shell of F_{27}-thiol molecules [17, 18]. Attractive F···F interactions among F_{27}-SH molecules were able to drive the crystal packing of superfluorinated 20 kDa gold nanoclusters, with an Au_{25} core coated by a shell of 18 $F_{27}SH$ molecules resulting in almost 500 fluorine atoms. In response to the oxidation state of the gold core, the $[Au_{25}(SF_{27})_{18}]$ clusters showed a switchable solubility in different fluorous media. Due to their high hydrophobicity, the use of such fully fluorine-coated particles for bioimaging purposes will require the development of an effective protocol for water dispersion.

A slight modification of this synthetic pathway allowed the synthesis of the related carboxylic acid (F_{27}-CO_2H) derivative, which was successfully esterified with cellulose [19]. The resulting innovative materials showed tunable mechanical and wettability properties and appeared extremely promising for their potential use as waterproof-breathable films or coatings in textiles. The presence of a carboxylic group might be also useful for the functionalization of different types of nanomaterials of biomedical interest, as well as for its potential fluorosurfactant properties.

The alcohol and mesylate intermediates (F_{27}-OH and F_{27}-OSO_2CH_3, respectively) were conjugated to different BODIPY dyes to create fluorescent probes with either 27 or 54 equivalent ^{19}F atoms, that used for complementary ^{19}F MR and fluorescence imaging modalities [20]. The different degree of conjugation of the BODIPY core enabled an emission into the far-red region, which is more suitable for *in vivo* imaging. Moreover, the presence of two iodine atoms on the BODIPY core provides sites for further chemical modifications and may promote efficient singlet oxygen generation for phototherapeutic purposes. However, it is also known that iodine (and generally heavy atoms) can display a strong fluorescence quenching effect. Similarly, Jung *et al.* covalently bound small, fluorinated alcohols to BODIPY dye, with no effect on its fluorescent properties [21]. As expected, all ^{19}F NMR spectra from these compounds showed a strong singlet from chemically and magnetically equivalent CF_3 groups, together with a weak quartet due to BF_2 group. Noteworthy, the F_{27}-dye was visualized in a mouse post-mortem, by both MRI and fluorescence imaging, thus showing potential for biological and medical applications.

In a very recent work, azide F_{27}-N_3 was used as building block in the synthesis of fluorinated Janus-type polyester dendrimers of different generations [22]. Increasing size, flexibility, and number of peripheral hydroxyl groups allowed for tuning of self-assembly behavior in aqueous media. Low generation dendrimers formed micelles with a high ^{19}F NMR signal-to-noise ratio and signal stability over time, whereas the highest generation derivatives displayed an interesting morphological transition from multilamellar dendrimersomes to tubules, with a subsequent reduction of CF_3 local mobility and switching-off of the NMR signal.

The covalent binding of one or more F_{27} moieties to existing molecular scaffolds was also applied to a family of 4-anilinoquinazoline analogues, which are potent and selective ATP-competitive inhibitors of the epidermal growth factor receptor tyrosine kinase, and thus represent significant molecular targets for anticancer drugs. Effective conjugation of the F_{27} group on their 6-position (65–93% yields) was reported, as well as the synthesis of the fluorinated analogues of gefitinib and erlotinib, two drugs used in the treatment of certain lung and colon cancers, in 93% and 90% yields, respectively [23].

For cell tracking applications, the low sensitivity of ^{19}F MRI can potentially be improved by paramagnetic relaxation enhancement. An interesting approach was based on the conjugation of the F_{27} group to a macrocyclic chelator able to form complexes with trivalent metal cations, namely DOTA (*viz* 1,4,7,10-tetraazacyclododecane-1,4,7,10-tetraacetate, Fig. 3.7c), which is already widely used in radionuclide therapy and ^{1}H MRI [24]. Insertion of a flexible oligo-oxyethylene linker between the fluorocarbon moiety and the macrocycle effectively modulated hydrophilicity and ^{19}F relaxation behavior of the resulting chelators, making them promising contrast agents for ^{1}H-^{19}F dual MRI.

Significant improvements in terms of stability, size controllability, biocompatibility, cellular uptake, and imaging sensitivity were achieved by Jiang *et al.* through the simultaneous use of a fluorinated Fe^{3+}-chelator and a ^{19}F MRI agent [25]. Encapsulation of a BODIPY dye in the lecithin-based emulsion of these two fluorinated probes allowed additional fluorescence imaging. In the presence of paramagnetic Fe^{3+} ions, significant T_1 and T_2 reduction was observed, and good-quality images were obtained *in vitro* at a ^{19}F concentration of 8 mM after 64 seconds of acquisition. Cytotoxicity assays on human lung adenocarcinoma cells (A549), mouse leukemic monocyte–macrophage cells (RAW264.7), and mouse fibroblast cells (L929) indicated good biocompatibility of the emulsion, and a proof-of-concept *in vivo* study of labeled RAW264.7 cells delivered to Balb/c nude mice was also carried out. In a following study, the same authors developed fluorinated theranostic emulsions with significantly enhanced *in vitro* and *in vivo* ^{129}Xe hyper-CEST MRI, ^{19}F MRI and fluorescence imaging signals, for the selective detection of integrin $\alpha_v\beta_3$ over-expressing cancer cells, and NIR-activated photodynamic therapy (PDT) [26]. The reported emulsions contained a F_{27}-dendron in the core (F_{27}-FG with FG = CH_3), fluorinated cryptophane for the capture of hyperpolarized ^{129}Xe, and a fluorescent tetrabenzylporphyrin on the surface, together with monodisperse polyethylene glycols (M-PEGs) as solubility and biocompatibility enhancers. The cryptophane would capture hyperpolarized ^{129}Xe in the system, whereas the porphyrin would both emit fluorescence and generate reactive oxygen species (ROS) for PDT.

In vivo tracking of fluorinated drugs by means of ^{19}F MRI is another challenge for medicine, but most often the ^{19}F content is too low for detection. Promising results were recently achieved in mice, using a liposomal drug delivery platform in which the antitumoral antibiotic doxorubicin and a dendritic amphiphilic contrast agent bearing three F_{27} groups per molecule (Fig. 3.7d) were simultaneously encapsulated [27].

Finally, it is worth to cite the work by Kikuchi and collaborators, who developed a series of multifunctional silica nanoparticles made up of a liquid perfluorocarbon core and a stable silica shell [28]. The combined use of NPs containing different PFCs allowed to perform multicolor ^{19}F MRI, a technique of great importance for revealing molecular interactions in living cells or animals. Moreover, the surface of these NPs could be modified with several functional groups, including small molecules, fluorescent dyes, PEG chains, and peptides. Surface functionalization may be useful for many biomedical applications, such as the detection of enzyme activities, cancer imaging, or drug delivery. Importantly, core-shell F_{27}-FG@ SiO_2 nanoparticles (FG = CH_3 in Fig. 3.7a) were also obtained and visualized in a living mouse after subcutaneous injection. Their covalent binding to fluorescent sulfo-cyanine 5 dye enabled dual-mode magnetic resonance and optical imaging.

3.4 Conclusive Remarks and Perspectives

Over the past 15 years, the need for more sustainable and effective polyfluorinated compounds, tailored for sensitive ^{19}F MRI, has stimulated the synthesis of several derivatives based on the use of PFTB as starting building block. In this regard, PERFECTA with its 36 magnetically equivalent ^{19}F atoms and symmetrical structure, composed of four PFTB units, has been widely exploited in different biocompatible formulations mainly for cell labelling purpose. In this contribution, we summarized the main results obtained using PERFECTA formulation as ^{19}F MRI probes. Despite high fluorine payload and its solid state, PERFECTA has shown excellent magnetic properties and solubility in aqueous solutions at high concentration using conventional biosurfactants. Moreover, PERFECTA formulations showed stability in the biological environment, high labelling efficiency and low cytotoxicity.

One of the main limitations of the use of ^{19}F MRI probes, mostly at pre-clinical level, is their difficult intracellular localization and fate, as ^{19}F MRI does not provide the necessary spatial resolution. This is usually overcome either through co-encapsulation of a fluorescent dye in the formulation or by covalent linkage of a fluorescent group to the fluorinated molecule. Neither of these solutions are optimal as we explained above. The possibility of using Raman imaging for direct visualization of fluorinated probes at cellular level in the cytoplasm represents an excellent alternative, avoiding any alteration of the formulation. In fact, polyfluorinated molecules, due to their strong C–F vibrational modes, provides specificity and avoids overlap with the cellular Raman signatures, and can be directly visualized intracellularly by Raman microscopy, thus enabling detection at tissue level after *in vivo* ^{19}F MRI localization.

Due to the presence of many C–F bonds in the PERFECTA structure, its molecular design permits replacement of a PFTB arm with different functional groups producing a series of F_{27}-derivatives, which may be useful building blocks for further functionalization and conjugation steps. Many examples have been reported highlighting the possibility of using F_{27}-tags for labelling biomolecules, substrates and drug delivery systems to enable *in vivo* ^{19}F MRI detection. In this regard, it is important to consider however that the functionalization should not impair the mobility of the fluorinated chains resulting in T_2 broadening. Additionally, synergic efforts from material scientists and radiologists, who have been working on the optimization of MRI pulse sequences and methods, have to be pursued to reap the full potential of emerging imaging probes.

References

1. Tirotta, I., Dichiarante, V., Pigliacelli, C., Cavallo, G., Terraneo, G., Baldelli Bombelli, F., Metrangolo, P., Resnati, G. (2015). 19F magnetic resonance imaging (MRI): from design of materials to clinical applications. *Chem. Rev.*, **115**, 1106–1129.

2. Srinivas, M., Cruz, L. J., Bonetto, F., Heerschap, A., Figdor, C. G., de Vries, I. J. M. (2010). Customizable, multi-functional fluorocarbon nanoparticles for quantitative *in vivo* imaging using 19F MRI and optical imaging. *Biomaterials*, **31**, 7070–7077.

3. Ahrens, E. T., Flores, R., Xu, H., Morel, P. A. (2005). *In vivo* imaging platform for tracking immunotherapeutic cells. *Nat. Biotechnol.*, **23**, 983–987.

4. Wu, T., Li, A., Chen, K., Peng, X., Zhang, J., Jiang, M., Chen, S., Zheng, X., Zhou, X., Jiang, Z. X. (2021). Perfluoro-tert-butanol: a cornerstone for high performance fluorine-19 magnetic resonance imaging. *Chem. Commun.*, **57**, 7743–7757.

5. Tirotta, I., Mastropietro, A., Cordiglieri, C., Gazzera, L., Baggi, F., Baselli, G., Bruzzone, M. G., Zucca, I., Cavallo, G., Terraneo, G., Baldelli Bombelli, F., Metrangolo, P., Resnati, G. (2014). A superfluorinated molecular probe for highly sensitive in vivo19F-MRI. *J. Am. Chem. Soc.*, **136**, 621–624.

6. Krafft, M. P., Riess, J. G. (2015). Per- and polyfluorinated substances (PFASs): environmental challenges. *Curr. Opin. Colloid Interface Sci.*, **20**, 192–212.
7. Chirizzi, C., Morasso, C., Caldarone, A., Tommasini, M., Corsi, F., Chaabane, L., Vanna, R., Baldelli Bombelli, F., Metrangolo, P. (2021). A bioorthogonal probe for multiscale imaging by 19 F-MRI and Raman microscopy: from whole body to single cells. *J. Am. Chem. Soc.*, **143**, 12253–12260.
8. Chirizzi, C., De Battista, D., Tirotta, I., Metrangolo, P., Comi, G., Baldelli Bombelli, F., Chaabane, L. (2019). Multispectral MRI with dual fluorinated probes to track mononuclear cell activity in mice. *Radiology*, **291**, 351–357.
9. Koshkina, O., White, P. B., Staal, A. H. J., Schweins, R., Swider, E., Tirotta, I., Tinnemans, P., Fokkink, R., Veltien, A., van Riessen, N. K., van Eck, E. R. H., Heerschap, A., Metrangolo, P., Baldelli Bombelli, F., Srinivas, M. (2020). Nanoparticles for "two color" 19F magnetic resonance imaging: towards combined imaging of biodistribution and degradation. *J. Colloid Interface Sci.*, **565**, 278–287.
10. Ayaz, N., Dichiarante, V., Pigliacelli, C., Repossi, J., Gazzera, L., Borreggio, M., Maiolo, D., Chirizzi, C., Bergamaschi, G., Chaabane, L., Fasoli, E., Metrangolo, P., Baldelli Bombelli, F. (2021). Hydrophobin-coated solid fluorinated nanoparticles for 19F-MRI. *Adv. Mater. Interfaces*, 2101677.
11. Jamgotchian, L., Vaillant, S., Selingue, E., Doerflinger, A., Belime, A., Vandamme, M., Pinna, G., Ling, W. L., Gravel, E., Mériaux, S., Doris, E. (2021). Tumor-targeted superfluorinated micellar probe for sensitive in vivo 19 F-MRI. *Nanoscale*, **13**, 2373–2377.
12. Jiang, Z.-X., Yu, Y. B. (2007). The design and synthesis of highly branched and spherically symmetric fluorinated oils and amphiles. *Tetrahedron*, **63**, 3982–3988.
13. Jiang, Z.-X., Liu, X., Jeong, E.-K., Yu, Y. B. (2009). Symmetry-guided design and fluorous synthesis of a stable and rapidly excreted imaging tracer for ^{19}F MRI. *Angew. Chem. Int. Ed.*, **48**, 4755–4758.
14. Jiang, Z.-X., Yu, Y. B. (2010). Fluorous mixture synthesis of asymmetric dendrimers. *J. Org. Chem.*, **75**, 2044–2049.
15. Yue, X., Taraban, M. B., Hyland, L. L., Yu, Y. B. (2012). Avoiding steric congestion in dendrimer growth through proportionate branching: a twist on da Vinci's rule of tree branching. *J. Org. Chem.*, **77**, 8879–8887.
16. Dichiarante, V., Martinez Espinoza, M. I., Gazzera, L., Vuckovac, M., Latikka, M., Cavallo, G., Raffaini, G., Oropesa-Nuñez, R., Canale, C., Dante,

S., Marras, S., Carzino, R., Prato, M., Ras, R. H. A., Metrangolo, P. (2018). A short-chain multibranched perfluoroalkyl thiol for more sustainable hydrophobic coatings, *ACS Sustainable Chem. Eng.*, **6**, 9734–9743.

17. Dichiarante, V., Tirotta, I., Catalano, L., Terraneo, G., Raffaini, G., Chierotti, M. R., Gobetto, R., Baldelli Bombelli, F., Metrangolo, P. (2017). Superfluorinated and NIR-luminescent gold nanoclusters. *Chem. Commun.*, **53**, 621–624.
18. Pigliacelli, C., Acocella, A., Díez, I., Moretti, L., Dichiarante, V., Demitri, N., Jiang, H., Maiuri, M., Ras, R. H. A., Baldelli Bombelli, F., Cerullo, G., Zerbetto, F., Metrangolo, P., Terraneo, G. (2022). High-resolution crystal structure of a 20 kDa Superfluorinated Gold Nanocluster. *Nature Commun.*, **13**, 2607. https://doi.org/10.1038/s41467-022-29966-2
19. Tedeschi, G., Guzman-Puyol, S., Ceseracciu, L., Benitez, J. J., Goldoni, L., Koschella, A., Heinze, T., Cavallo, G., Dichiarante, V., Terraneo, G., Athanassiou, A., Metrangolo, P., Heredia-Guerrero, J. A. (2021). Waterproof-breathable films from multi-branched fluorinated cellulose esters. *Carbohydrate Polym.*, **271**, 118031.
20. Martinez Espinoza, M. I., Sori, L., Pizzi, A., Terraneo, G., Moggio, I., Arias, E., Pozzi, G., Orlandi, S., Dichiarante, V., Metrangolo, P., Cavazzini, M., Baldelli Bombelli, F. (2019). BODIPY dyes bearing multibranched fluorinated chains: synthesis, structural, and spectroscopic studies. *Chem. Eur. J.*, **25**, 9078–9087.
21. Huynh, A. M., Mìller, A., Kessler, S. M., Henrikus, S., Hoffmann, C., Kiemer, A. K., Bìcker, A., Jung, G. (2016). Small BODIPY probes for combined dual ^{19}F MRI and fluorescence imaging. *ChemMedChem.*, **11**, 1568–1575.
22. Rosati, M., Acocella, A., Pizzi, A., Turtu, G., Neri, G., Demitri, N., Nonappa, Raffaini, G., Donnio, B., Zerbetto, F., Baldelli Bombelli, F., Cavallo, G., Metrangolo, P. (2022). Janus-type dendrimers based on highly branched fluorinated chains with tunable self-assembly and ^{19}F nuclear magnetic resonance properties, *Macromolecules*, **55**, 2486–2496.
23. Shi, H., Lai, B., Chen, S., Zhou, X., Nie, J., Ma, J.-A. (2017). Facile synthesis of novel perfluorocarbon-modulated 4-anilinoquinazoline analogues. *Chin. J. Chem.*, **35**, 1693–1700.
24. Jiang, Z.-X., Yu, Y. B. (2008). The design and synthesis of highly branched and spherically symmetric fluorinated macrocyclic chelators. *Synthesis*, **2**, 215–220.
25. Peng, Q., Li, Y., Bo, S., Yuan, Y., Yang, Z., Chen, S., Zhou, X., Jiang, Z.-X. (2018). Paramagnetic nanoemulsions with unified signals for sensitive ^{19}F MRI cell tracking. *Chem. Commun.*, **54**, 6000–6003.

26. Zhang, H., Yu, Q., Li, Y., Yang, Z., Zhou, X., Chen, S., Jiang, Z.-X. (2020). Fluorinated cryptophane-A and porphyrin-based theranostics for multimodal imaging-guided photodynamic therapy. *Chem. Commun.*, **56**, 3617–3620.
27. Bo, S., Yuan, Y., Chen, Y., Yang, Z., Chen, S., Zhou, X., Jiang, Z.-X. (2018). In vivo drug tracking with ^{19}F MRI at therapeutic dose. *Chem. Commun.*, **54**, 3875–3878.
28. Akazawa, K., Sugihara, F., Nakamura, T., Matsushita, H., Mukai, H., Akimoto, R., Minoshima, M., Mizukami, S., Kikuchi, K. (2018). Perfluorocarbon-based ^{19}F MRI nanoprobes for in vivo multicolor imaging. *Angew. Chem. Int. Ed.*, **57**, 16742–16747.

Chapter 4

Development of Perfluorocarbon-Encapsulated Silica Nanoparticle as ^{19}F MRI Tracer Agent

Masafumi Minoshima,[a] Yuki Konishi,[a] Yue Wu[a] and Kazuya Kikuchi[a,b,c]
[a]*Graduate School of Engineering, Osaka University, Suita, Osaka, Japan*
[b]*Immunology Frontier Research Center, Osaka University, Suita, Osaka, Japan*
[c]*Quantum Information and Quantum Biology Osaka University, Suita, Osaka, Japan*
kkikuchi@mls.eng.osaka-u.ac.jp

4.1 Introduction

^{19}F magnetic resonance imaging (MRI) is a powerful method for *in vivo* imaging of biomolecules, cells, and target tissues in animals, taking advantage of its negligible background signal [1–4]. To obtain

Fluorine Magnetic Resonance Imaging: Methods and Applications in Biomedicine
Edited by Eric T. Ahrens and Ulrich Flögel
Copyright © 2025 Jenny Stanford Publishing Pte. Ltd.
ISBN 978-981-5129-37-3 (Hardcover), 978-1-003-53004-6 (eBook)
www.jennystanford.com

high-quality images using ^{19}F MRI, fluorinated tracers need to be delivered with a high density per voxel. To address the sensitivity challenge, increasing the number of fluorine atoms in MRI probes is essential [5]. A simple increase in the fluorine atoms per molecule may not provide a useful approach towards higher sensitivity due to the following two reasons: (1) increasing the number of fluorine atoms decreases the aqueous solubility of the molecular probes [6] and (2) suppression of the molecular mobility in the large molecular size shortens the transverse relaxation time (T_2), resulting in attenuation of the MRI signal [7]. The use of perfluorocarbon (PFC)-containing nanoemulsions is an attractive strategy to provide a high ^{19}F density in the tracer thereby overcoming these challenges. The emulsification of PFCs with detergents, such as amphiphilic phospholipids and polymers, can generate useful nanoemulsions. Perfluoro-15-crown-5-ether (PFCE) is widely used as a ^{19}F tracer in nanoemulsions because of its attractive MRI properties. Firstly, PFCE has 20 equivalent fluorine atoms, which show a sharp single ^{19}F NMR peak yielding a strong signal. Secondly, PFCE is a liquid and can maintain high molecular mobility, even at high concentrations. Considering these advantages, PFCE nanoemulsions have been used for *in vivo* cell tracking [8–10]. However, nanoemulsions can be unstable over time under physiological buffered conditions by flocculation, coalescence, or Ostwald ripening [11, 12]. Moreover, PFCE nanoemulsions cannot be handled in organic solvents, which may limit the available avenues for surface functionalization with organic reagents. To overcome these limitations, we developed a multifunctional ^{19}F MRI tracer agent, fluorine-accumulated silica nanoparticles for MRI signal enhancement (FLAME) [13]. The encapsulation of PFC nanoemulsions with a robust silica shell has superior properties, such as surface modifiability, biocompatibility, and sufficient *in vivo* stability, while retaining high sensitivity in ^{19}F MRI. In this chapter, we describe the design of multifunctional ^{19}F MRI nanoprobes for applications to theranostic agents with drugs, multicolor ^{19}F MRI, and detection of enzyme activity *in vivo*.

4.2 Development of Fluorine Accumulated MRI Signal Enhancement (FLAME)

4.2.1 Preparation and Characterization of FLAME

Introducing a silica shell onto the surface of the nanoemulsion appears to be appealing for combining the advantages of silica nanoparticles and PFCE nanoemulsions (Fig. 4.1a). Silica nanoparticles possess

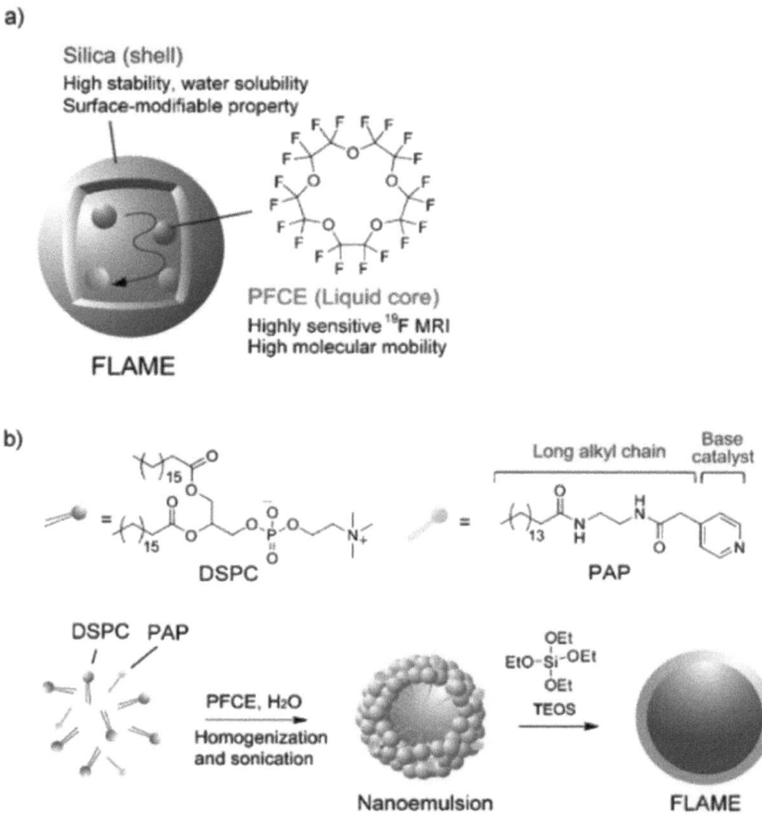

Figure 4.1 Design and synthetic strategy of FLAME. (a) The components of FLAME. (b) The chemical structure of PAP and synthesis of FLAME. Reprinted with permission from Ref. [13].

attractive features such as bio-inertness, favorable colloidal properties, and high surface modification ability [14]. The synthetic protocol for FLAME is presented in Fig. 4.1b. Distearoyl-sn-glycero-3-phosphocholine (DSPC) was used to form the PFCE nanoemulsion. To cover the PFCE-phospholipid nanoemulsion with silica gel, we developed a novel surfactant, PAP, which consists of an alkyl moiety that interacts with phospholipids and a basic pyridinyl group. When template nanoemulsions are formed in the presence of PAP [15], the basic site of PAP was expected to be displayed on the nanoemulsion surface and to initiate the sol–gel process of tetraethyl orthosilicate (TEOS), a silica precursor [16]. As a result, silica condensation occurred only on the surface of the nanoemulsion, and silica coating of the PFCE was achieved.

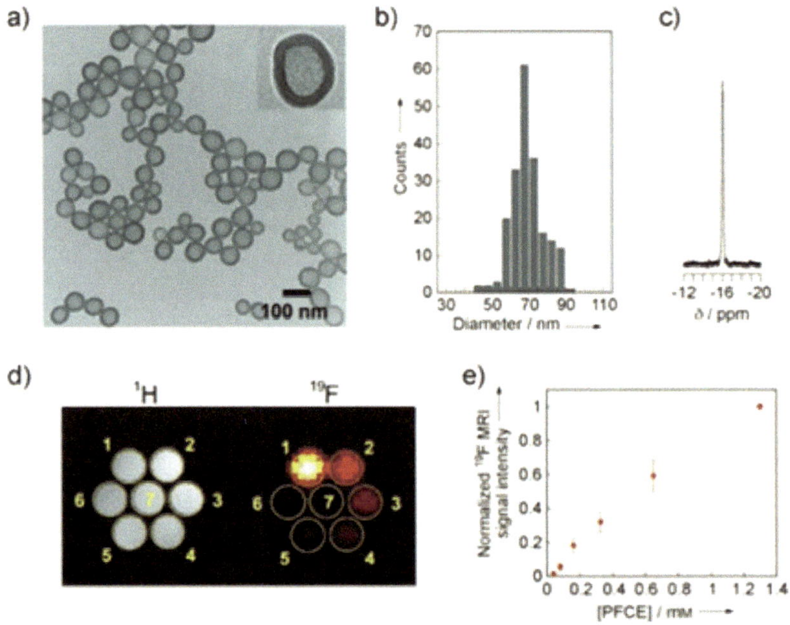

Figure 4.2 Characterization of FLAME. (a) TEM image of FLAME. (b) Particle size distribution histogram calculated from TEM images. (c) ^{19}F NMR spectrum of FLAME in PBS (pH 7.4). (d) MRI of phantoms (1–6) filled with FLAME in PBS (500 μL). PFCE concentrations: 1.30, 0.650, 0.325, 0.163, 0.0813, and 0.0406 mM for phantoms 1–6, respectively. Phantom 7 contains PBS. (e) Plot of the normalized ^{19}F MRI signal intensity versus PFCE concentration. Reprinted with permission from Ref. [13].

The core–shell structures of FLAME were clearly observed by transmission electron microscopy (TEM; Fig. 4.2a). The average diameter of the particles was 76 ± 9 nm (Fig. 4.2b). Dynamic light scattering (DLS) measurements reflected the formation of the silica coating and showed that the hydrodynamic diameter changed from 80 nm of the nanoemulsion to 131 nm for FLAME. The inclusion of PFCE in the particles was confirmed by NMR spectroscopy. The ^{19}F NMR spectrum of FLAME dispersed in phosphate-buffered saline (PBS) solution showed a PFCE-derived single sharp peak at −16.4 ppm (Fig. 4.2c). In the ^{19}F MRI measurements, strong ^{19}F MRI signals were observed using FLAME (Fig. 4.2d). The ^{19}F MRI signal intensity was proportional to PFCE concentration (Fig. 4.2e).

4.2.2 FLAME-PEG for *In Vivo* Targeting

The potential of FLAMEs for *in vivo* targeting was evaluated in tumor-bearing mice. The nanomaterials were delivered to the tumor using the enhanced permeability and retention (EPR) effect, which promotes nanoparticle accumulation in cancerous tissues by a combination of leaky angiogenic blood vessels and deficient tumor lymphatics [17]. However, naked nanoparticles should be trapped immediately by the reticuloendothelial system (RES) and show a decreased circulation time. Therefore, the surface of FLAME was modified with polyethylene glycol (PEG) for effective delivery to the tumors (Fig. 4.3a). PEGylation reduces uptake by increasing the electrostatic interactions between proteins and small molecules [18].

Passive targeting and accumulation of FLAME-PEG using the EPR effect were demonstrated by MRI. Mice bearing a tumor were administered intravenous injections of FLAME (Fig. 4.3b). Strong ^{19}F MRI signals of FLAME-PEG at the tumor site indicated passive targeting of the nanoparticles. The ^{19}F NMR measurements of homogenized tissue samples showed that NMR signals were detected in the liver (C_{PFCE} = 113 nmol g^{-1}) and tumor (C_{PFCE} = 28.8 nmol g^{-1}). However, the ^{19}F MRI signals of non-PEGylated FLAME-COOH were detected only in the liver, indicating that FLAME-COOH was immediately trapped by the RES. These results indicate that the ^{19}F MRI signals of FLAMEs are sufficiently strong for *in vivo* studies and demonstrate that ^{19}F MRI signal localization reflects the actual accumulation position of FLAMEs *in vivo*.

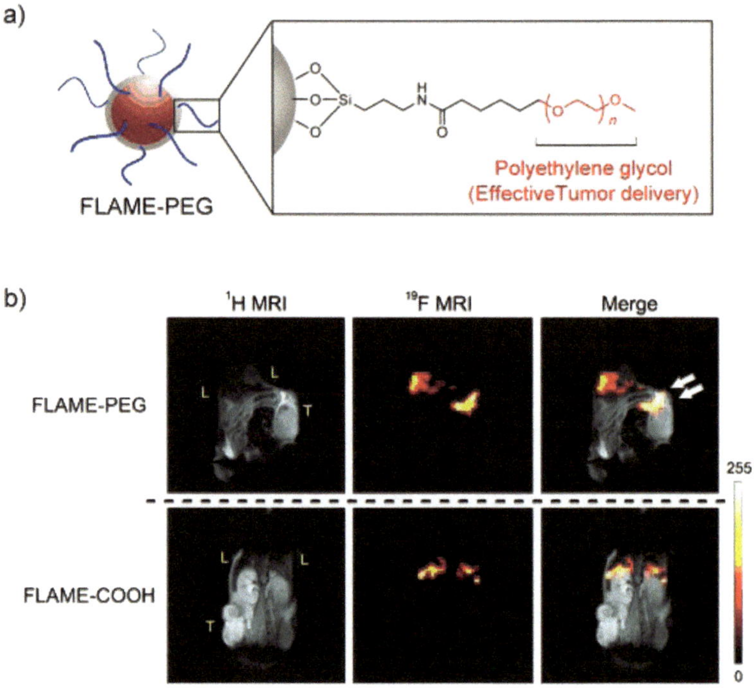

Figure 4.3 *In vivo* accumulation of FLAME at tumor site. (a) Structure of FLAME-PEG and FLAME-COOH (b) *In vivo* MRI of FLAME-PEG and FLAME-COOH in tumor-bearing mice. The positions of the liver and tumor are represented as L and T, respectively. Reprinted with permission from Ref. [13].

4.3 Development of Perfluorocarbon-Encapsulated Mesoporous Silica Nanoparticles

4.3.1 Design, Preparation, and Physical Properties of ^{19}F MRI- and Fluorescence-Traceable Drug Delivery Carrier

Mesoporous silica nanoparticles (MSNs) have attracted significant attention due to their attractive properties, such as extremely large surface areas (1000 m^2 g^{-1}), tunable pore sizes (1.5–10 nm), and ease of functionalization *via* various synthetic approaches [19, 20]. Because the controlled release of drugs from the pores of MSNs results in prolonged drug efficacy, MSNs serve as ideal materials for drug delivery [21, 22]. To assess drug efficacy and toxicity of drug carriers, it is essential to monitor the localization of the drug carrier. MRI-traceable MSNs have also attracted attention in the field of drug delivery due to their deep tissue penetration capability.

Based on the excellent performance of FLAME, the idea of combining MSN with FLAME emerged as an *in vivo* traceable drug carrier [23]. By modifying the silica coverage of the PFC core, we designed a novel drug carrier with dual modality imaging capabilities (NIR/^{19}F MRI), termed mFLAME (mesoporous FLAME, Fig. 4.4a). mFLAME consisted of a PFCE core and mesoporous silica shell. Mesoporous silica shells are stable in aqueous solutions, and drugs can be loaded into their pores. Furthermore, active targeting to foci can be achieved by modifying the targeting ligands on the mFLAME surface. To impart fluorescence imaging capabilities, Cy5 dye was covalently modified on a mesoporous silica shell by silica polymerization in the presence of Cy5-conjugated 3-aminopropyltriethoxysilane (APTES).

The procedure used to prepare mFLAME was similar to that used for FLAME. DSPC was used to form a PFCE nanoemulsion owing to its extremely low water solubility. In addition, we found that *n*-cetyltrimethylammonium bromide (CTAB), which is commonly used for the synthesis of MSNs, was also capable of dispersing PFCE in water. Furthermore, core–shell-type nanoparticles that constitute the PFCE core and that of the mesoporous silica shell could be produced from the PFCE emulsions by a sol–gel process.

The nanomaterials were characterized using DLS. The ζ-potential and hydrodynamic diameter of the PFCE emulsions were +51.0 mV and 78 nm, respectively. In contrast, those of mFLAME were −21.1 mV and 165 nm, respectively. The DLS data revealed that mFLAME did not form aggregates in an aqueous solution. The increase in size and decrease in ζ-potential as compared to those of the PFCE emulsions were due to the formation of the silica shell. TEM revealed that mFLAME had mesopore and core–shell structures, and the average diameter of the particles was 79 ± 20 nm (Fig. 4.4b). The N_2 adsorption/desorption isotherms of mFLAME revealed a typical mesoporous structure with a Brunauer–Emmett–Teller (BET) surface area of 715 $m^2\ g^{-1}$, pore volume of 1.21 $cm^3\ g^{-1}$, and pore width of 6.7 nm (Fig. 4.4c). Fluorescence measurements confirmed that mFLAME could be traced by NIR fluorescence (Fig. 4.4d). Experiments also showed that mFLAME has MRI properties similar to FLAME, confirming that mFLAME has sufficient sensitivity for *in vivo* ^{19}F MRI applications.

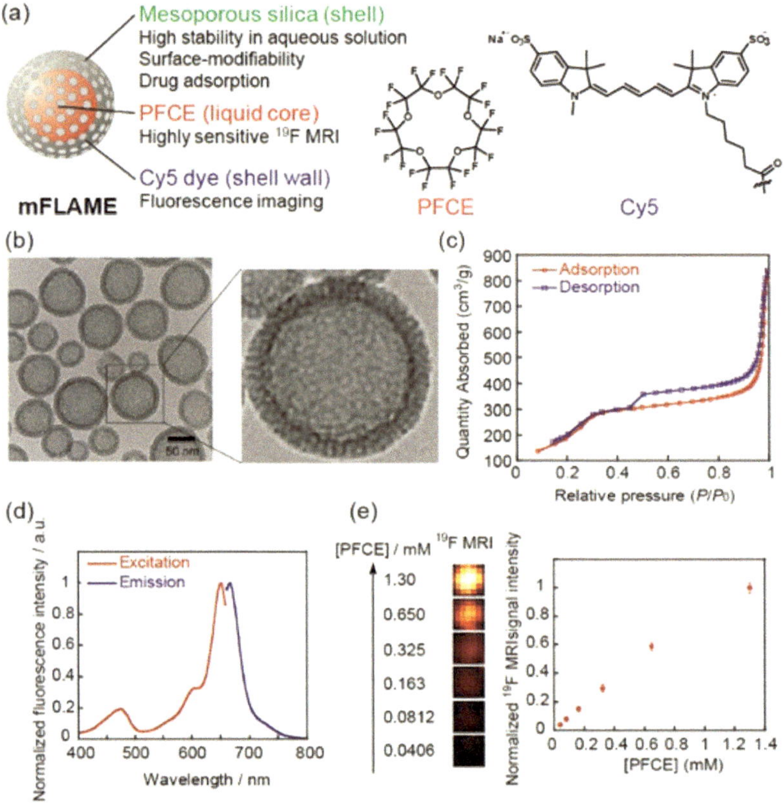

Figure 4.4 (a) The components of mFLAME. (b) Transmission electron microscopy (TEM) image of mFLAME. (c) N_2 adsorption/desorption isotherms of mFLAME. (d) Excitation and fluorescence spectra of mFLAME in phosphate buffered saline (pH 7.4). (e) (Left) ^{19}F MRI of mFLAME in phosphate buffered saline (500 μL); (right) plot of normalized ^{19}F MRI signal intensity versus PFCE concentration. Adapted with permission from Ref. [23].

4.3.2 Cellular Uptake of Folate-Functionalized mFLAME

For efficient delivery of mFLAMEs to tumors, we focused on the folate receptor, which is a 38 kDa glycophosphatidylinositol-linked membrane protein found on the surface of most solid tumors [24]. It was recently reported that folate receptor-mediated uptake could be exploited to facilitate the entry of nanomaterials into cells [25]. Hence, a folate-functionalized mFLAME nanoparticle (mFLAME–FA) was synthesized. The ζ-potential value of mFLAME–FA was -45.1 mV, while that of the amino-functionalized mFLAME (mFLAME–NH_2) was +18.4 mV. PEGylated mFLAME (mFLAME–PEG) was also prepared as a control. The absorption spectra of mFLAME–PEG and mFLAME–FA revealed that only mFLAME–FA had absorption peaks at 280 and 362 nm, which were derived from the folate moiety.

Since Cy5 fluorophores are incorporated into a mesoporous silica shell, the nanoparticle uptake into cells was visualized using fluorescence confocal laser scanning microscopy (Fig. 4.5a). mFLAME–FA was internalized in KB cells that highly express folate receptors [26]. Furthermore, pre-incubation with free folic acid inhibited the uptake of mFLAME–FA. In contrast, uptake of mFLAME–PEG was not observed in KB cells.

Flow cytometry was also carried out to investigate the cellular uptake of mFLAME by KB cells (Fig. 4.5b). KB cells took up approximately 6.5-fold more mFLAME–FA than mFLAME–PEG. Upon treatment with free folic acid, endocytosis of mFLAME–FA decreased by more than 75%.

The nanoparticle uptake into cells was also examined using ^{19}F MRI. After mFLAME–FA was incubated with KB cells, the cells were transferred to a well of a microtiter plate, and ^{1}H/^{19}F MRI experiments were performed (Fig. 4.5c). Figure 4.5d shows MR images of the microtiter plate, including those of KB cells treated with mFLAME. ^{19}F MRI signals were observed from mFLAME–FA, while no ^{19}F MRI signal was observed in the control sample. These results demonstrated that mFLAME could be used as a multimodal probe for ^{19}F MRI and fluorescence imaging.

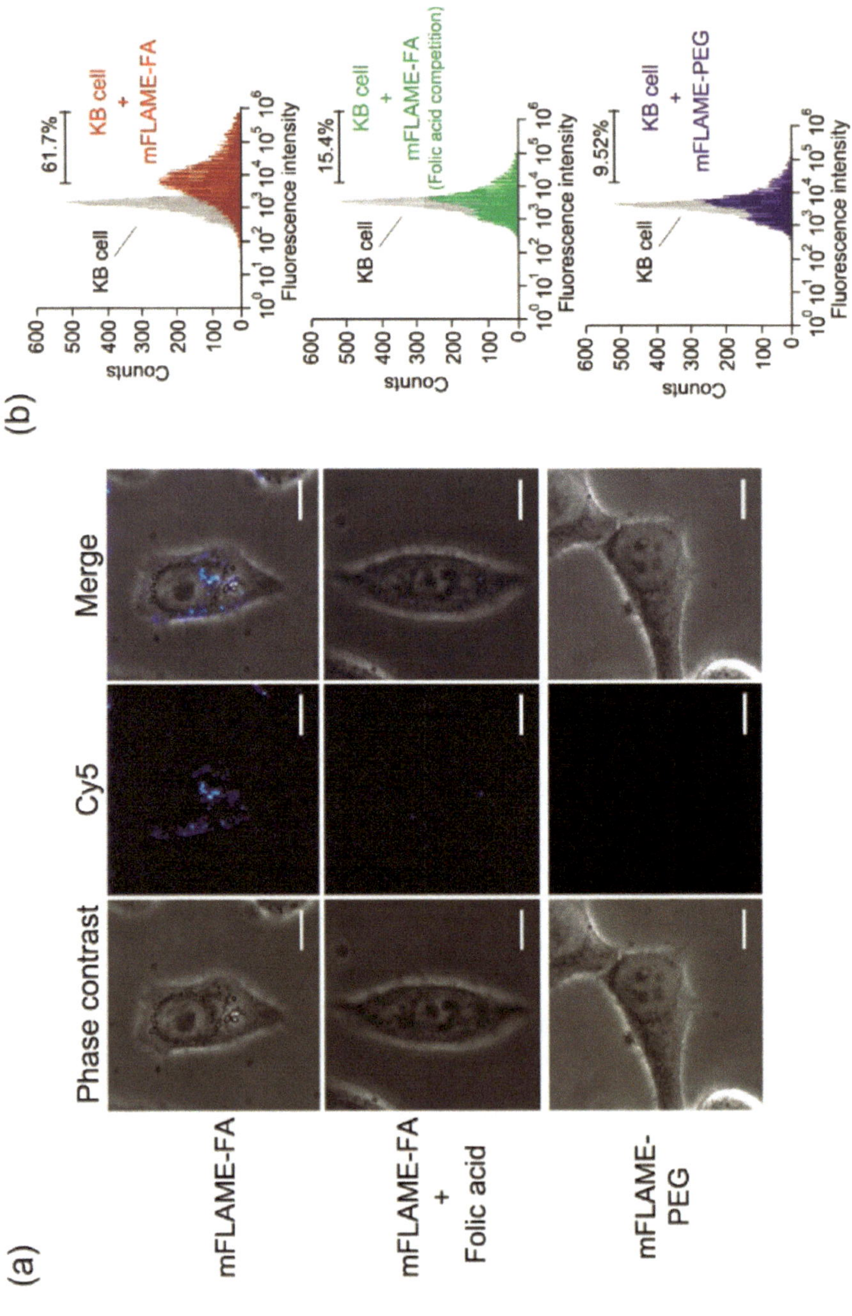
(a)
mFLAME-FA
mFLAME-FA + Folic acid
mFLAME-PEG
Phase contrast
Cy5
Merge
(b)
Counts
Fluorescence intensity
KB cell
61.7%
KB cell + mFLAME-FA
15.4%
KB cell + mFLAME-FA (Folic acid competition)
9.52%
KB cell + mFLAME-PEG

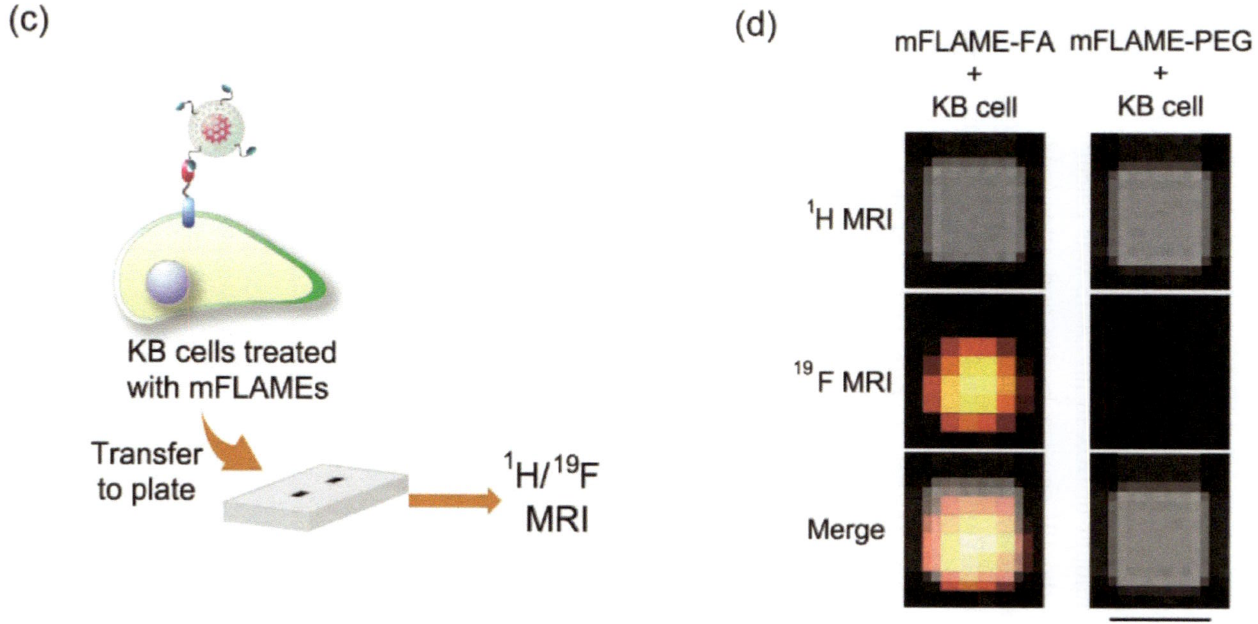

Figure 4.5 (a) Confocal laser scanning microscopy images of KB cells treated with mFLAME–FA or mFLAME–PEG with or without folic acid for 4 h. Measurement conditions: Cy5 was excited at 635 nm and detected at 660–760 nm. Scale bar: 10 μm. (b) Flow cytometric analysis of cellular uptake. The gray histograms show the distribution of non-treated cells. (c) Illustration of the experimental procedure for the MRI detection of mFLAME–FA in KB cells. Scale bar indicates 0.3 mm. Adapted with permission from Ref. [23].

4.3.3 Drug Encapsulation and Cellular Toxicity of Drug-Loaded mFLAMEs

To examine the potential of mFLAMEs for drug delivery, a chemotherapeutic agent and well-known anti-cancer drug, doxorubicin (DOX) [27], was loaded into the pores of mFLAME. The loading amount of DOX in mFLAME was calculated using the difference in the UV spectra before and after loading. We then evaluated the release profiles of the encapsulated DOX at different pH values using UV-visible absorption spectroscopy. Encapsulated-DOX was released gradually from mFLAME in a time-dependent manner (Fig. 4.6a). The release rates from mFLAME at pH 5.0, were faster than those at pH 7.5, which suggested that the surrounding pH affects the electrostatic interactions between mFLAME and DOX. The results indicated that mFLAME may have useful drug release abilities in the endosome and lysosome (pH 5.0–5.5), which have lower pHs than other cellular components, such as the cytosol. mFLAME may also be useful for the treatment of cancer because the pH of tumors is lower than that of normal tissues.

Next, DOX-loaded mFLAME were evaluated by assessing the viability of KB cells. Figure 4.6b shows that DOX-loaded mFLAME–FA had a greater cytotoxic effect on KB cells as compared to that of mFLAME–PEG. In addition, the cytotoxicity of mFLAME–FA was greater than that of free DOX because mFLAME–FA was effectively internalized into KB cells owing to the folate ligand. The cellular uptake of DOX-loaded mFLAME was analyzed by confocal laser scanning microscopy (Fig. 4.6c). The fluorescence of DOX was similar in KB cells incubated with DOX-loaded mFLAME–FA and in cells treated with free DOX. In contrast, a weaker fluorescence intensity was observed in cells incubated with DOX-loaded mFLAME–PEG, indicating that DOX-loaded mFLAME–FA exhibited strong cytotoxic effects on KB cells owing to efficient cellular internalization and drug release in the cells.

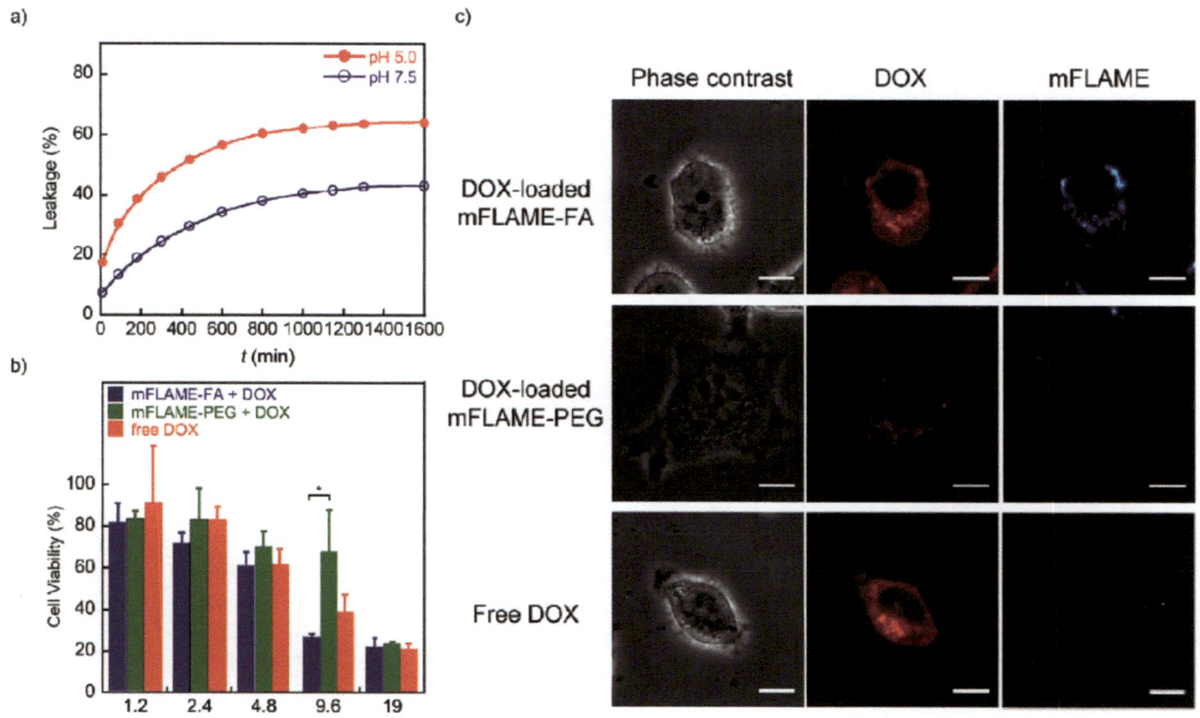

Figure 4.6 (a) Release profile of DOX-loaded mFLAME–FA in citric acid buffer (pH 7.5 and 5.0) at 37 °C [DOX] = 19 mg mL^{-1}. (b) Concentration dependent cell survival data of folate receptor positive-KB cells treated with free DOX or DOX-loaded mFLAME for 1 day. Values are expressed as the mean ± SE (error bars; $n = 3$, $*P < 0.05$). (c) Fluorescence imaging of KB cells treated with DOX-loaded mFLAME–FA, DOX-loaded mFLAME–PEG, and free DOX for 6 h [DOX] = 9.6 mg mL^{-1}. Scale bar represents 10 μm. Reprinted with permission from Ref. [23].

4.4 Multicolor Imaging by Perfluorocarbon-Encapsulated Silica Nanoparticles

4.4.1 Development of PFC-Encapsulated Silica Nanoparticles for Multicolor ^{19}F MRI

Multicolor imaging is a useful technique for visualizing the biodistribution and interactions of molecules, nanoparticles, and cells using different trackers. Multicolor MRI can detect different trackers in deep tissues. For multicolor imaging, ^{19}F MRI has advantages over ^{1}H MRI because a broad chemical shift range (>350 ppm) of ^{19}F MRI tracers is available without interference from endogenous background signals. Efforts to develop small-molecule, polymer, or inorganic nanocrystal-based ^{19}F MRI probes that show different ^{19}F chemical shift peaks have resulted in multicolor ^{19}F MRI images of phantoms [28–30]. Different PFC-encapsulated nanoemulsions have also been employed as multicolor ^{19}F MRI tracer agents for the imaging of stem cells and neuroinflammation in living mice [31–33]. Recently, different fluorine nanocrystals (CaF_2 and SrF_2) have been used for multicolor ^{19}F MRI to map immune specificity in real time [34]. Our silica nanoparticles (FLAME) can contain different liquid PFCs with different chemical shifts in the core for multicolor ^{19}F MRI [35]. FLAME has the advantages of high stability and facile surface functionalization. Moreover, the T_2 values of liquid PFCs inside the particles are much longer than those of solid-state substances, such as nanocrystals.

We selected three different perfluorocarbons to encapsulate in silica nanoparticles (Fig. 4.7a). The first is PFCE, which has a strong singlet peak derived from 20 equivalent fluorine atoms (δ = –16.4 ppm [TFANa is used as internal standard]). The second is TPFBME (1,1,1-tris(perfluoro-tert-butoxymethyl) ethane), which has 27 equivalent fluorine atoms (δ = 3.3 ppm). TPFBME was synthesized from perfluoro-tert-butanol in only one step. The third is perfluorotributylamine (PFTBA), which shows a single peak at –53.0 ppm (6 fluorine atoms). Three PFC-encapsulated silica nanoparticles (PFCE@SiO_2, TPFBME@SiO_2, and PFTBA@SiO_2) were prepared. In the TEM images, all nanoparticles showed a core-shell structure. The longitudinal relaxation time (T_1) and T_2 of PFCs in

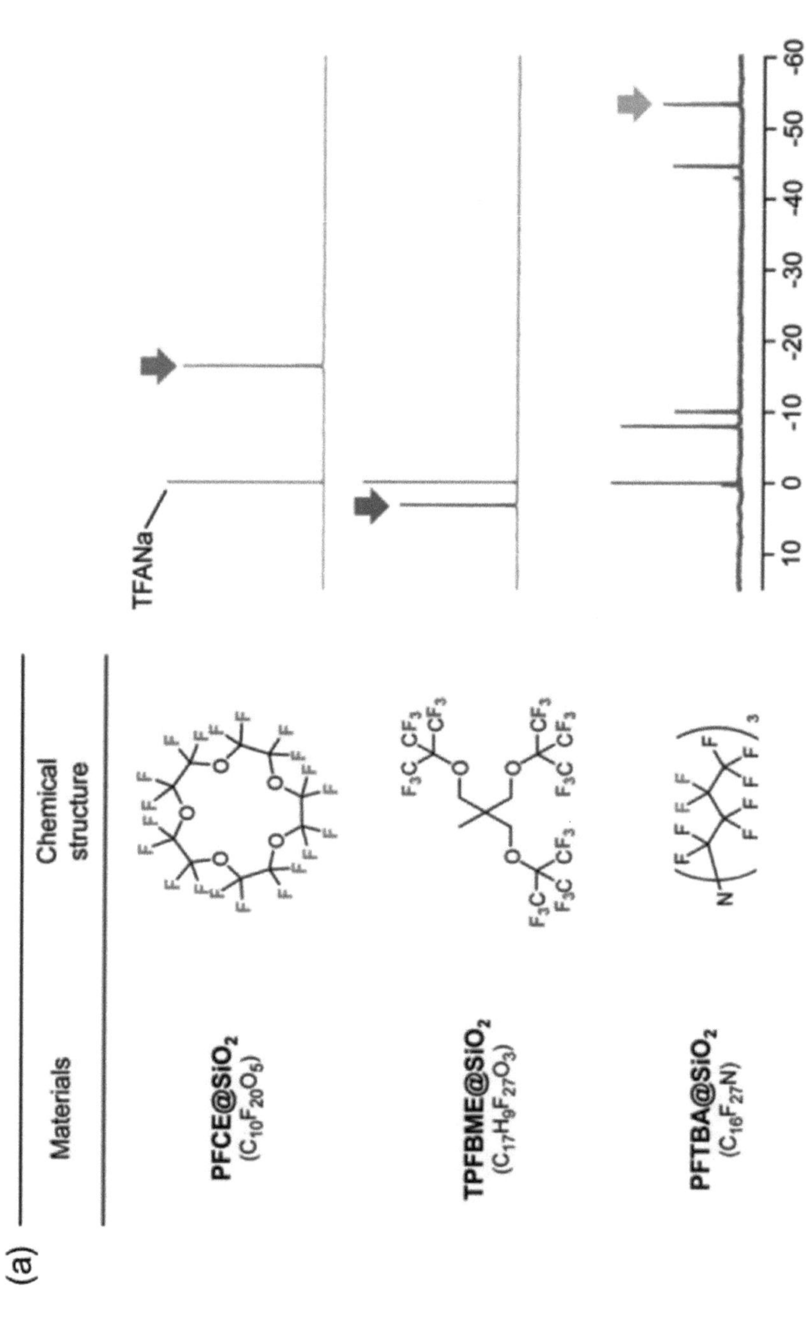
(a)
Materials
Chemical structure
PFCE@SiO2
(C10F20O5)
TPFBME@SiO2
(C17H9F27O3)
PFTBA@SiO2
(C16F27N)
TFANa
δ / ppm
10
0
-10
-20
-30
-40
-50
-60

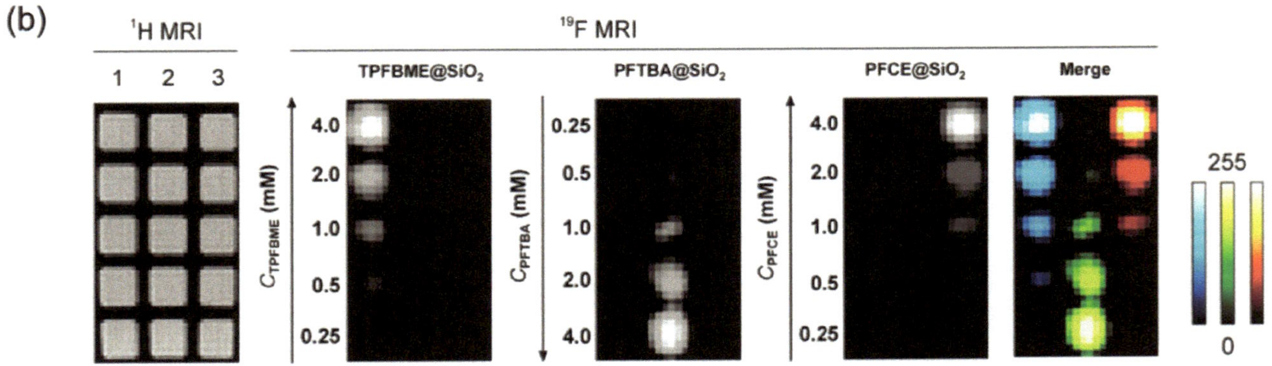

Figure 4.7 (a) Chemical structures of PFCs and ^{19}F NMR spectra of PFC@SiO_2. The peaks of colored fluorine atoms in the chemical structures are indicated by arrows in each spectrum. (b) ^{19}F MRI phantom images of TPFBME@SiO_2, PFTBA@SiO_2, and PFCE@SiO_2. The nanoprobes were placed in vertical lanes 1 (TPFBME@SiO_2), 2 (PFTBA@SiO_2), and 3 (PFCE@SiO_2) on ^1H MRI, respectively. The center frequencies of the TPFBME peak (at approximately $\delta = 3.3$ ppm), PFTBA peak (at approximately $\delta = -53.0$ ppm), and PFCE peak (at approximately $\delta = -16.4$ ppm) were chosen as the frequencies for radiofrequency (RF) output. Adapted with permission from Ref. [35].

silica nanoparticles were comparable to values measured in the unencapsulated oils. For multicolor ^{19}F MRI, nanoparticles were added into a 384-well microplate at a series of PFC concentrations for quantification analysis using ^{19}F MRI phantom images. ^{19}F MRI phantom images were acquired by excitation of the peaks at different chemical shift values. As shown in Fig. 4.7b, the signal of the three silica nanoparticles increased in proportion to the concentration of nanoparticles in multi-color ^{19}F MRI phantom images.

4.4.2 Triple Color ^{19}F MRI with PFC@SiO_2 Nanoparticles

To confirm the sensitivity of PFC-encapsulated silica nanoparticles *in vivo*, we simultaneously administered nanoparticles (C_{PFC} = 10 Mm) to a living mouse *via* subcutaneous injection. ^{19}F images were acquired using the same RF excitation frequencies as used in the multicolor ^{19}F phantom studies. Distinct MRI signals of the silica nanoparticles were detected at each injection site, demonstrating the feasibility of *in vivo* multicolor ^{19}F MRI with spectrally distinct PFC-encapsulated silica nanoparticles.

Next, we applied an *in vivo* multicolor ^{19}F MRI system to evaluate the liver uptake of surface-functionalized PFC@SiO_2 nanoparticles. Three types of surface-modified silica nanoparticles were prepared with different PFCs and surface functional groups, namely PFCE@SiO_2-PEG (PFCE@SiO_2 modified with PEG on the surface), TPFBME@SiO_2-COOH (TPFBME@SiO_2 modified with carboxylic acid), and PFTBA@SiO_2-OH (unmodified PFTBA@SiO_2) (Fig. 4.8a). Prior to the evaluation with *in vivo* ^{19}F MRI, each ^{19}F MRI signal of the nanoparticles in MRI tubes was measured to normalize the signal intensity because each nanoparticle type had different ^{19}F content. Thereafter, nanoparticles were administered into a healthy mouse *via* intravenous injection, and ^{19}F MRI was acquired at 3, 12, and 24 h. In all nanoparticles, ^{19}F MRI signals were found in the liver at every time point (Fig. 4.8b). Quantification of the normalized signal to noise (S/N) ratio of the ^{19}F MRI signals revealed the relative amount of nanoparticles in the liver (Fig. 4.8c). The S/N ratio of PFCE@SiO_2-PEG was reduced by approximately 50% compared to that of TPFBME@SiO_2-COOH and PFTBA@SiO_2-OH. This result indicates that the accumulation of PFCE@SiO_2-PEG by liver was suppressed. This multicolor ^{19}F MRI imaging system can be utilized to evaluate the liver uptake of surface-functionalized nanoparticles in a single mouse.

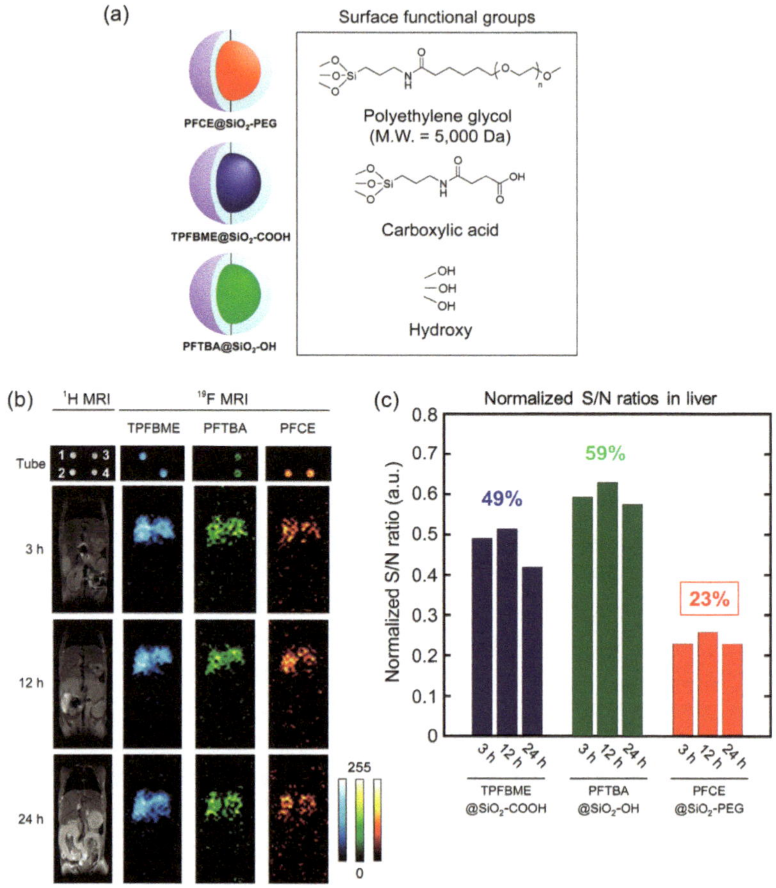

Figure 4.8 (a) Nanoprobe design and the chemical structures of surface functional groups. (b) ^{19}F MRI images of PFCE@SiO_2-PEG, TPFBME@SiO_2-COOH, and PFTBA@SiO_2-OH (C_{PFC} = 3.3 mm) in tube (1. TPFBME@SiO_2-COOH, 2. PFCE@SiO_2-PEG, 3. PFTBA@SiO_2-OH, and 4. Mixture of all tracer agents) and a live mouse at 3, 12, and 24 h after injection. (c) The S/N ratios in the tube and the liver. Adapted with permission from Ref. [35].

4.5 Activatable ^{19}F MRI Nanoprobes

Activatable MRI probes change signals in response to external stimuli. MRI probes for metal ions, pH, redox states, and enzymatic activity have been developed by switching the relaxivity or chemical exchange saturation transfer effect [36–42]. Meade *et al.* developed an activatable Gd^{3+}-based ^{1}H MRI contrast agent for β-galactosidase enzyme activity and demonstrated the use of this contrast agent to track β-gal expression in living *Xenopus laevis* embryos [38].

Activatable ^{19}F MRI probes are more desirable for *in vivo* sensing applications because they show signals without endogenous background ^{1}H signals and accurately detect the target biological analytes [43–45]. Our group has developed an activatable ^{19}F MRI probe based on the paramagnetic relaxation enhancement (PRE) effect that shortens T_2 of the MR-observable nuclei by paramagnetic molecules [46, 47]. The initial activatable ^{19}F MRI probe comprises a paramagnetic Gd^{3+} complex, a cleavable enzyme substrate, and a ^{19}F-containing group. This probe shows weak MRI signals due to the intramolecular PRE effect between the Gd^{3+} complex and the ^{19}F-containing group. The PRE effect is effective over short distances between the nuclei and a paramagnetic center because of the distance dependency (proportional to r^{-6}). Cleavage of the linker causes an increase in T_2 and recovery of the ^{19}F MRI signal. To expand this switching OFF/ON-system to perfluorocarbon-encapsulated silica nanoparticles, activatable ^{19}F MRI nanoprobes were prepared by modifying the surface of FLAME with a paramagnetic Gd^{3+} complex *via* a cleavable linker [48].

4.5.1 Switching ^{19}F MRI Signal of FLAME by Paramagnetic Relaxation Enhancement (PRE) Effect

Grüll *et al.* observed the PRE of PFCE in Gd^{3+}-modified nanoemulsions [49]. In these nanoemulsions, the distance between the Gd^{3+} complexes and the fluorine core was less than 22 Å. In contrast, PFCE molecules in the core of FLAME are more than 50 Å away from the surface-modified paramagnetic metal ion complexes, considering the thickness of the silica shell. Thus, potentially the

PRE effect would not sufficiently attenuate the ^{19}F NMR/MRI signals. Nonetheless, we prepared FLAME-DTPA-Gd, FLAME modified with Gd^{3+} diethylenetriaminepentaacetate (DTPA) complexes on the surface and examined the ^{19}F NMR/MRI signal (Fig. 4.9a). The ^{19}F NMR spectrum of FLAME-DTPA-Gd showed a broad peak (T_2 = 40 ms), whereas that of FLAME-DTPA without Gd^{3+} exhibited a single sharp peak (T_2 = 420 ms) (Fig. 4.9b). Thus, despite our concerns, the ^{19}F NMR/MRI signals of PFCE in FLAME were significantly affected

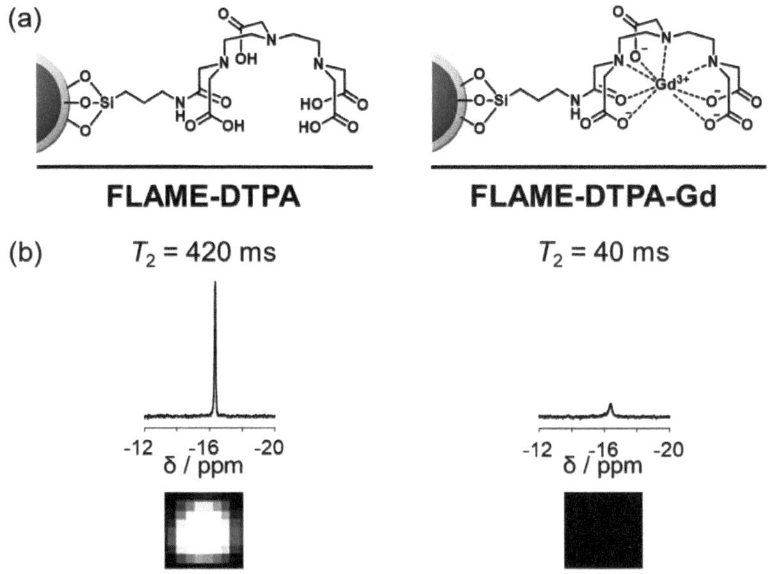

Figure 4.9 (a) Chemical structure of FLAME-DTPA and FLAME-DTPA-Gd. (b) ^{19}F NMR spectra and ^{19}F MRI phantom images of FLAME-DTPA and FLAME-DTPA-Gd. Adapted with permission from Ref. [48].

by PRE from surface-modified Gd^{3+} complexes. These results can be explained by the molecular mobility on the NMR/MRI measurement timescale. Iwahara and Clore reported that the PRE effect was efficient despite the long average distance when MR-observable nuclei can occasionally enter the effective range of the PRE effect [50, 51]. The long T_2 indicates that the PFCE in FLAME maintains high molecular mobility even in the nanoparticle structure. Although the PFCE at the center of the FLAME core is approximately 250 Å away from the surface Gd^{3+} complexes (where PRE is not efficient), the fluorine compounds stochastically access the inner shell of FLAME on the measurement time scale. Near the inner shell, the fluorine nuclei in PFCE are impacted by the PRE effect from multiple Gd^{3+} complexes on the surface. Although the contribution of one Gd^{3+} complex to the PRE effect is small, the combination of many paramagnetic molecules efficiently decreases T_2.

4.5.2 Development of Thiol-Activatable FLAMEs

Based on this finding, we designed activatable FLAMEs, a series of FLAME-SS-Gd^{3+} (FSG), to image reducing environments. Redox reactions play crucial roles in biological processes, and aberrant redox reactions are implicated in inflammation, neurodegenerative diseases, and tumors [52]. FSGs have Gd^{3+} complexes through disulfide linkers on the surface of FLAME. FSGs also reduce the T_2 of the PFCE in the nanoparticle core *via* the PRE effect. When the disulfide bonds of FSG are reduced, the Gd^{3+} complexes are cleaved away from the FLAME surface, and the baseline T_2 of the encapsulated PFCE and ^{19}F NMR/MRI signal recovers. To optimize the amount of surface Gd^{3+} complex required, three FSGs (FSG1-3) were prepared, and their NMR properties were analyzed. The ratio of the number of fluorine atoms and Gd^{3+} ions per nanoparticle was calculated as 1.8×10^3, 7.7×10^2, and 5.3×10^2 for FSG1-3. ^{19}F NMR spectra of FSGs after treatment with a reducing agent, tris(2-carboxyethyl)-phosphine (TCEP), showed that the ^{19}F NMR peaks became sharper and increased compared to those without TCEP treatment. The addition of TCEP also recovered the T_2 values of FSG1-3 within 2 h to baseline.

Next, ^{19}F MR phantom images of FSGs in buffer with or without TCEP were acquired with various effective echo times ($T_{E,eff}$). Similar to the ^{19}F NMR results, TCEP treatment induced a noticeable ^{19}F MRI signal enhancement in FSG1-3 at any $T_{E,eff}$. At $T_{E,eff}$ = 12 ms, the ^{19}F MRI signals were increased by approximately 60- and 40-fold in FSG2 and FSG3, respectively (Fig. 4.10). These results demonstrate that our activatable FSGs enable off-on switching of ^{19}F MR images by reductive stimulation. FSGs have the advantage of high sensitivity due to the large number of fluorine atoms in the core. The ratio of fluorine atoms to Gd^{3+} complexes was over 500, which is much higher than that of known PRE-based probes.

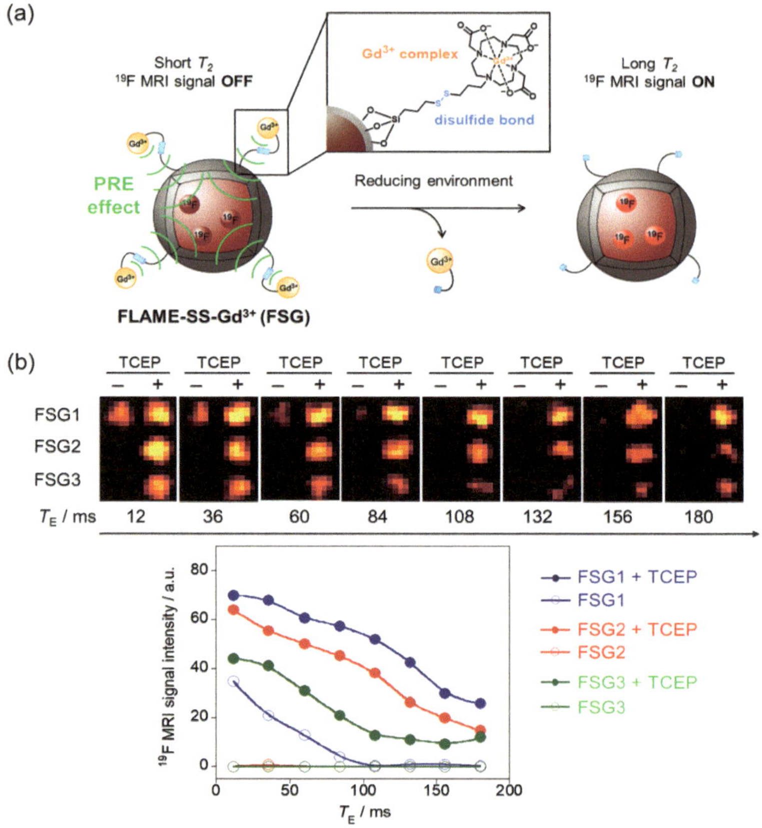

Figure 4.10 (a) Chemical structure of FLAME-SS-Gd^{3+} (FSG). (b) ^{19}F MRI signal enhancement of FSGs by TCEP. (top) ^{19}F MRI phantom images of FSG1-3 with or without TCEP. (bottom) Plot of ^{19}F MRI signal intensity of FSG1-3 at different $T_{E,eff}$ with (filled circles) or without (open circles). Adapted with permission from Ref. [48].

4.5.3 Development of Caspase-3/7-Activatable FLAME

Based on the PRE effect between surface-functionalized Gd^{3+} and the PFCE core in the nanoparticle, we developed ^{19}F MRI nanoprobes for detecting enzyme activity *in vivo* [53]. As the target enzymes, caspase-3 and -7 were selected because they are involved in the maintenance of tissue homeostasis [54]. This activity is used as an apoptosis marker for the development of anticancer drugs [55]. Caspase-3/7-responsive ^{19}F MRI nanoprobes [FLAME-DEVD X (X = 1, 2)] were designed by conjugating a surface-carboxylated FLAME with caspase-3/7 substrate peptides (DEVD) bearing Gd^{3+}-DOTA complexes (Fig. 4.11). Similar to FSGs, the cleavage of substrate peptides by caspase-3/7 increases T_2 and recovers the ^{19}F MRI signal owing to the dissociation of the Gd^{3+} complexes from the FLAME surface. Moreover, the substrate peptide was tandemly repeated in FLAME-DEVD 2 to facilitate access to the surface of the nanoprobe and to increase the number of substrates that can be cleaved.

Caspase-3/7 enzymatic reaction was performed using FLAME-DEVD X and caspase-3. After co-incubation at 37 °C for 12 h, the cleaved peptide fragment (Gd^{3+}-DOTA-Deg-DEVD) was detected in the supernatant of reactions containing FLAME-DEVD X and caspase-3, indicating cleavage of the substrate peptides on the FLAME surface by caspase-3. Next, ^{19}F MRI phantom images of FLAME-DEVD X were measured at the indicated time points. The ^{19}F MRI signal was markedly enhanced after incubation of FLAME-DEVD 1 or FLAME-DEVD 2 with caspase-3. This corresponds with the increase in the recovered T_2 values of FLAME-DEVD 1 (230 ms) and FLAME-DEVD 2 (276 ms), respectively. In contrast, weak ^{19}F MRI signals were observed in the absence of caspase-3. The normalized ^{19}F MRI signal intensity of FLAME-DEVD 2 was 1.7-fold higher than that of FLAME-DEVD-1 after incubation for 12 h. The surface density of Gd^{3+} complexes on the nanoparticle was higher for FLAME-DEVD 2 than for FLAME-DEVD 1.

To test agents in cells, we treated RAW264.7 cells with tumor necrosis factor (TNF)-α and cycloheximide to induce apoptosis, and the cell lysate was then co-incubated with FLAME-DEVE 2. ^{19}F MRI showed that the signals of FLAME-DEVD 2 increased by 3.5-fold after incubation with apoptotic cell lysate. The addition of the caspase-3/7 inhibitor suppressed the ^{19}F MRI signal. These results indicate ^{19}F MRI detection of caspase-3/7 activity in an apoptotic cell lysate with FLAME-DEVD 2.

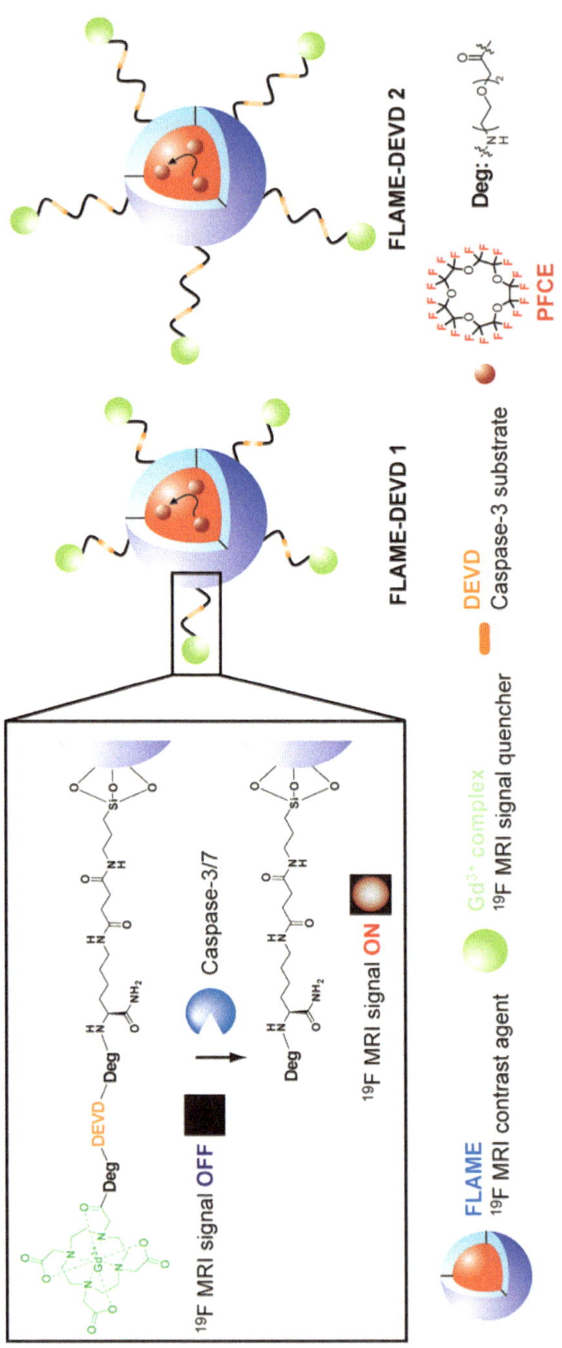

Figure 4.11 Schematic of FLAME-DEVD X (X = 1, 2), enzyme-responsive ^{19}F MRI nanoprobes for detecting caspase-3/7 activity. Reprinted with permission from Ref. [53].

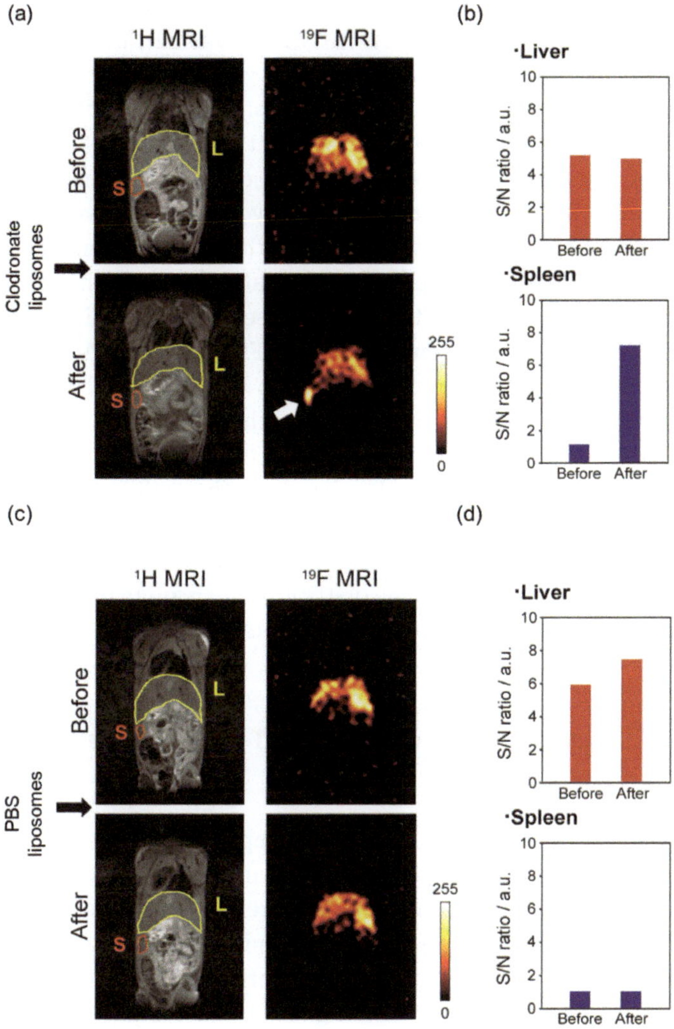

Figure 4.12 (a and c) ^{19}F MRI of caspase-3/7 activity in mouse spleen. FLAME-DEVD 2 (C_{PFCE} = 10 mM, 200 μL) was intravenously injected into a mouse, and ^{1}H/^{19}F MR images were acquired (Before). Then, clodronate (a) or PBS (c) liposomes (100 μL) were intravenously injected. After 24 h, ^{1}H/^{19}F MR images were acquired (After). The positions of the liver (L) and spleen (S) are indicated on the ^{1}H MRI. (b and d) S/N ratio of the liver and spleen before and after injection of clodronate or PBS liposomes. Adapted with permission from Ref. [53].

Next, caspase-3/7 activity of apoptotic macrophages in the liver and spleen was visualized by ^{19}F MRI *in vivo*. To induce apoptosis *in situ*, clodronate liposomes [56, 57] were administered to mice. FLAME-DEVD 2 was then intravenously injected, and the mouse was imaged using ^{1}H/^{19}F MRI. ^{19}F MRI signals for FLAMEDEVD 2 were observed in the liver regardless of the clodronate liposome treatment, likely due to nonspecific hydrolysis by liver proteases. In contrast, no signal was detected in the spleen. After injection of clodronate liposomes, the ^{19}F MRI signals in the spleen were clearly enhanced, and the S/N ratios in the spleen were 6.2-fold higher than before clodronate liposome injection (Fig. 4.12a, b). After injection of control liposomes containing PBS, no ^{19}F MRI signal was detected in the spleen, and the S/N ratio in the spleen did not increase (Fig. 4.12c, d). Fluorescence analysis of the tissue sections indicated that FLAME-DEVD 2 was localized in the spleen regardless of apoptosis induction. Hence, the enhancement of the ^{19}F MRI signal for FLAME-DEVD 2 in the spleen was attributed to caspase-3/7 activity, rather than the probe distribution. Through these strategies, caspase-3/7 activity at a targeted diseased site could potentially be visualized by ^{19}F MRI.

4.5.4 Development of Caspase-1-Activatable ^{19}F MRI Nanoprobes

This design strategy of tandemly repeating substrates can be used to improve the reactivity of another enzyme, caspase-1 [58]. Caspase-1 is a family of cysteine proteases involved in the maturation and secretion of pro-inflammatory cytokines in response to an inflammatory response [59]. Therefore, imaging of caspase-1 activity facilitates understanding of the mechanisms of various inflammatory diseases and evaluating the therapeutic efficacy of anti-inflammatory drugs. To prepare the caspase-1 activatable nanoprobe, Gd^{3+} complex-conjugated peptides containing the substrate sequence, WEHD [60], were conjugated to the FLAME surface. To test whether the tandem repetition of the substrate peptide could improve the cleavage efficiency of caspase-1 on the nanoparticle surface, we prepared three nanoprobes, FLAME-WEHD X (X = 1–3). X indicates the number of repeats of the substrate peptide. After incubation with caspase-1, clear ^{19}F MRI signals were observed in FLAME-WEHD 2 and 3. The cleavage efficiencies of the substrate peptides were enhanced by increasing the number of repeats (Fig. 4.13). These results support the validity of the tandem repetition of the substrate peptide sequence to improve the response of ^{19}F MRI signal upon enzyme reaction. Using FLAME-WEHD 3, we visualized the immune response in living mice by ^{19}F MRI. Our probes led to outstanding sensitivity in visualizing endogenous enzymatic activity *in vivo*.

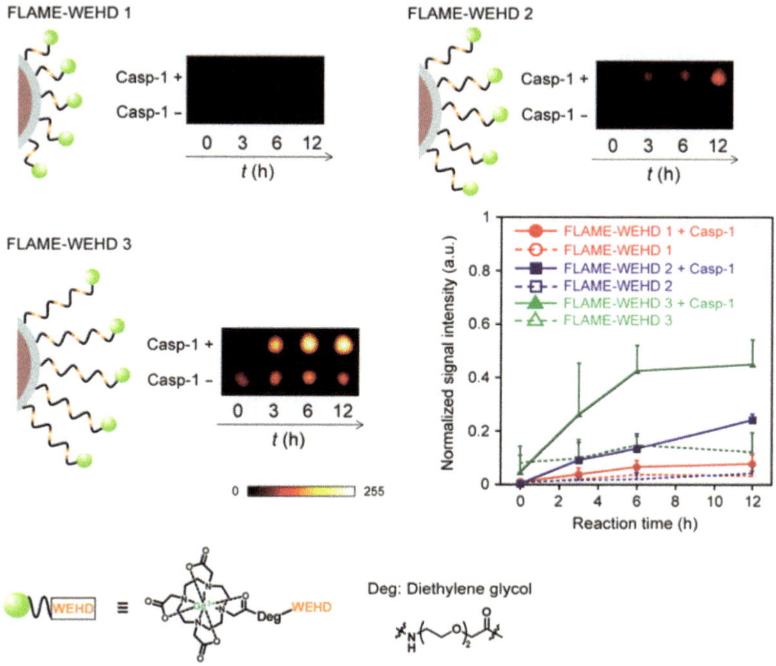

Figure 4.13 ^{19}F MRI phantom images and time course of ^{19}F MRI signal intensity of FLAME-WEHD 1–3 with (filled) or without (open) caspase-1. Data are presented as means ± SD (n = 3). Adapted with permission from Ref. [58].

4.5.5 Development of Cathepsin K-Activated ^{19}F MRI Nanoprobes

Cysteine cathepsins are essential proteolytic enzymes involved in various physiological processes, such as protein processing and protein degradation. Cathepsin K is a cysteine protease that is highly expressed in osteoclasts and plays an important role in bone tissue resorption [61]. The enzyme is secreted extracellularly from the lysosomal compartment of activated osteoclasts to degrade type I/II collagen fibers in the bone matrix. Excessive osteoclast activity is implicated in bone tissue diseases, such as osteoporosis, rheumatoid arthritis, and bone metastasis. A class of small molecules have been developed as selective cathepsin K inhibitors, some of which are employed in preclinical and clinical studies. Thus, the detection of cathepsin K activity *in vivo* is required to develop these drugs and to understand the mechanisms of these diseases in detail.

Based on the design strategy of the activatable ^{19}F MRI probe, ^{19}F MRI nanoparticle probes (FLAME-Gd-Acp and FLAME-Gd-Deg) were developed to detect the activity of cathepsin K (Fig. 4.14a) [62]. This nanoparticle-based probe consists of a PFCE-encapsulated nanoparticle, a Gd^{3+} complex, a cathepsin K substrate, and a linker. The HPGGPQ peptide was selected as a cathepsin K substrate because of its high reactivity and selectivity against cathepsin K [63]. Hydrophobic (aminocaproic acid: Acp) and hydrophilic (diethylene glycol: Deg) linkers were incorporated into each end of the substrate to evaluate the effect of the linkers. The amount of modified Gd^{3+} complexes in the nanoparticles was 10^6 per silica nanoparticle, indicating that there was sufficient Gd^{3+} to attenuate ^{19}F MRI signals.

^{19}F NMR spectra of FLAME-(Gd-Acp) and FLAME-(Gd-Deg) showed ^{19}F NMR peak broadening compared to FLAME-COOH without Gd^{3+} complexes. T_2 values of FLAME-(Gd-Acp) and FLAME-(Gd-Deg) were also shortened to 41 and 43 ms, respectively, from 432 ms (unmodified FLAME-COOH). These results indicate that ^{19}F NMR signals from these nanoprobes were largely suppressed because of T_2 shortening by the PRE effect.

To detect cathepsin K activity by ^{19}F MRI, FLAME-(Gd-X) (X: Acp or Deg) was incubated with cathepsin K in a reaction buffer, and ^{19}F images were acquired at various time points. After the enzyme reaction, the ^{19}F MRI signals of both FLAME-(Gd-Acp) and FLAME-(Gd-Deg) were recovered and almost saturated after 3 h. In contrast, ^{19}F MRI signals were not detected after incubation without cathepsin K (Fig. 4.14b). To examine the selectivity, ^{19}F images were acquired after the reaction of FLAME-(Gd-Acp) and FLAME-(Gd-Deg) probes with a series of cathepsins (Fig. 4.14c). ^{19}F MRI signals of these probes were detected in samples incubated with cathepsin

K, whereas almost no signal was detected in other cathepsins (cathepsins B, S, and L). Consequently, both FLAME-(Gd-Acp) and FLAME-(Gd-Deg) have high selectivity for the imaging of cathepsin K activity by ^{19}F MRI. Our OFF/ON-switching strategy can be broadly applied to detect the activity of various enzymes by varying the amino acid sequence of the substrate.

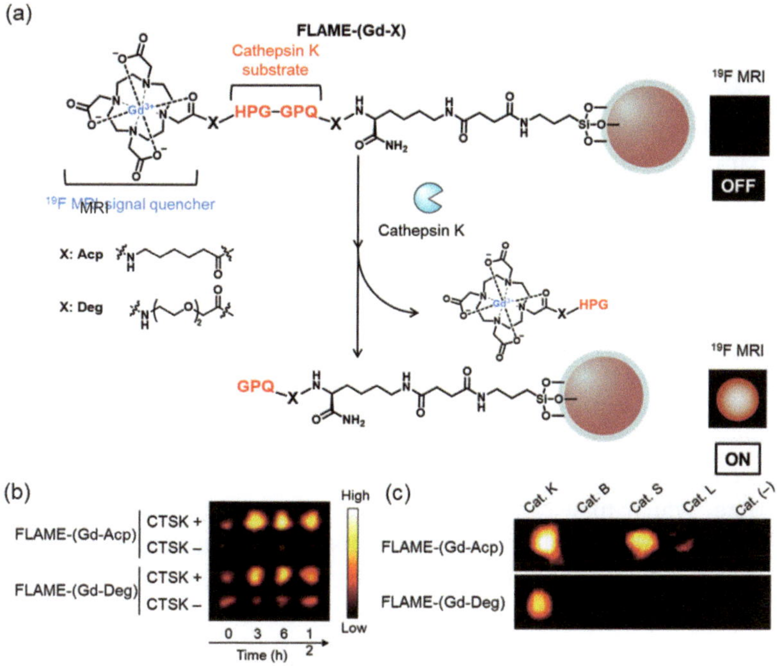

Figure 4.14 ^{19}F MRI measurement of FLAME-(Gd-X) (X = Acp, Deg). (a) Chemical structure of FLAME-(Gd-Acp) and FLAME-(Gd-Deg) and principle of off/on switching. (b) ^{19}F MRI phantom image of FLAME-(Gd-X) with or without cathepsin K. (c) ^{19}F MRI phantom images of FLAME-(Gd-Acp) and FLAME-(Gd-Deg) after incubation with cathepsin K, B, S, and L.

4.6 Summary

In this chapter, we describe the design and preparation of perfluorocarbon-encapsulated silica nanoparticles (FLAME) and their use as *in vivo* ^{19}F MRI tracer agents. FLAME contains an F-dense liquid core enabling strong ^{19}F signals *in vivo*. Our silica nanoprobes can have a variety of molecules as a component of the core, shell, and surface, rendering multiple functions of drug release, targeted delivery, and biomolecular sensing. Encapsulation of different perfluorocarbons enables multicolor ^{19}F MRI. Using mesoporous silica as a shell, nanoparticles can be used as drug carriers by incorporating the drugs into the pores. In addition, fluorophores can be conjugated with the silica shell of FLAME for fluorescent/MRI dual-mode imaging. Furthermore, the surface of FLAME can be modified with various functional groups in organic solvents. PEG-modified FLAME reduced liver uptake and accumulated in tumors in living mice. As in the case of small-molecule ^{19}F MRI probes, the paramagnetic Gd^{3+}-DOTA complex-modified FLAME attenuated the ^{19}F MRI signals by the PRE effect. This enables the design of activatable ^{19}F MRI nanoprobes for detecting reducing molecules and enzyme activities of caspase and cathepsin. The caspase-3 activatable nanoprobe can detect the apoptotic response of spleen macrophages upon anticancer drug treatment.

For future challenges, the biodistribution of silica nanoparticles needs to be improved. Silica nanoparticles tend to accumulate in the liver and spleen despite surface PEG modifications. Moreover, the accumulated silica nanoparticles remain in these tissues for a long time, which may cause undesirable effects such as inflammation and carcinogenesis. As the *in vivo* distribution of the nanoparticles is dependent on the size and surface functional groups [64], the preparation of FLAME will be optimized for suitable biodistribution. The development nanoparticles with biodegradability or rapid excretion would also be beneficial for further *in vivo* diagnostic applications. Moreover, by combining targeted delivery of OFF-ON activatable switches and drug release systems, FLAME-based nanoprobes may enable the detection of target biomolecules and controlled drug release at local sites *in vivo*. FLAME-based activatable/targetable nanoprobes may also be used with multicolor ^{19}F MRI, which may provide distinctive images of the dynamics of biomolecules or cells in an individual subject. Overall, the perfluorocarbon-encapsulated silica-nanoparticle platform should be useful for various *in vivo* diagnostic and therapeutic applications in the biomedical field.

References

1. Higuchi, M., Iwata, N., Matsuba, Y., Sato, K., Sasamoto, K., and Saido, T. C. (2005). ^{19}F and ^{1}H MRI detection of amyloid beta plaques *in vivo. Nat. Neurosci.*, **8**, 527–533.
2. Srinivas, M., Heerschap, A., Ahrens, E. T., Figdor, C. G., and de Vries, I. J. M. (2010). ^{19}F MRI for quantitative *in vivo* cell tracking. *Trends Biotechnol.*, **28**, 363–370.
3. Ruiz-Cabello, J., Barnett, B. P., Bottomley, P. A., and Bulte, J. W. M. (2011). Fluorine (^{19}F) MRS and MRI in biomedicine. *NMR Biomed.*, **24**, 114–129.
4. Tirotta, I., Dichiarante, V., Pigliacelli, C., Cavallo, G., Terraneo, G., Bombelli, F. B., Metrangolo, P., and Resnati, G. (2015). ^{19}F magnetic resonance imaging (MRI): from design of materials to clinical applications. *Chem. Rev.*, **115**, 1106–1129.
5. Liu, J., and Lee, L. S. (2007). Effect of fluorotelomer alcohol chain length on aqueous solubility and sorption by soils. *Environ. Sci. Technol.*, **41**, 5357–5362.
6. Tanaka, K., Kitamura, N., Naka, K., and Chujo, Y. (2008). Multi-modal ^{19}F NMR probe using perfluorinated cubic silsesquioxane-coated silica nanoparticles for monitoring enzymatic activity. *Chem. Commun.*, 6176–6178.
7. Ahrens, E. T., Flores, R., Xu, H. Y., and Morel, P. A. (2005). *In vivo* imaging platform for tracking immunotherapeutic cells. *Nat. Biotechnol.*, **23**, 983–987.
8. Ruiz-Cabello, J., Walczak, P., Kedziorek, D. A., Chacko, V. P., Schmieder, A. H., Wickline, S. A., Lanza, G. M., and Bulte, J. W. (2008). *In vivo* "hot spot" MR imaging of neural stem cells using fluorinated nanoparticles. *Magn. Reason. Med.*, **60**, 1506–1511.
9. Kislukhin, A. A., Xu, H. Y., Adams, S. R., Narsinh, K. H., Tsien, R. Y., and Ahrens, E. T. (2016). Paramagnetic fluorinated nanoemulsions for sensitive cellular fluorine-19 magnetic resonance imaging. *Nat. Mater.*, **15**, 662–668.
10. Tirotta, I., Mastropietro, A., Cordiglieri, C., Gazzera, L., Baggi, F., Baselli, G., Bruzzone, M. G., Zucca, I., Cavallo, G., Terraneo, G., Bombelli, F. B., Metrangolo, P., and Resnati, G. (2014). A superfluorinated molecular probe for highly sensitive *in vivo* ^{19}F-MRI. *J. Am. Chem. Soc.*, **136**, 8524–8527.

11. Riess, J. G., and Postel, M. (1992). Stability and stabilization of fluorocarbon emulsions destined for injection. *Biomater. Artif. Cells Immobilization Biotechnol.*, **20**, 819–830.
12. Freire, M. G., Dias, A. M. A., Coelho, M. A. Z., Countinho, J. A. P., and Marrucho, I. M. (2005). Aging mechanisms of perfluorocarbon emulsions using image analysis. *J. Colloid Interface Sci.*, **286**, 224–232.
13. Matsushita, H., Mizukami, S., Sugihara, F., Nakanishi, Y., Yoshioka, Y., and Kikuchi, K. (2014). Multifunctional core-shell silica nanoparticles for highly sensitive ^{19}F magnetic resonance imaging. *Angew. Chem., Int. Ed.*, **53**, 1008–1011.
14. Vivero-Escoto, J. L., Huxford-Phillips, R. C., and Lin, W. (2012). Silica-based nanoprobes for biomedical imaging and theranostic applications. *Chem. Soc. Rev.*, **41**, 2673–2685.
15. Watanabe, S., Ozaki, H., Mitsuhashi, K., Nakahama, S., and Yamaguchi, K. (1992). Emulsion polymerization of styrene using phospholipids with polymerizable chain as an emulsifier. *Makromol. Chem.*, **193**, 2781–2792.
16. Wang, S., Zhang, M., Wang, D., and Liu, S. (2011). Synthesis of hollow mesoporous silica microspheres through surface sol–gel process on polystyrene-co-poly(4-vinylpyridine) core–shell microspheres. *Microporous Mesoporous Mater.*, **139**, 1–7.
17. Matsumura, Y., and Maeda, H. (1986). A new concept for macromolecular therapeutics in cancer chemotherapy: mechanism of tumoritropic accumulation of proteins and the antitumor agent smancs. *Cancer Res.*, **46**, 6387–6392.
18. Li, S., and Huang, L. (2009). Nanoparticles evading the reticuloendothelial system: role of the supported bilayer. *Biochim. Biophys. Acta Biomembr.*, **1788**, 2259–2266.
19. Li, Z., Barnes, J. C., Bosoy, A., Stoddart, J. F., and Zink, J. I. (2012). Mesoporous silica nanoparticles in biomedical applications. *Chem. Soc. Rev.*, **41**, 2590–2605.
20. Tang, F., Li, L., and Chen, D. (2012). Mesoporous silica nanoparticles: synthesis, biocompatibility and drug delivery. *Adv. Mater.*, **24**, 1504–1534.
21. Lai, C.-Y., Trewyn, B. G., Jeftinija, D. M., Jeftinija, K., Xu, S., Jeftinija S., and Lin, V. S.-Y. (2003). A mesoporous silica nanosphere-based carrier system with chemically removable CdS nanoparticle caps for stimuli-responsive controlled release of neurotransmitters and drug molecules. *J. Am. Chem. Soc.*, **125**, 4451–4459.

22. Giri, S., Trewyn, B. G., Stellmarker M. P., and Lin, V. S.-Y. (2005). Stimuli-responsive controlled-release delivery system based on mesoporous silica nanorods capped with magnetic nanoparticles. *Angew. Chem., Int. Ed.*, **44**, 5038–5044.

23. Nakamura, T., Sugihara, F., Matsushita, H., Yoshioka, Y., Mizukami, S., and Kikuchi, K. (2015). Mesoporous silica nanoparticles for ^{19}F magnetic resonance imaging, fluorescence imaging, and drug delivery. *Chem. Sci.* **6**, 1986–1990.

24. Sega E. I., and Low, P. S. (2008). Tumor detection using folate receptor-targeted imaging agents. *Cancer Metastasis Rev.*, **27**, 655–664.

25. Ke, C.-Y., Mathias, C. J., and Green, M. A. (2003). The folate receptor as a molecular target for tumor-selective radionuclide delivery. *Nucl. Med. Biol.*, **30**, 811–817.

26. Choi, H., Choi, S. R., Zhou, R., Kung H. F., and Chen, I.-W. (2004). Iron oxide nanoparticles as magnetic resonance contrast agent for tumor imaging via folate receptor-targeted delivery. *Acad. Radiol.*, **11**, 996–1004.

27. Hickman, J. A. (1992). Apoptosis induced by anticancer drugs. *Cancer Metastasis Rev.*, **11**, 121–139.

28. Huang, X. N., Huang, G., Zhang, S. R., Sagiyama, K., Togao, O., Ma, X. P., Wang, Y. G., Li, Y., Soesbe, T. C., Sumer, B. D., Takahashi, M., Sherry, A. D., and Gao, J. M. (2013). Multi-chromatic pH-activatable ^{19}F-MRI nanoprobes with binary ON/OFF pH transitions and chemical-shift barcodes. *Angew. Chem. Int. Ed.*, **52**, 8074–8078.

29. Bar-Shir, A., Yadav, N. N., Gilad, A. A., van Zijl, P. C. M., McMahon, M. T., and Bulte, J. W. M. (2015). Single ^{19}F probe for simultaneous detection of multiple metal ions using miCEST MRI. *J. Am. Chem. Soc.*, **137**, 78–81.

30. Ashur, I., Allouche-Arnon, H., and Bar-Shir, A. (2018). Calcium fluoride nanocrystals: tracers for *in vivo* ^{19}F magnetic resonance imaging. *Angew. Chem. Int. Ed.*, **57**, 7478–7482.

31. Partlow, K. C., Chen, J. J., Brant, J. A., Neubauer, A. M., Meyerrose, T. E., Creer, M. H., Nolta, J. A., Caruthers, S. D., Lanza, G. M., and Wickline, S. A. (2007). ^{19}F magnetic resonance imaging for stem/progenitor cell tracking with multiple unique perfluorocarbon nanobeacons. *FASEB J.*, **21**, 1647–1654.

32. Lim, Y. T., Noh, Y. W., Cho, J. H., Han, J. H., Choi, B. S., Kwon, J., Hong, K. S., Gokarna, A., Cho, Y. H., and Chung, B. H. (2009). Multiplexed imaging of therapeutic cells with multispectrally encoded magnetofluorescent nanocomposite emulsions. *J. Am. Chem. Soc.*, **131**, 17145–17154.

33. Weise, G., Basse-Lusebrink, T. C., Kleinschnitz, C., Kampf, T., Jakob, P. M., and Stoll, G. (2011). *In vivo* imaging of stepwise vessel occlusion in cerebral photothrombosis of mice by ^{19}F MRI. *PLoS One*, **6**, e28143.
34. Cohen, D., Mashiach, R., Houben, L., Galisova, A., Addadi, Y., Kain, D., Lubart, A., Blinder, P., Allouche-Arnon, H., and Bar-Shir, A. (2021). Glyconanofluorides as immunotracers with a tunable core composition for sensitive hotspot magnetic resonance imaging of inflammatory activity. *ACS Nano*, **15**, 7563–7574.
35. Akazawa, K., Sugihara, F., Nakamura, T., Matsushita, H., Mukai, H., Akimoto, R., Minoshima, M., Mizukami, S., and Kikuchi, K. (2018). Perfluorocarbon-based ^{19}F MRI nanoprobes for *in vivo* multicolor imaging. *Angew. Chem. Int. Ed.*, **57**, 16742–16747.
36. Zhang, S., Kuangcong, W., and Sherry, A. D. (1999). A novel pH-sensitive MRI contrast agent. *Angew. Chem., Int. Ed.*, **38**, 3192–3194.
37. Aime, S., Barge, A., Castelli, D. D., Fedeli, F., Mortillaro, A., Nielsen, F. U., and Terreno, E. (2002). Paramagnetic lanthanide(III) complexes as pH-sensitive chemical exchange saturation transfer (CEST) contrast agents for MRI applications. *Magn. Reson. Med.*, **47**, 639–648.
38. Louie, A. Y., Hüber, M. M., Ahrens, E. T., Rothbacher, U., Moats, R., Jacobs, R. E., Fraser, S. E., and Meade, T. J. (2000). *In vivo* visualization of gene expression using magnetic resonance imaging. *Nat. Biotechnol.* **18**, 321–325.
39. Perez, J. M., Josephson, L., O'Loughlin, T., Högemann, D., and Weissleder, R. (2002). Magnetic relaxation switches capable of sensing molecular interactions. *Nat. Biotechnol.* **20**, 816–820.
40. Yoo, B., and Pagel, M. D. (2006). A PARACEST MRI contrast agent to detect enzyme activity. *J. Am. Chem. Soc.* **128**, 14032–14033.
41. Okada, S., Bartelle, B. B., Li, N., Breton-Provencher, V. Lee, J., Rodriguez, E., Melican, J., Sur, M., and Jasanoff, A. (2018). Calcium-dependent molecular fMRI using a magnetic nanosensor. *Nat Nanotechnol.*, **13**, 473–477.
42. Ye, D., Shuhendler, A. J., Pandit, P., Brewer, K. D., Tee, S. S., Cui, L., Tikhomirov, G., Rutt, B., and Rao, J. (2014). Caspase-responsive smart gadolinium-based contrast agent for magnetic resonance imaging of drug-induced apoptosis. *Chem. Sci.* **5**, 3845–3852.
43. Takaoka, Y., Sakamoto, T., Tsukiji, S., Narazaki, M., Matsuda, T., Tochio, H., Shirakawa, M., and Hamachi, I. (2009). Self-assembling nanoprobes that display off/on ^{19}F nuclear magnetic resonance signals for protein detection and imaging. *Nat. Chem.*, **1**, 557–561.

44. Matsuo, K., Kamada, R., Mizusawa, K., Imai, H., Takayama, Y., Narazaki, M., Matsuda, T., Takaoka, Y., and Hamachi, I. (2013). Specific detection and imaging of enzyme activity by signal amplifiable self-assembling ^{19}F MRI probes. *Chem. Eur. J.* **19**, 12875–12883.

45. Yuan, Y., Sun, H. B., Ge, S. C., Wang, M. J., Zhao, H. X., Wang, L., An, L. N., Zhang, J., Zhang, H. F., Hu, B., Wang, J., and Liang, G. (2015). Controlled intracellular self-assembly and disassembly of ^{19}F nanoparticles for MR imaging of caspase 3/7 in zebrafish. *ACS Nano,* **9**, 761–768.

46. Mizukami, S., Takikawa, R., Sugihara, F., Hori, Y., Tochio, H., Wälchli, M., Shirakawa, M., and Kikuchi, K. (2008). Paramagnetic relaxation-based ^{19}F MRI probe to detect protease activity. *J. Am. Chem. Soc.* **130**, 794–795.

47. Mizukami, S., Matsushita, H., Takikawa, R., Sugihara, F., Shirakawa, M., and Kikuchi, K. (2011). ^{19}F MRI detection of β-galactosidase activity for imaging of gene expression. *Chem. Sci.*, **2**, 1151–1155.

48. Nakamura, T., Matsushita, H., Sugihara, F., Yoshioka, Y., Mizukami, S., and Kikuchi, K. (2015). Activatable ^{19}F MRI nanoparticle probes for the detection of reducing environments. *Angew. Chem., Int. Ed.,* **54**, 1007–1010.

49. De Vries, A., Moonen, R., Yildirim, M., Langereis, S., Lamerichs, R., Pikkemaat, J. A., Baroni, S., Terreno, E., Nicolay, K., Strijkers, G. J., and Grüll, H. (2014). Relaxometric studies of gadolinium-functionalized perfluorocarbon nanoparticles for MR imaging. *Contrast Media Mol. Imaging,* **9**, 83–91.

50. Clore, G. M., and Iwahara, J. (2009). Theory, practice, and applications of paramagnetic relaxation enhancement for the characterization of transient low-population states of biological macromolecules and their complexes. *Chem. Rev.*, **109**, 4108–4139.

51. Iwahara, J., and Clore, G. M. (2006). Detecting transient intermediates in macromolecular binding by paramagnetic NMR. *Nature,* **440**, 1227–1230.

52. Townsend, D. M., Tew, K. D., and Tapiero, H. (2003). The importance of glutathione in human disease. *Biomed. Pharmacother.*, **57**, 145–155.

53. Akazawa, K., Sugihara, F., Nakamura, T., Mizukami, S., and Kikuchi, K. (2018). Highly sensitive detection of caspase-3/7 activity in living mice using enzyme-responsive ^{19}F MRI nanoprobes. *Bioconjugate Chem.*, **29**, 1720–1728.

54. Porter, A. G., and Janicke, R. U. (1999). Emerging roles of caspase-3 in apoptosis. *Cell Death Differ.*, **6**, 99–104.

55. Blankenberg, F. G. (2008). *In vivo* detection of apoptosis. *J. Nucl. Med.* **49**, 81–95.

56. Lehenkari, P. P., Kellinsalmi, M., Napankangas, J. P., Ylitalo, K. V., Monkkonen, J., Rogers, M. J., Azhayev, A., Vaananen, H. K., and Hassinen, I. E. (2002). Further insight into mechanism of action of clodronate: inhibition of mitochondrial ADP/ATP translocase by a nonhydrolyzable, adenine-containing metabolite. *Mol. Pharmacol.* **61**, 1255–1262.

57. Buiting, A. M. J., Zhou, F., Bakker, J. A. J., van Rooijen, N., and Huang, L. (1996). Biodistribution of clodronate and liposomes used in the liposome mediated macrophage 'suicide' approach. *J. Immunol. Methods* **192**, 55–62.

58. Akazawa, K., Sugihara, F., Minoshima, M., Mizukami, S., and Kikuchi, K. (2018). Sensing caspase-1 activity using activatable 19 F MRI nanoprobes with improved turn-on kinetics. *Chem. Commun.*, **54**, 11485–11788.

59. Franchi, L., Eigenbrod, T., Munoz-Planillo R., and Nunez, G. (2009). The inflammasome: a caspase-1-activation platform that regulates immune responses and disease pathogenesis. *Nat. Immunol.*, **10**, 241–247.

60. Thornberry, N. A., Rano, T. A., Peterson, E. P., Rasper, D. M., Timkey, T., Garcia-Calvo, M., Houtzager, V. M., Nordstrom, P. A., Roy, S., Vaillancourt, J. P., Chapman, K. T., and Nicholson, D. W. (1997). A combinatorial approach defines specificities of members of the caspase family and granzyme B. Functional relationships established for key mediators of apoptosis. *J. Biol. Chem.*, **272**, 17907–17911.

61. Palermo, C., and Joyce, J. A. (2007). Cysteine cathepsin proteases as pharmacological targets in cancer. *Trends Pharmacol. Sci.*, **29**, 22–28.

62. Konishi, Y., Okunishi, A., Sugihara, F., Nakamura, T., Akazawa, K., Minoshima, M., and Kikuchi, K. (2021). Development of off-on switching ^{19}F MRI probes for cathepsin K activity detection. *Bull. Chem. Soc. Jpn.*, **94**, 1690–1694.

63. Lecaille, F., Weidauer, E., Juliano, M. A., Brömme, D., and Lalmanach, G. (2003). Probing cathepsin K activity with a selective substrate spanning its active site. *Biochem. J.*, **375**, 307–312.

64. Blanco, E., Shen, H., and Ferrari, M. (2015). Principles of nanoparticle design for overcoming biological barriers to drug delivery. *Nat. Biotechnol.*, **33**, 941–951.

Chapter 5

Perfluorocarbon Theranostic Nanoemulsions: From Imaging to Treatment of Inflammatory Diseases and Pain

Jelena M. Janjic,[a] Riddhi Vichare[a] and Sravan K. Patel[b,c]

[a]*Graduate School of Pharmaceutical Sciences, School of Pharmacy, Duquesne University, Pittsburgh, PA, USA*

[b]*School of Pharmacy, University of Pittsburgh, PA, USA*

[c]*Magee-Women's Research Institute, Pittsburgh, PA, USA*

janjicj@duq.edu

5.1 Theranostic Nanomedicines as Future Personalized Medicines

Theranostic nanomedicine, as a concept, aims to fully personalize treatment by combining therapeutics with diagnostics into one nanosystem. The goal is to create a tailored intervention for every individual that is adaptable to the patient's true state; this enables the delivery of only the treatment needed, no more and no less.

Fluorine Magnetic Resonance Imaging: Methods and Applications in Biomedicine
Edited by Eric T. Ahrens and Ulrich Flögel
Copyright © 2025 Jenny Stanford Publishing Pte. Ltd.
ISBN 978-981-5129-37-3 (Hardcover), 978-1-003-53004-6 (eBook)
www.jennystanford.com

Furthermore, theranostic nanomedicines deliver the additional benefit of full targeting to specific cells and/or tissues, which avoids off-site side-effects and improves efficacy and safety. Imaging specific cells/tissues and responses to treatment could contribute to a better understanding of the disease and the effect of the drug in a specific patient, and as an additional benefit, could give insights into the disease pathology and provide mechanistic insights. Ideally, a diagnostic agent in a theranostic nanomedicine works in synergy with the therapeutic moiety (drug) by providing input on amount of drug reaching its anatomical, cellular, and/or molecular target, feedback on drug effects on the underlying pathology, and therapeutic efficacy and/or side effects. Furthermore, by imaging directly the cellular-molecular changes in real time, the drug treatment can be fine-tuned for maximal efficacy. The diagnostic moiety in the theranostic nanomedicine can also provide clues to the mechanisms of the drug target and indirectly help fine-tune both the dose and dosing regimen. Therefore, an ideal theranostic nanomedicine provides an opportunity to personalize treatment during the entire course of therapy in an individual patient.

To achieve the above goals, theranostic nanomedicine must grow beyond smart engineering and into pharmaceutical development. Adding an imaging agent or drug to an existing drug delivery system or imaging platform is only a partial solution and does not offer personalized treatment. The two have to work in synergy, which can be achieved if the whole theranostic nanosystem is viewed from the lens of drug design and that vision is maintained from the inception of the nanosystem through all steps of development until reaching a patient. For a pharmaceutical scientist, theranostic nanomedicines are the future, but they must meet the same standards of effectiveness and safety as all other clinically approved treatments. Figure 5.1 summarizes the key structural and functional features of theranostic nanomedicines (nanotheranostics) as future personalized medicines [1].

This chapter discusses perfluorocarbon (PFC) based materials that offer unique opportunities for advancing theranostic nanomedicines to clinically viable therapeutic and diagnostic solutions for inflammatory diseases, cancer, and post-trauma/surgical pain and inflammation management. While a recent review [3] surveys diverse structural nanosystems that can be imaged by ^{19}F magnetic resonance imaging (^{19}F MRI), here we will focus specifically on PFC nanoemulsions (PFC-Nes) as the theranostic nanomedicine platform for imaging and treatment of inflammation in inflammatory diseases (*e.g.*, rheumatoid arthritis, inflammatory bowel disease), cancer associated inflammation, neuroinflammation following trauma, infection, and tissue (*e.g.*, muscle, skin) injury and associated pain, both acute and chronic.

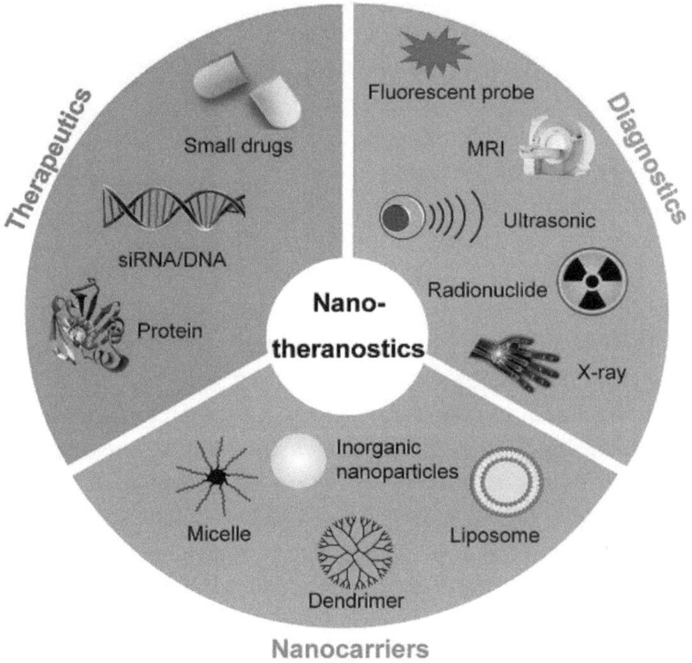

Figure 5.1 Nanotheranostic critical features, functions and capabilities. Figure reproduced from Ref. [2].

Theranostic nanomedicine development is still challenged primarily by the inherent complexity of combining a therapeutic entity (*i.e.*, small molecule drug, protein, RNA or DNA) with a diagnostic into one nanosystem. To address this, we aim to discuss some novel design strategies and the utilization of modern pharmaceutical dosage form development methods as applied to theranostic PFC-Nes. We will present examples where application of quality by design (QbD) methodologies led to macrophage targeted PFC-Nes that showed both therapeutic effects and imaging functionality in rodent models of inflammation and pain [4–9].

Typically, nanosystems, including theranostics, are targeted to affected organs and tissues either passively or actively for both imaging and drug delivery. The biodistribudion of nanoparticles is influenced by their composition and structure [10]. The pharmacokinetic and functional profiles of each component of a nanosystem, therapeutic and diagnostic, contribute to the overall theranostic nanosystem individually and in combination. For example, if the aim is to achieve high tumor targeting with an antiproliferative agent using a nanosystem, consideration must be made for drug loading and release from the nanosystem as well as the bioditribution and targeting of the nanosystem. If the goal is to simply image the biodistribution of the nanosystem, it may not be desirable for the imaging agent to be released along with the drug, or a completely different release profile may be needed for individual entities. In an example from our work, we introduced an imaging agent to a drug delivery nanosystem because the drug carrier biodistribution is what we are assessing. This approach to theranostics development [11] provides very useful information on the drug delivery system performance with respect to stability, biodistribution, and target tissue/organ retention. Although it may be useful to image biodistribution of the drug carrier, does this in itself truly define a theranostic? Biodistribution of a nanosystem alone still does not provide clues on how well the drug is working or what the changes are in the tissue or an organ in a particular pathophysiological state of the disease in response to drug delivery. Furthermore, dosing balance between imaging agent and the drug in a theranostic nanosystem is crucial. In a recent review, Janib *et al.* [12] pointed out that for some types of diagnostics to achieve their full imaging efficacy, the required dose may lead to toxicity and confoud understanding the pathology it aims to investigate. On the other hand, to counterbalance toxicity associated with imaging agents, one may need to adjust the overall dose of nanotheranostic, which could lead to subtherapeutic levels of the drug delivered by the nanosystem. One way to solve this problem would be to increase drug loading. However, doing so may lead to loading barriers for both the imaging and therapeutic moiety. By approaching the design of theranostic nanosystems with full appreciation of these challenges and making smart material choices, one may achieve the goal of truly personalized nanomedicine.

In the drug discovery field, fragment-based drug design has been used to overcome similar barriers to balance the efficacy and safety of new drugs. Fragment-based drug design considers optimization of fragments of a drug molecule and subsequent incorporation of these fragments into a functional whole [13]. If we consider the same principles for the rational design of nanotheranostics, it is possible to see solutions for balancing imaging with drug delivery. Taking this further, we can consider viewing each component of the nanosystem (*e.g.*, surfactants, shells, etc.) as fulfilling more than one role. For example, the drug carrier could also serve as an imaging agent.

5.2 Perfluorocarbon Nanoemulsions as Model Nanotheranostics

An advantage of nanotheranostics incorporating PFCs is background free imaging of *in vivo* biodistribution, which can provide indirect insight into the therapeutic effect of the drug delivered. For example, inflammation changes are observed by fluorine-19 MRI with targeted theranostic PFC nanoemuslion in response to the anti-inflammatory drug delivered at the site of nerve injury [9] or inflammatory insult [14]. Overall, PFCs designed for ^{19}F MRI can be divided into three general groups: (1) perfluorinated small molecules (MW < 1000 Da) [15], (2) perfluorinated polymers [16–20], and (3) perfluorocarbon-hydrocarbon conjugates (*e.g.*, perfluorinated surfactants, polyethyleneglycol (PEG) conjugates) [21–23]. PFCs are biologically and metabolically inert, as well as chemically stable in vastly changing environments and do not interact with biological fluids. Hence, PFCs can be used to build nanotheranostics with potentially increased safety profiles compared to other materials, such as those incorporating metals and radioactive tracers. As ^{19}F MRI lacks tissue depth detection limitations, this imaging technique allows for monitoring of theranostic nanosystem biodistribution throughout the body non-invasively, quantitatively, and repeatedly over time. One concern that arises from using PFCs is their slow elmination, which is depended on the PFC structure [15, 24]. However, in preclinical studies, this long half-life can be an advantage to allow for long-term follow up using ^{19}F MRI. Further, PFCs can be conjugated to other molecules which enables a combination of MRI with other imaging modalities such as fluorescence imaging [20, 25, 26] and PET [27, 28], or can enhance cell targeting efficacy, for example, by conjugation to a cell penetrating peptide [29]. Versatile PFC conjugation chemistries are available for functionalization [20, 25, 28–31] of PFC-NEs, which further expands their utility as nanotheranostic platforms.

Kadayyakara *et al.* [32] demonstrated the utility of PFCs as oxygen sensors inside cancer cells using ^{19}F MRI in rats. The changes in ^{19}F relaxation times measured by MRI were directly correlated to the effects of chemotherapeutic drug bischloroethylnitrosourea (BCNU) delivered to rats with growing tumors in the brain. Though the drug was delivered separately as a solution and not with the PFC as a unique nanosystem, the oxygen sensing properties of PFCs demonstrate their theranostic potential. The biosensing properties of PFCs can also be achieved by conjugation to pH sensing fluorescent probes [25]. Nanotheranostics built with these conjugates could provide insight into drug delivery into specific cellular compartments and indirectly serve as a measure of drug payload availability inside the targeted cells. In a recent study, a combination of near-infrared fluorescence (NIRF) imaging and ^{19}F MRI provided a previously unknown relationship between macrophage infiltration levels at the site of nerve injury and associated pain. Here, PFC-NEs labeled with NIRF labeled dye were used to track macrophages in a rat model of neuropathic pain [7, 9, 33]. As we will discuss later in this chapter, loading PFC-NEs with nonsteroidal anti-inflammatory drugs (NSAIDs) led to the supression of pain behaviour in these animals, which was further correlated to a reduction in NIRF signal at the site of injury [9].

Overall, PFCs are effective materials for developing nanotheranostics as they can fulfil multiple roles, both diagnostic and therapeutic carrier.

5.3 Biphasic vs. Triphasic Perfluorocarbon Nanoemulsions

Nanoemulsions are kinetically stable colloidal systems with sub-micron sized droplets and high oil content relative to low amount of surfactants, typically between 1 and 10% [34]. Nanoemulsions have wide applications in pharmaceutical industry where they are typically used to increase tissue penetration of poorly soluble drugs (transdermal delivery), and improve drug solubility and bioavailability (parenteral and oral) [35–39]. They are also easily incorporated into other dosage forms, such as capsules and gels, and can be produced on an industrial scale [39–41]. A variety of processes are used to produce pharmaceutical nanoemulsions that can be divided into two main groups, low energy and high energy methods [42–44]. Phase inversion-based low energy methods were developed for nanoemulsion production on scales amenable to large-scale commercial production [45–47]. High energy processes include high pressure homogenization, high shear processing (*e.g.*, microfluidization), rotor-stator high speed stirring [48] and acoustic cavitation processsing [49]. These processing methods are less sensitive to composition changes and varying drug types, oils and surfactants. In our work, microfludization was the method of choice for theranostic nanoemulsion preparation due to ease of operation, predictable nanoemulsion properties and high colloidal stability of produced nanoemulsions with varied payload and composition [17, 26, 30, 50–52]. What distinguishes most nanoemulsions in the pharmaceutical literature from theranostic nanoemulsions prepared with PFCs is the nature of the internal (oil) phase. Most nanoemulsions for drug delivery purposes are made with hydrocarbon (HC) oils and surfactants, whereas PFC is the major component of the internal phase in nanoemulsions developed for imaging and theranostic applications.

Classical nanoemulsions are composed of two phases: oil and water, where oil droplets are dispersed in water and stabilized by surfactants. Triphasic PFC nanoemulsions are composed of three distinct phases, where both fluorous phase (PFC) and organic phase (HC oil) are combined and the droplets are stabilized in water (aqueous phase) by surfactants, as shown in Fig. 5.2 [50, 52]. Due to the hydrophobic and lipophobic nature of PFCs, the incorporation of hydrocarbon oil into the nanoemulsion internal phase generates a triphasic nanoemulsion system [53].

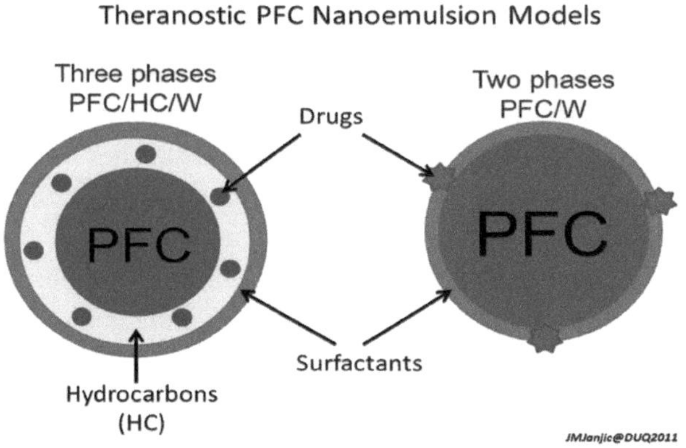

Figure 5.2 Two models of PFC nanotherapeutic design.

Early development of PFC nanoemulsions followed the existing hydrocarbon oil emulsification strategies. PFCs, used as the internal oil phase, were dispersed in the aqueous phase and stabilized by a variety of surfactants, from lipids to non-ionic polymers [54, 55]. These nanoemulsions were first developed as blood substitutes and quickly their potential was realized as ^{19}F MRI agents. An expansive list of PFC colloidal systems (nano and microbubbles and droplets, including nanoemulsions) and their utilization in biomedical research is discussed in depth in a review article by Craig and Krafft [56]. The potential of PFC nanoemulsions for anti-cancer drug delivery was demonstrated in several preclinical models [57–59], including by reducing angiogenesis [60]. Table 5.1 provides an updated list of examples of biphasic (traditional) PFC nanoemulsions. In most cases (Table 5.1), the drug carrying function of the PFC nanoemulsion is based on the entrapment or anchoring of the therapeutic molecules to the surfactant layer of the nanoemulsion droplet, with PFC oil as the internal phase. In one such example, Wickline *et al.* [61] incorporated a cytolitic peptide (melittin) as an anticancer agent into the lipid surfactant layer of the PFC nanoemulsion droplet. They showed that tumor targeting of incorporated melittin lead to a dramatic reduction in tumor size [62]. At the same time, the nanoemulsion yielded ^{19}F MRI signals in tumor in mouse. In another example, paclitaxel was incorporated into a block-polymer non-ionic surfactant stabilizing a perfluoropentane-based nanoemulsion

Table 5.1 Summary of preclinical studies highlighting colloidal and therapeutic/diagnostic properties of perfluorocarbon nanosystems with anti-cancer potential

	Disease	Drug/Gene/Imaging Agent	Perflurocarbon	Mechanism of Action	Size (nm) and Zeta Potential	Route	Biological Model (Cells or Animals)	Therapeutical and Pharmacological Outcomes	Ref.
1.	Breast cancer	PD-1/DsRed (fusion protein) sinoporphyrin sodium	Perfluorotributylamine (PFTBA)	Immune checkpoint blockade (ICB) Anti-tumor activity	150 nm −14.5 mV	I.V.	Subcutaneous 4T1 tumor-bearing mouse model	Rejuvenate T cell response Oxygen delivery to the hypoxic tissue	[70]
2.	Cancer	Benzoporphyrin derivative (BPD), indocyanine green (ICG)	Perfluoropentane	Increased oxygenation improves the efficiency of Photodynamic therapy ROS generation results in necrosis in tumor tissue	204.5 ± 17.3 nm nm	I.V.	Human squamous cell carcinoma (FaDu cell) Swiss nu/nu mice infected with FaDu cells	↑ necrotic area in the tumor ↑ oxygen capacity ↑ contrast for photoacoustic imaging	[71]

(*Continued*)

Table 5.1 (*Continued*)

	Disease	Drug/Gene/Imaging Agent	Perflurocarbon	Mechanism of Action	Size (nm) and Zeta Potential	Route	Biological Model (Cells or Animals)	Therapeutical and Pharmacological Outcomes	Ref.
3.	Bladder cancer	Indocyanine green (ICG), Mitomycin C (MMC), and Anti-EGFR-mAbs	Perfluorooctyl bromide (PFOB)	DNA cross-linking/breaking, glutathione alkylation and inhibition of thioredoxin reductase	257 ± 19.4 nm −12.3 ± 2.33 mV		T24 cells, a cell line established from a human urinary bladder cancer cell	↑Hyperthermic effect with production of singlet oxygen ↑ tumor cell uptake	[72]
4.	Colon cancer	5-Fluorouracil (antimetabolite analogue of pyrimidine) Epidermal growth factor functionalized	Perflurocarbon	5-Fu inhibits nucleoside metabolism and DNA synthesis, resulting in cell apoptosis	200 ± 10.84 nm − 23.7 ± 1.4 mV		Human SW620 colon cancer cells SW620 tumor-bearing female BALB/c mice	↑ binding to EGF receptor-overexpressing solid tumors ↑oxygen diffusion in tumor	[73]
5.	Breast Cancer	Doxorubicin, Indocyanine green, Anti-HER2-monocoloal antibody	Perfluorooctyl bromide (PFOB)	DNA intercalation and free radical generation	340.6 ± 4.5 nm and −57.1 ± 5.6 mV		MDA-MB-453 cells line	↑Resistance to macrophage clearance, ↑ HER2 binding specificity	[74]

	Disease	Drug/Gene/Imaging Agent	Perflurocarbon	Mechanism of Action	Size (nm) and Zeta Potential	Route	Biological Model (Cells or Animals)	Therapeutical and Pharmacological Outcomes	Ref.
6.	Breast cancer	Indocyanine green (ICG), Camptothecin (CPT), Anti-HER2-monocoloal antibody	Perfluorooctyl bromide (PFOB)	Binds to HER2-expressing cells, CPT can block the synthesis of topoisomerase I	292.2 ± 12 nm −13.3 ± 3.1 mV	Intra-tumoral	MDA-MB-453 cell line (HER2(+) human breast meta-static carcinoma cell line), Nude mice bearing MDA-MB-231 tumors	↓Tumorigenic cells by DNA damage, FDOT for tumor detection ↑Hyperthermic effect with production of singlet oxygen	[75]
7.	Cancer	Superparamagnetic Fe_3O_4, 10-hydroxycamptothecin (HCPT), FA conjugated	Perfluoropentane	Active and passive targeting by selectively bind to FA-positive tumor cells	309 nm −25.8 mV	I.V.	SKOV3 ovarian cancer cell line, Tumor-bearing mice	↑ Inhibitory effect on tumor growth ↑contrast enhancement in MRI and PA imaging	[76]

(*Continued*)

Table 5.1 (*Continued*)

	Disease	Drug/Gene/Imaging Agent	Perflurocarbon	Mechanism of Action	Size (nm) and Zeta Potential	Route	Biological Model (Cells or Animals)	Therapeutical and Pharmacological Outcomes	Ref.
8.	Breast cancer	c-MYC (MYC) inhibitor prodrug (MI3-PD), Labeled with αvβ3-PEG-mimetic antagonist	Perfuorooctylbromide (PFOB)	Reprograms M2 (pro-tumor macrophages), Regulates the expression of tumor-promoting genes by inhibiting MYC, αvβ3-Integrins targeted	262 nm –20 mV	I.V.	PyMT-Bo1 β3KO cell line, 4T1-GFP-FL and PyMT-Bo1 β3KO bearing orthotopic mammary fat pad (MFP) tumors	↓M2 polarization, ↓ M2-like tumor-associated macrophages (TAMS), ↓Mammary fat pad TAMS	[77]
9.	Cancer	Cell penetrating peptide (CPP) from the Transactivator of transcription (TAT) peptide, Cyanine-5 (Cy5)	Perfluoropolyether (PFPE)	TAT facilitates endocytosis, Immunotherapy against cancer	180 nm	Intra-tumoral	Mice bearing flank glioma tumors.	↑ Cellular uptake due to labelling CAR T receptor ↑ ^{19}F signal by 8 times	[78]

	Disease	Drug/Gene/Imaging Agent	Perflurocarbon	Mechanism of Action	Size (nm) and Zeta Potential	Route	Biological Model (Cells or Animals)	Therapeutical and Pharmacological Outcomes	Ref.
10.	Cancer	Melittin	PFOB	Compromises cell permeability by pore formation causing cell death	227 nm		Xenograft (MDA-MB-435 human breast cancer); and precancerous lesions in K14-HPV16 mice with squamous carcinoma	↑delivery of a potent model cytolytic peptide (melittin) resulting in apoptosis ↓ Solid tumors and precancerous lesions.	[79]
11.	Pancreatic cancer	siRNA silencing nerve growth factor (NGF), Polymeric C-X-C chemokine receptor type 4 (CXCR4) antagonist (PCX)	Perfuorooctylbromide (PFOB)	Silencing the expression of NGF which contributes to cancer cell survival. CXCR4 inhibition induces apoptosis of cancer cells.	~147 nm and ~19 mV	Intraperi-toneal	KPC8060, COLO357, and PANC-1 cell lines, Human PANC-1 bearing athymic nude mice.	↓ Cancer cell proliferation, ↓Colony formation and metastasis, ↑ Cancer cell apoptosis ↑ Tumor penetration	[80]

designed for image guided delivery using both ultrasound and MRI in a mouse tumor model [63]. Lanza *et al.* [64] also reported targeted delivery of anti-proliferative drugs to smooth muscle cells using perfluorooctyl bromide (PFOB) nanoemulsions. PFC nanoemulsions were also reported to deliver antigen to dendritic cells (DC) with the goal of boosting the immune response in DC-based vaccines [65].

Additionally, PFCs are used in biomedicine to carry oxygen, for example as artifical oxygen carriers and in transplantation and organ preservation [66, 67]. The oxygen carrying capacity of PFCs has been utilized to improve photodynamic therapy (PDT) in solid tumors in preclinical models. In malignant solid tumors, as the distance between cancer cells and vasculature increases, there is a decreased oxygen diffusion. A tumor-specific immune response is triggered by PDT as it targets tumor cells by generating singlet oxygen ($^1O_2$) leading to the release of tumor-associated antigens (TAAs). Due to a lack of oxygen, PDT's effectiveness in hypoxic solid tumors is restricted [68, 69]. Zang *et al.* [70] boosted the efficacy of PDT by using PFC nanoemulsion as an oxygen carrier, as well as a delivery vehicle for sinoporphyrin sodium (DVDMS), a potent photosensitizer. In similar studies by Xavierselvan *et al.* [71], a 9.1 fold increase in oxygen levels was observed inside tumors following administration of a perfluropentane nanoemulsion, subsequently improving PDT efficacy. The effect of photo- and chemo-therapeutics for bladder cancer was demonstrated by Lee *et al.* [72] by encapsulating Mitomycin C (MMC), a chemotherapeutic drug, and indocyanine green (ICG).

Wu *et al.* [73] encapsulated 5-flurouracil (5-Fu) in nanoparticles (NPs) using low energy emulsification technique for the purpose of treatment of colon cancer in an animal model. The NPs were further functionalized with epidermal growth factor (EGF), that selectively binds to the overexpressed EGF-receptors on the solid tumors in colon cancer. This modification promoted cellular uptake *via* EGF-receptor mediated endocytosis, resulting in a higher rate of apoptosis and antitumor effects. The incorporation of PFC also modulated the hypoxic tumor microenvironment (TME) by transporting oxygen, thereby improving therapeutic outcomes.

Fluorescence diffuse optical tomography (FDOT) is an emerging imaging technique that relies on the quantity and species of fluorescent probes and is used to locate and quantify fluorescent molecules within deep tissues. ICG near-infrared dye can serve as an exogenous fluorophore in FDOT to enable cancer cell detection [81]. However, this NIR dye can rapidly degrade leading to loss of fluoresence [82]. NIR dyes are commonly incorporated in PFC nanoemulsions by integrating them into the surfactant layer [83], a hydrocardon oil core [84], or by droplet surface conjugation [85]. Lee and colleagues [74] reported a doxorubicin (DOX) and ICG containing anti-human epidermal growth factor receptor 2 (anti-HER2) surface conjugated double nanoemulsion. HER2 is overexpressed in aggressive breast cancer. The PFC aided in retarding degradation and loss of fluorescence of ICG used for diagnostics. This approach eschews the off-target effects of DOX while protecting ICG from the tumor microenvironment and external stimuli like pH, heat, and light. The same group also co-administered PFC nanoemulsion harboring Campothecin (CPT), an anti-cancer drug blocking the activity of topoisomerase I, along with ICG in (HER+) breast cancer cells [75].

PFC nanoemulsions have been used to bind multiple types of proteins and peptides for the use in oncology applciations. Liu *et al.* [76] investigated folate-targeted nanoemulsions loaded with 10-hydroxycamptothecin (HCPT) and superparamagnetic Fe_3O_4. The ability of perfluoropentane to undergo liquid-to-gas phase transition is achieved through low-intensity focused ultrasound (LIFU) sonication. The application of LIFU stimulates acoustic droplet vaporization of liquid perfluoropentane. This in turn enhances the ultrasound imaging capability at tumor location and releases HCPT for anti-cancer activity. In the studies of Esser *et al.* [77], a PCF nanoemulsion was used to deliver a MYC inhibitor prodrug (MI3-PD) to M2 (pro-tumor) macrophages, consequently disrupting M2 polarization. To preferentially deliver payload to M2 macrophages, the surface of nanoemulsion droplets was coated with integrin $\alpha v\beta 3$. Hingorani *et al.* [78] engineered peptide-PFC nanoemulsion by chemically conjugating TAT peptide on the surface. This significantly improved *ex vivo*, intracellular uptake of the PFC agent in therapeutic chimeric antigen receptor (CAR) T cells that were subsequently injected and imaged by ^{19}F MRI in a solid tumor model. Soman *et al.* [86] studied the delivery of Melittin, a cytolytic peptide, that cannot be incorporated into conventional bilayer nanoparticle scaffolds (*e.g.*, liposomes) as it exhibits membrane disruption by pore formation activity. Use of PFC as a nanovehicle retained structural integrity even after adding Melittin. This is attributed to the hydrophobic and lipophobic nature of the PFC core which imparts stability and slows the dissociation of Melittin from nanoemulsion droplets. This study demonstrates the applicability of PFC containing nanoemulsions to overcome the incorporation and delivery challenges of cytolytic peptides [79]. Additionally, PFC nanoemulsions as a delivery vehicle for small interfering RNA (siRNA) was investigated by Ding *et al.* [80] to modulate tumor-neuronal interactions in TME by silencing nerve growth factor (NGF) expression in pancreatic cancer.

All of the examples listed in Table 5.1 utilize so-called two-phasic systems, where the internal phase is PFC (oil phase), which is stabilized with surfactants in water (external phase). In these examples, the therapeutic entity was required in relatively low doses at the targeted organs and tissues to achieve efficacy. For low concentrations, drug loading in the surfactant layer of the PFC nanoemulsion is adequate. However, this strategy has limited mass loading capacity for less potent drugs, and has limited control of the drug release. To improve loading of poorly soluble small molecule drugs (<1000 Da) drugs into PFC-based nanoemulsions, we employed triphasic systems. Figure 5.2 shows a side-by-side structural comparison depicting two-phase (PFC/water) and triphasic (PFC/hydrocarbon/water) nanoemulsions.

Increasing the HC oil fraction to values as high as 1:1 v/v ratio of HC to PFC oil provided increased capacity for carrying poorly water soluble drugs [17, 26, 51, 52] and water insoluble NIR dyes, such as quaterrylene dye [50]. Furthermore, a high HC oil content can yield increased stabilization of PFC nanoemulsions against Ostwald ripening [87]. Table 5.2 summarizes a few representative examples of poorly soluble drug and/or dye loaded triphasic nanoemulsions produced in our lab [17, 50, 87–89].

Recently, we developed several PFC nanoemulsions with perfluoro-15-crown-5 ether (PCE) for both cell tracking by MRI [16, 58, 90–94] and drug delivery [57, 58, 63, 95]. PCE symetrical structure offers 20 magnetically equivalent ^{19}F atoms per molecule with only one resonance peak in the NMR spectrum. This attractive feature for MRI is hindered by nanoemulsion instability arising from Ostwald ripening. The smaller the molecular weight of HC or PFC molecule in internal phase of the nanoemulsion, the faster the rate of Ostwald ripening [96]. However, when PCE nanoemulsion was formulated as a triphasic system by combining with a HC oil, Miglyol 812 in 1:1 v/v ratio, the resultant nanoemulsion remained stable for

182 days, showing no change in either droplet size or polydispersity [50]. Whereas, nanoemulsion with the same surfactant to oil ratio prepared only with PCE showed a steady increase in droplet size over 100 days follow-up *via* dynamic light scattering (DLS) measurements, indicating the superior stability of the triphasic system.

Table 5.2 Triphasic PFC nanoemulsions produced by microfluidization with varied PFCs and hydrocarbon oils

PFC	Hydrocarbon (HC) Oil	Size (nm)	PDI
PFPE	–	180	0.12
PFPE	Miglyol 812N	160	0.15
PFPE	Olive oil	145	0.16
PFPE	Capmul® PG-8 NF	103	0.19
PFPE	PG/Olive oil	138	0.15
PCE	–	158	0.1
PCE	–	107	0.12
PCE	–	121	0.08
PCE	Miglyol 812N	130	0.15
PFTE	Olive oil	185	0.15
PFTE	Olive oil	159	0.19
PFH	Miglyol 812N	145	0.18

Abbreviations: PFPE: perfluoropolyether; PCE: perfluoro-15-crown-5-ether; PFTE: perfluoro-terbutyl-ether; PFH: perfluorohexane.

Triphasic PFC/HC/Water nanoemulsions offer a dramatic improvement in fluorescent signal performance of lipid-soluble NIRF dyes. These dyes are readily incorporated at high volume fraction in HC oil of triphasic nanoemulsions compared to the surfactant layer of two-phase (PFC/Water) nanoemulsions. One such example pertains to exeptionally stable quaterrelene dyes. These dyes showed a dramatic improvement in fluorescence signal when incorporated into triphasic PFC nanoemulsions. The triphasic PFC nanoemulsions can be produced on a large scale (>0.5 L) while maintaining the same size distribution profile across manufactured batches (Fig. 5.3A–B), with confirmed triphasic structure (Fig. 5.3C) and stable colloidal properties (size distribution, zeta potential) upon storage, varied temperatures (Fig. 5.3D–F) and under stress (centrifugation, Fig. 5.3G and filtration, Fig. 5.3H), and maintain fluorescence upon storage (Fig. 5.3I).

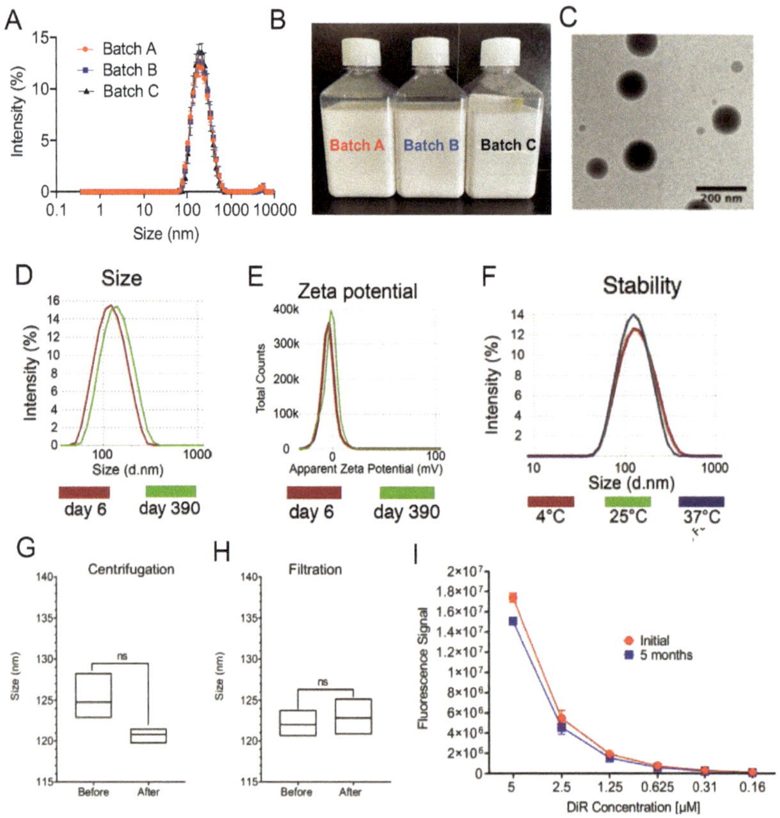

Figure 5.3 Scale-up production of triphasic PFC nanoemulsions and quality assessments *in vitro*. (A–B) Size distribution and visual comparison of three consecutive batches prepared on 600 mL scale. (C) Cryo-TEM image of an example triphasic nanoemulsion showing PFC core surrounded by hydrocarbon corona. (D) The long term follow up (>300 days) of size distribution. (E) Zeta potential, showing no significant changes after 390 days. (F) Size distribution comparison of nanoemulsions stored at three different temperatures show no change. (G) Centrifugation and (H) sterile filtration do not significantly impact droplet size distribution. (I) Triphasic nanoemulsions retain fluorescent properties upon storage for 5 months. Data sources for (A–C) Janjic Lab unpublished, (D–F) reproduced with permission from Ref. [9], and (G–I) reproduced with permission from Ref. [5].

Triphasic nanoemulsions can be easily prepared by microfluidization with varied lipophilic payload (dye and drug), as indicated in Table 5.2. The data show that changes in composition can be tolerated while maintaining small droplet size (100–200 nm), narrow size distribution, and low polydispersity index (PDI <0.2). This is advantageous for theranostic platform development because it allows for adaptations to imaging and drug compostions without significant changes in the formulation process parameters [8, 30, 50].

For Table 5.2, the general nanoemulsion formulation manufacturing process is as follows. The lipophilic payload, NIR dye (*e.g.*, DiR) and a drug (*e.g.*, Celecoxib), are dissolved at room temperature at desired concentration (*e.g.*, 5 mg/mL) in Miglyol 812N (HC oil). This oil solution is further combined with PFC and surfactant mixtures in a pre-emulsification step using high speed mixing on a planetary centrifugal mixer at room temperature under ambient pressure. The pre-emulsion is then processed using a Microfluidizer M110S (Microfludics) [51, 88]. When produced by microfluidization, the triphasic nanoemulsions showed a narrow size distribution, which is not affected by the presence of the lipophilic drug or fluorescent dye [8, 30, 50, 89, 97].

5.4 Perfluorocarbon Nanoemulsions Stability and Quality Control

Nanoemulsions are metastable systems, so given sufficient time they are prone to breakdown, at a rate that is proportional to the height of the energy barrier between the separated phases and the formed nanoemulsion. To formulate a nanoemulsion with prolonged stability, the height of the energy barrier should be greater than 20 kT as it imparts kinetic stability to the system [34]. The kinetic stability of the nanoemulsion is largely determined by the extent of interactions between the droplets which are primarily governed by Brownian motion, applied shear, or gravitational forces [34]. Nanoemulsion breakdown is driven by destabilizing mechanisms like Ostwald ripening, flocculation, gravity separation (creaming or sedimentation), and coalescence [98]. In particular, due to the large specific weight (1.5–1.8 g/ml) of PFCs, Ostwald ripening is a commonly reported problem during long-term storage. For PFC nanoemulsions, maintaining colloidal stability is critical for achieving reproducibility in MRI applications where cells are labeled (*e.g.*, cell tracking), as well as in overall biodistribution and drug delivery efficacy. Changes in internal conditions (*e.g.*, elevated temperature, pH changes, salts) or external conditions (*e.g.*, exposure to biological fluids, agitation) can lead to nanodroplet destabilization. However, we found that relatively small changes (<10%) in droplet size upon prolonged storage are inevitable and do not interfere with the imaging or drug delivery functionality of PFC nanoemulsions. Larger changes (>20%) in size and polydispersity could modify drug loading and release characteristics, alter biodistribution and/or imaging detectability. Moreover, a faster rate of sedimentation during storage has an impact on product homogeneity, and can potentially lead to error in theranostic dosing.

A combination of surfactants was found to be effective in suppressing droplet size growth in PFC nanoemulsions, along with the addition of water-immiscible oil [17, 20, 51, 54, 96, 99, 100]. Once injected *in vivo*, nanoemulsion is exposed to the complex biological environment in the circulatory system, which could further accelerate the nanodroplet degradation. Therefore, we developed stress tests that can evaluate PFC nanoemulsion stability under biologically relevant conditions *in vitro* [101]. Representative data on storage stability, size distribution, and zeta potential of our earlier reported PFC nanoemulsions are shown in Fig. 5.3.

We recently established critical quality attributes (CQAs) for PFC nanoemulsions used as nanotheranostics [8]. Extensive analytical studies are needed due to the multifunctional nature of the formulations to assure quality throughout the lifetime of the product, that is from time of manufacture, through storage, and use. Colloidal stability and imaging properties (*e.g.*, fluorescence) need to be continuously tested over the life-time of the product [17, 25, 50, 51, 102]. DLS instrumenatation (*e.g.*, Zetasizer Nano ZS Malvern, UK) can be used for stability monitoring of nanoemulsion droplet size. Typically, the nanoemulsion samples are diluted in deionized water or other media at optimized volume-to-volume ratios (typically 1/40 or 1/80 v/v). All size measurements are run at 25 °C with a scattering angle of 173° on the Zetasizer Nano ZS [51, 103] and complemented by microscopy. Drug-loading and drug release studies are typically performed using the vertical Franz diffusion cell [104–108].

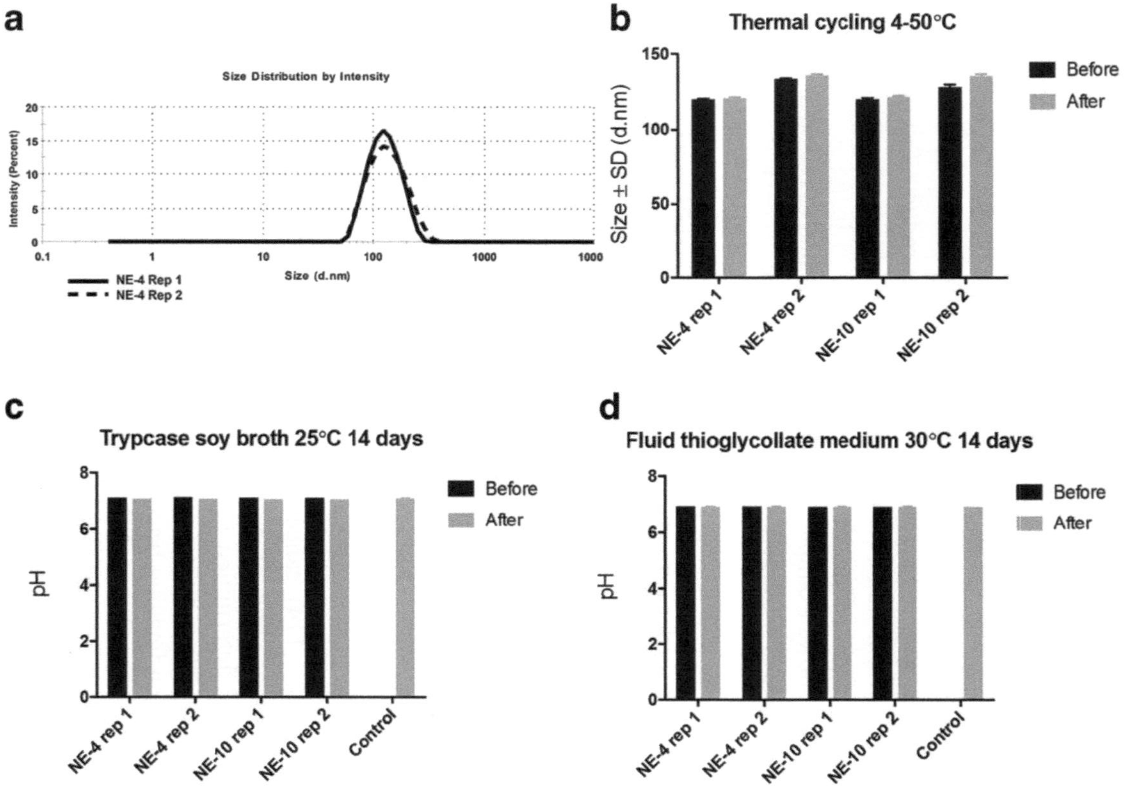

Figure 5.4 PFC nanoemulsion *in vitro* assessments. (a) Size distribution overlays of the selected PCE nanoemulsion (NE-4, NE-4 replicate 1, and NE-4 replicate 2). (b) Thermal cycling study on selected nanoemulsions. (c–d) Sterility test. Select nanoemulsions were diluted 1:80 v/v and incubated for 14 days in trypcase soy broth at 25 °C or thioglycolate medium at 35 °C. pH measurements were taken before and after the incubation period. Reproduced with permission from Ref. [8].

Depending on the purpose of prepared PFC nanoemulsions, whether they are intended for *ex vivo* cell labeling or *in vivo* delivery, quality assessments need to include additional tests that can be representative of the PFC nanoemulsion droplet exposure. For example, when they are intended to be used for *ex vivo* cell labeling, PFC nanoemulsions need to remain stable upon mixing with complex biological media and cells, centrifugation, agitation, elevated temperatures, and cooling (thermal cycling). Also, it is imperative to test PFC nanoemulsions for sterility. PFC nanoemulsions must be free of pathogens and pathogen microorganism products. Our group reported several studies where PFC nanoemulsions were evaluated under biological conditions and with the stressors listed above [8]. Figure 5.4 shows examples of stability tests needed to characterize and optimize PFC nanoemulsions for *in vitro* and *in vivo* use.

5.5 Macrophage Targeted Perfluorocarbon Theranostic Nanoemulsions

Passive targeting of circulating inflammatory cells by PFC nanoemulsions has been extensively investigated for the purpose of detecting inflammation using ^{19}F MRI [109–112]. Macrophages have emerged as the most accessible cellular targets due to their ability to take up nanoparticles injected into the blood stream. Multiple studies show preferred labeling of macrophages with PFC nanoemulsions, which is used to quantify inflammation by ^{19}F MRI in transplantation [91], abscess [113], inflammatory bowel disease (IBD) [114], cancer [115], rheumatoid arthritis (RA) [116], and pulmonary inflammation [117]. In a rat model of RA, Balducci *et al.* [116] determined severity of RA by measuring ^{19}F content in arthritic joints, thus showing that PFC can be used both as an inflammation imaging tracer as well as a quantitative indicator for disease progression and remission. Furthermore, treatment with conventional corticosteroid subsequently resulted in lowering of ^{19}F content indicating resolution of inflammation. ^{19}F MRI has been used in numerous other inflammation models where the macrophage burden is reduced with therapeutic intervention concomitant with reduced ^{19}F signal [114, 118]. In other studies, Balducci *et al.* [119] reported a NIRF labeled PFC nanoemulsion that could be used to detect breast cancer tumor-associated inflammation by ^{19}F MRI and NIRF. A linear correlation between these two imaging signals was observed in excised organs (Fig. 5.5) as measured by ^{19}F NMR. The presence of a fluorescent dye incorporated in the nanoemulsions facilitated assessment of nanoemulsion accumulation in specific cell types using flow cytometry and histology [110].

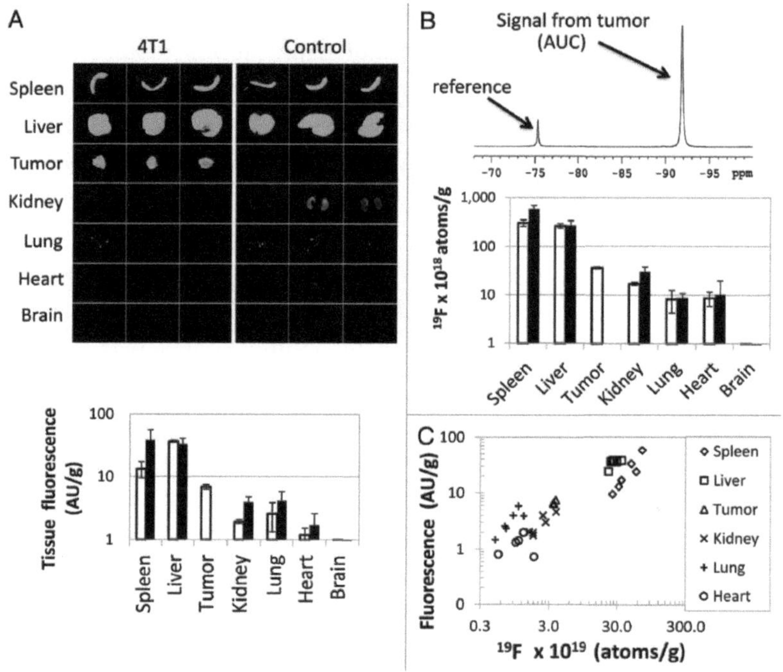

Figure 5.5 ^{19}F NMR and fluorescence correlation of PFC-NE tissue distributions in a mouse tumor model: (A) NIRF images of excised tumors and other organs were obtained 72 h post-administration of the PFC-NE agent. (B) A representative NMR spectrum of an excised tumor with a chemically shifted reference compound (trifluoroacetic acid at –76.00 ppm). The data is normalized by tissue weight (n = 3 per tissue, data represents an average with standard deviation). Error bars represent the standard deviation. (C) Correlation of fluorescent and ^{19}F NMR signals of individual tissues (obtained from both control and tumor-bearing animals), plotted on a log scale. Data reproduced with permission from Ref. [119].

In another example, triphasic nanoemulsion are able to incorporate NIR fluorescent dyes into the hydrocarbon shell surrounding a PCE core. The two imaging reporters, NIRF dye and ^{19}F MR tracer, tag macrophages following intravenous injection and enable multimodal imaging of inflammation in nerve injury, as well as in acute inflammation rodent models [9, 97]. Therefore, this provides unique opportunities to assess both superficial and deep tissues and quantify macrophage burden present in regions of interest using both NIRF and ^{19}F MRI (Fig. 5.6).

To demonstrate the potential of triphasic PFC nanoemulsions as macrophage targeted theranostic, we loaded the hydrocarbon core with a COX-2 inhibitor, celecoxib (CXB), and a lipid soluble NIRF dye, DiD. These nanoemulsions have dual mode (NIR/^{19}F MRI) properties; when nanoemulsion was taken up by activated macrophages in culture, a reduced production of pro-inflammatory mediator PgE2 was noted [9, 26, 51, 52]. The nanoemulsions, with celecoxib (CXBNE and without, drug-free nanoemulsion DFNE), were tested at different doses and no viability change was observed in macrophages upon 24 h exposure (Fig. 5.7A). In order to confirm that these nanoemulsions do not produce any pro-inflammatory effects, NO release and CD86 expression changes were monitored after 24 h of exposure. A non-toxic concentration of LPS was used, and cells were activated for 18 h. We found that CD86 expression and NO release were not altered by the nanoemulsions (Fig. 5.7B–C). CD86 and NO levels were comparable to untreated cells and significantly lower than LPS-activated cells, which clearly indicate that nanoemulsions do not stimulate a pro-inflammatory phenotype in macrophages.

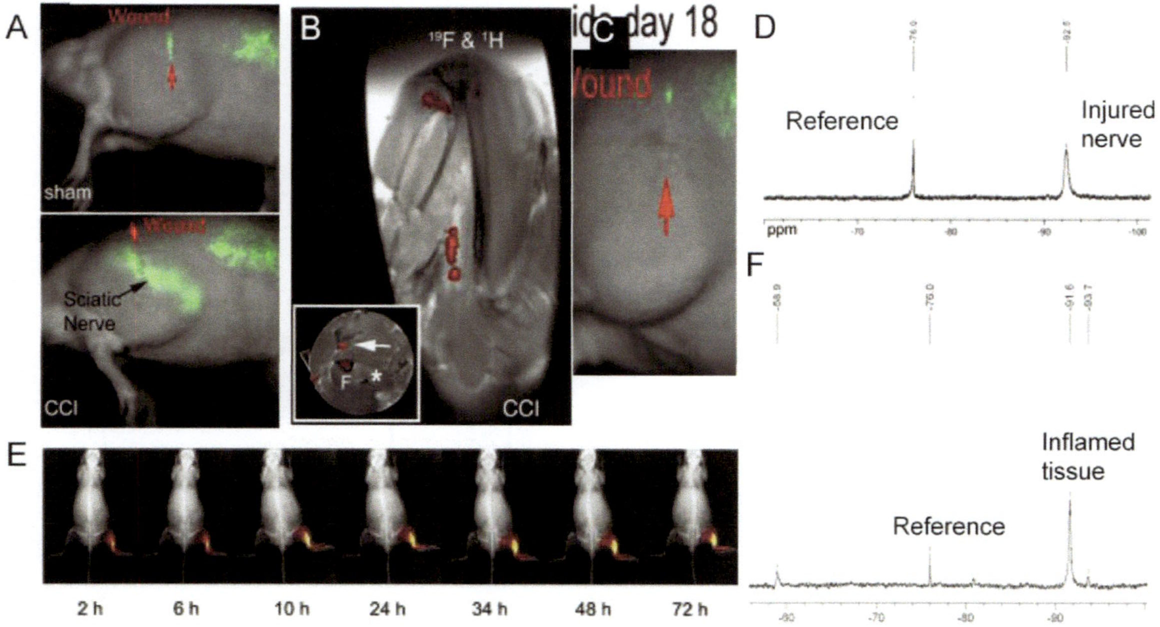

Figure 5.6 NIRF PFC nanoemulsion labeled macrophage detection *in vivo* in rodent models of nerve injury and inflammation. (A) NIRF imaging of macrophage infiltration patterns in sham and chronic constriction injury (CCI) of sciatic nerve in rats. (B) End-point ^{19}F MRI of macrophage accumulation at CCI. (C) End-point NIRF imaging of excised CCI nerve. (D) ^{19}F NMR spectrum of excised CCI nerve populated with labeled macrophages. (E) Live sequential NIRF imaging in complete Freund's adjuvant (CFA) treated mice. Increasing NIRF signal correlates with labeled macrophage accumulation at the site of CFA injection. (F) ^{19}F NMR spectrum of excised CFA treated paw tissue infiltrated with PFC-NE labeled macrophages. Data reproduced from Refs. [9, 97].

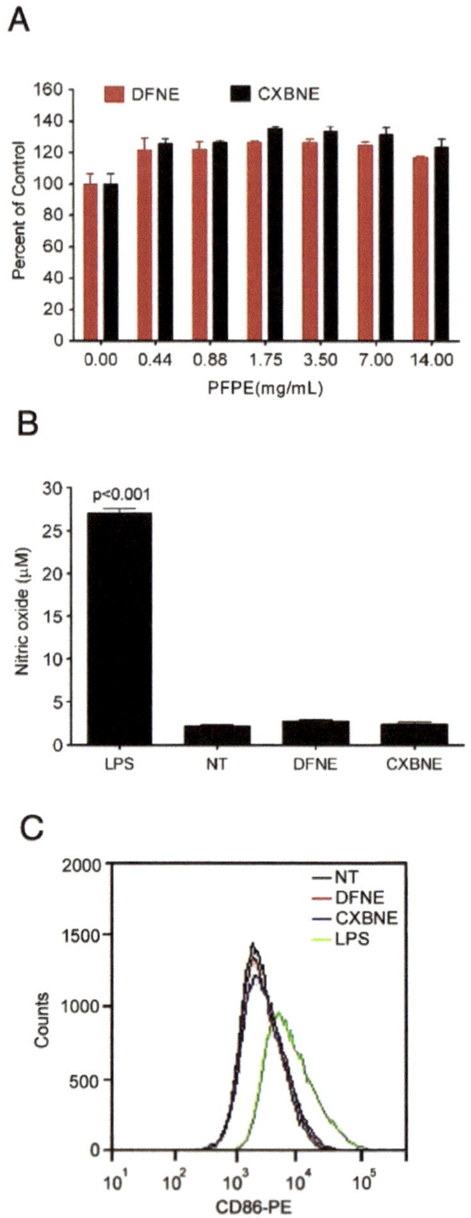

Figure 5.7 Effect of nanoemulsion on macrophage viability and activation. (A) Viability of RAW 264.7 macrophages exposed to DFNE and CXBNE for 24 h. Data is shown as mean of percent unexposed cells ± SD (n = 4). (B) Nitric oxide release from macrophages exposed to nanoemulsions for 24 h. (C) CD86 expression in macrophages exposed to nanoemulsions for 24 h. Cells not exposed to any treatment are designated as NT (no treatment). Figure reproduced from Ref. [97].

To understand the cellular uptake and intracellular fate of nanoemulsions, cellular uptake studies were conducted in RAW 264.7 macrophages. Fluorescence analysis of cells exposed to nanoemulsion showed an identical emission profile (Fig. 5.8A) to that of dye in solution and nanoemulsion, whereas untreated cells did not display fluorescence (Fig. 5.8). This result confirms that the fluorescent properties of the theranostics are not altered and can be detected in cells using fluorescence microscopy. Confocal microscopy showed that the nanoemulsions were readily internalized by macrophages and localized to acidic endosomal/lysosomal compartments (Fig. 5.8).

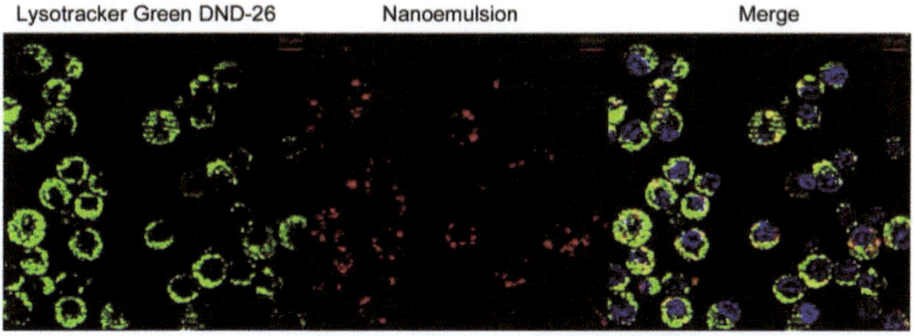

Figure 5.8 Nanoemulsion uptake in macrophages using fluorescence methods. Confocal images showing co-localization of nanoemulsion (red) with acidic compartments (green) labeled with Lysotracker Green-DND. Merged image shows the co-localized pixels (yellow). Nuclei are stained with DAPI (blue). Figure reproduced from Ref. [97].

Further confirmation of the potential of a celecoxib-loaded triphasic nanoemulsion (CXBNE) for treatment of inflammatory diseases was obtained in a paw inflammation mouse model, while using free drug celecoxib (CXB) and drug-free nanoemulsion (DFNE) as treatment controls. In this study, we demonstrated the feasibility of fluorescent imaging of inflammation with dye-loaded nanoemulsions in a CFA mouse model, with follow up *ex vivo* tissue analyses from the excised sites of inflammation (Fig. 5.9). Nanoemulsions were injected and mice imaged at multiple time points (Fig. 5.9A–B), showing significant differences in fluorescence signal in inflamed paws between drug-loaded nanoemulsion and drug-free nanoemulsions. Histology confirmed that nanoemulsions (red) associate with COX-2 expressing macrophages in inflamed

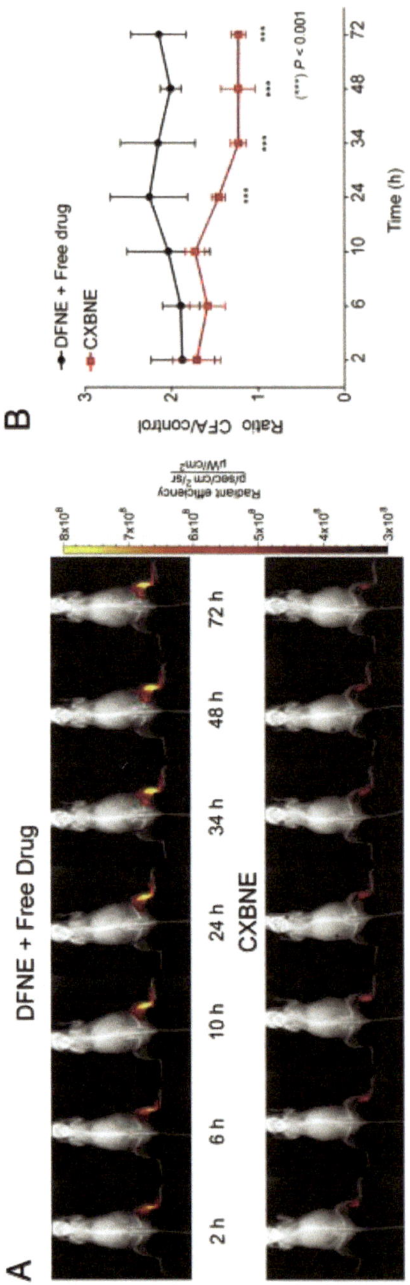
A
DFNE + Free Drug
2 h
6 h
10 h
24 h
34 h
48 h
72 h
CXBNE
Radiant efficiency
B
DFNE + Free drug
CXBNE
Ratio CFA/control
Time (h)
(***) P < 0.001

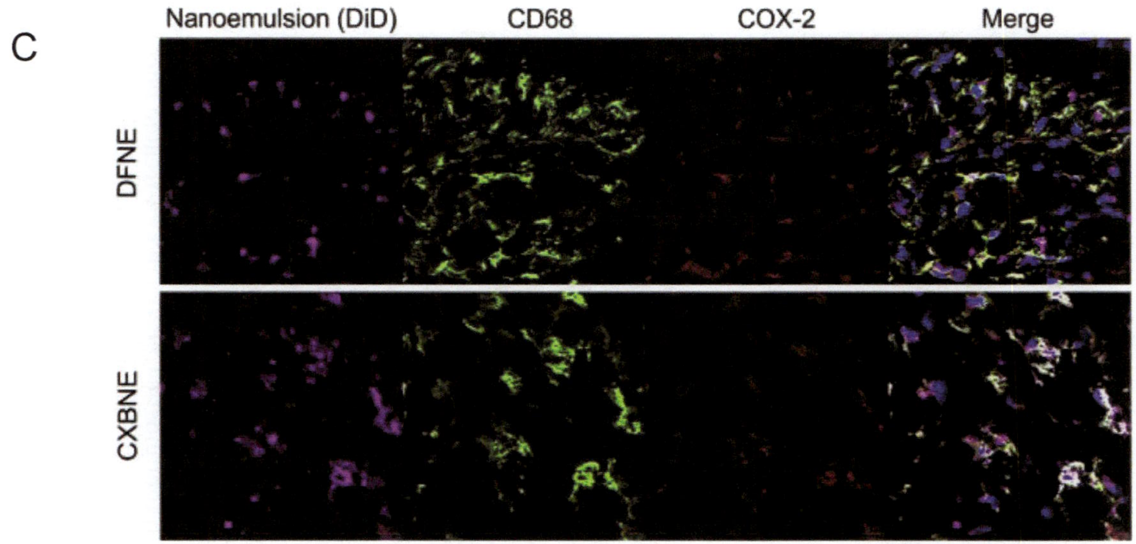

Figure 5.9 Theranostic PFC nanoemulsion (CXBNE) as an anti-inflammatory therapeutic and a macrophage tracking agent in an inflammation mouse model. (A) Representative time-course images showing overlay of fluorescence (red) and anatomical X-ray (grey) in CFA-mouse obtained on an IVIS Lumina XR scanner. The nanoemulsion was I.V. injected 12 h prior to the right paw being injected with CFA and left paw with phosphate-buffered saline (control). (B) Fluorescence imaging of CFA injected paws show significantly less signal when celecoxib nanoemulsion was injected as compared to drug-free nanoemulsions. (C) Immunofluorescence indicates that nanoemulsions, both CXBNE and DFNE were selectively associated with macrophages expressing COX-2. Figure reproduced from Ref. [97].

tissues (Fig. 5.9C). These results indicate that drug delivery and macrophage imaging can be performed together. Furthermore, imaging can be utilized not just to visualize macrophages in inflammation, but also assess the changes in accumulation as an indicator of therapeutic effect. Essentially, features of combined therapy and therapy diagnosis can be integrated using drug-loaded PFC nanoemulsions.

In another example of the theranostic potential of COX-2 inhibiting triphasic PCE nanoemulsions, we followed their effect on inflammation and pain in the CFA model for 40 days in both male and female mice (Fig. 5.10) [6]. The von Frey assay indicated a significant and persistent reduction in mechanical hypersensitivity with CXB-NE over DF-NE and drug solution (CXB) controls in an inflammatory pain mouse model in both sexes (Fig. 5.10C–D). To resolve inflammation and pain, mice were injected IV with a single dose of NIRF labeled CXB-NE or controls (drug free nanoemulsion) or free celecoxib (CXB) solution, and followed for 40 days using von Frey behavior assay and NIRF live imaging. Overall, the CXB dose was kept at 1 mg/kg across all groups.

We found a persistent and statistically significant reduction in mechanical hypersensitivity with CXBNE over DFNE and drug solution (CXB) controls. In male mice, apparent pain control persists for up to 32 days, while it lasts up to 11 days in females (Fig. 5.10C) [6]. Importantly, extended pain relief is achieved in both male and female mice with CXBNE. Both males and females showed no improvement in mechanical hypersensitivity when treated with matched low concentrations of CXB delivered as a non-targeted, free drug solution. Thus, the key to the effectiveness of the low dose of CXB lies in targeted delivery to macrophages *via* nanoemulsion, which is in agreement with findings in rat injury models [9]. We also demonstrated the feasibility of live and *ex vivo* NIRF imaging of inflammation with theranostic CXBNEs in CFA-treated mice with follow-up *ex vivo* analyses from the excised site of inflammation. The hindpaw NIRF signal persists for up to 40 days in both males and females. Immunofluorescent staining demonstrated that $CD68^+$ macrophages internalized both the CXBNE and DFNE controls. In a related study in a chronic nerve constriction injury (CCI) rat model, we also demonstrated that CXBNE suppresses pain hypersensitivity at extremely low drug doses (0.24 mg/kg, 100-fold lower than

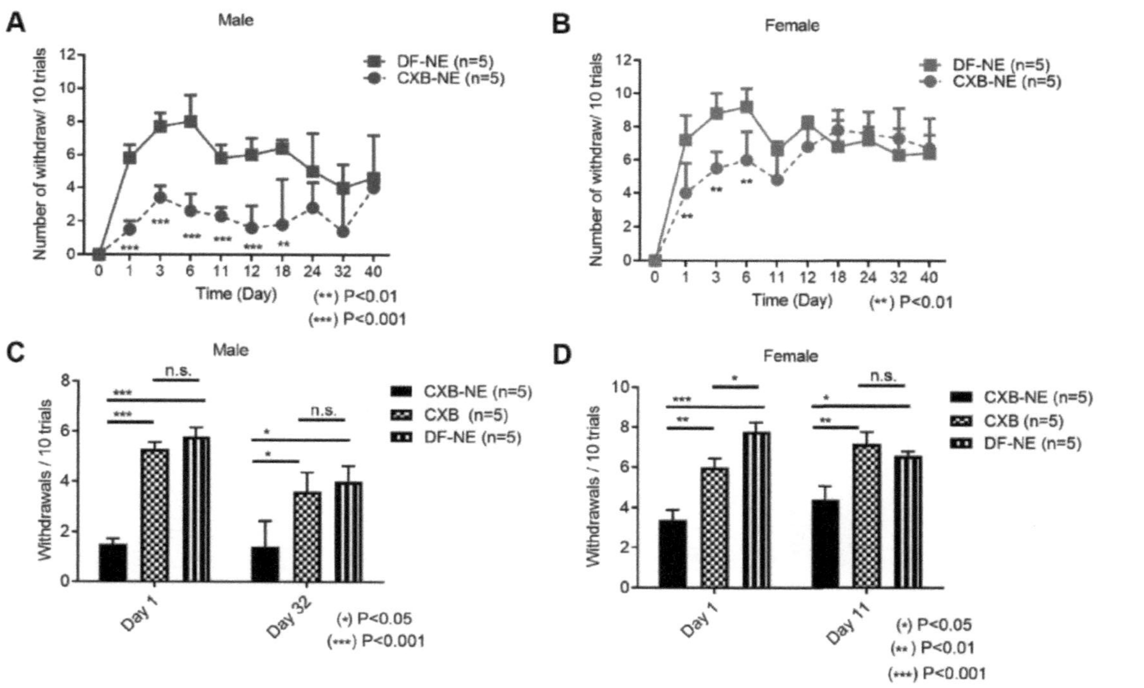

Figure 5.10 Long-lasting pain relief in CFA mice with a single dose of macrophage immunomodulating nanoemulsion (CXBNE). (A) Long term follow up of males and (B) females injected with CXBNE or drug free control (DFNE). (C) Pain reduction in males lasts 32 days and (D) in females 11 days. Data reproduced from Ref. [6].

a typical systemic dose in animals) [9]. Pain relief was associated with an intriguing trend of decreased NIRF signal in CXBNE treated animals versus controls, corresponding to a reduction in macrophage infiltration, which was confirmed by immunofluorescent histology [9]. Taken together, the above studies in inflammation and pain rodent models demonstrate highly effective nanotheranostic functionality for triphaisic PFC nanoemulsions.

5.6 Pharmaceutical Perspective on Perfluorcarbon Theranostics

For successful translation of PFC nanoemulsions into the clinic as MRI agents and theranostics, pharmaceutical process development is of critical importance. Microfluidization, a high-pressure homogenization technique, is widely used in food and pharmaceutical industry. Microfluidization is fast, reliable and scalable emulsification process for kinetically stable nanoemulsions [120–123]. When compared with other high energy emulsification methods such as sonication and high pressure homogenization, microfluidization produces more stable nanoemulsions with smaller sizes and a narrower size distribution [122, 123]. The droplet size of oil in water nanoemulsions produced by microfluidization is influenced by both formulation variables (oil volume fraction, phase viscosity, surfactant concentration) and processing conditions (*e.g.*, pressure) [121, 124–126]. Lu *et al.* [103] reported large scale manufacturing of theranostic PFC nanoemulsions produced by microfluidization. This study compared two processes on two microfluidizers with different geometries and pressure control systems [103].

We posit that a Quality by Design (QbD) approach [127] can serve as a road map towards successful clinical translation of PFC nanoemulsions as imaging agents and as nanotheranostics. QbD is an intentional strategy that ensures the product quality by utlizing statistical, analytical and risk-management methodology in the design, development and manufacturing. This approach to pharmaceutical product design and manufacturing has been encouraged by worldwide regulatory authorities, including the U.S. Food and Drug Administration (FDA) and European Medicines Agency (EMA), and is widely adopted by the pharmaceutical industry.

The heart of a QbD approach is the rational development of a design space for manufacturing and it has recently been proposed as the key strategy for advancing nanomedicine into the clinic [127]. One of the key parameters that drives PFC nanoemulsions biodistribution and quality is droplet size. The droplet size of nanoemulsions is driven by composition variables (PFC and/or hydrocarbon oil type, surfactant structure and concentration, surfactant to oil ratio) as well as processing conditions (viscosity, temperature, pressure, etc.). A thorough investigation of the impact of these parameters on nanoemulsion performance and quality during development stage is critical to reduce risk of product failure during scale-up.

We developed PFC nanoemulsions and applied failure mode effects and criticality analysis (FMECA) [8, 128] which led to predictive mathematical models for establishing direct relationships between manufacturing process parameters and composition parameters with desirable quality atributes (*e.g.*, size, polydispresity, drug loading, etc.). This approach is more cost-effective over traditional one-factor-at-a time optimization approach. FMECA helps narrow the design space search and limits the number of PFC nanoemulsions that need to be produced and tested. Further, it provides a direct insight into colloidal behaviors of PFC nanoemulsions under relevant conditions to their use. For example, the QbD/FMECA approach aided in the development of a PFC nanoemulsion with ten times higher celecoxib loading, as compared to previous reports [97], and allowed us to establish shelf life without the need for long term follow ups.

5.7 Conclusions

Successful clinical realization of theranostic systems is possible with careful selection of agent building blocks and applying principles of fragment-based drug design. Overall, PFC nanoemulsions have characteristics making them advantageous for theranostics due to their non-toxic nature, and background-free imaging potential using ^{19}F MRI. The roadmap to successful clinical translation of PFC nanoemulsions is paved by QbD, which includes rational design, extensive optimization, risk assessment and intentional focus on final product quality.

Abbreviations

PFPE: Perfluoropolyether; PFTE: Perfluorotrimethylether; PFC: Perfluorocarbon; PFC-NE: Perfluorocarbon-nanoemulsion; PCE: Perfluoro-15-crown-5-ether; PFOB: Perfluorooctylbromide; PFDB: Perfluorodecylbromide; HC: Hydrocarbon; DF-NE: Drug-free nanoemulsion; CXB-NE: Celecoxib nanoemulsion.

References

1. Mura, S., and P. Couvreur. Nanotheranostics for personalized medicine. *Adv Drug Deliv Rev*, 2012, **64**(13): 1394–1416.
2. Wang, J., and H. Cui. Nanostructure-based theranostic systems. *Theranostics*, 2016, **6**(9): 1274–1276.
3. Li, Y., *et al.* ^{19}F MRI nanotheranostics for cancer management: progress and prospects. *ChemMedChem*, 2022, **17**(4): e202100701.
4. Deal, B., *et al.* Behavioral and inflammatory sex differences revealed by celecoxib nanotherapeutic treatment of peripheral neuroinflammation. *Sci Rep*, 2022, **12**(1): 8472.
5. Nichols, J.M., *et al.* Tracking macrophages in diabetic neuropathy with two-color nanoemulsions for near-infrared fluorescent imaging and microscopy. *J Neuroinflammation*, 2021, **18**(1): 299.
6. Liu, L., *et al.* Sex differences revealed in a mouse CFA inflammation model with macrophage targeted nanotheranostics. *Theranostics*, 2020, **10**(4): 1694.
7. Saleem, M., *et al.* Nanomedicine-driven neuropathic pain relief in a rat model is associated with macrophage polarity and mast cell activation. *Acta Neuropathol Commun*, 2019, **7**(1): 108.
8. Herneisey, M., *et al.* Development of theranostic perfluorocarbon nanoemulsions as a model non-opioid pain nanomedicine using a quality by design (QbD) approach. *AAPS PharmSciTech*, 2019, **20**(2): 65.
9. Janjic, J.M., *et al.* Low-dose NSAIDs reduce pain via macrophage targeted nanoemulsion delivery to neuroinflammation of the sciatic nerve in rat. *J Neuroimmunol*, 2018, **318**: 72–79.
10. Alexis, F., *et al.* Factors affecting the clearance and biodistribution of polymeric nanoparticles. *Mol Pharm*, 2008, **5**(4): 505–515.
11. Lammers, T., *et al.* Theranostic nanomedicine. *Acc Chem Res*, 2011, **44**(10): 1029–1038.

12. Janib, S.M., A.S. Moses, and J.A. MacKay. Imaging and drug delivery using theranostic nanoparticles. *Adv Drug Deliv Rev*, 2010, **62**(11): 1052–1063.

13. Hajduk, P.J., and J. Greer. A decade of fragment-based drug design: strategic advances and lessons learned. *Nat Rev Drug Discov*, 2007, **6**(3): 211–219.

14. Patel, S.K., and J.M. Janjic. Macrophage targeted theranostics as personalized nanomedicine strategies for inflammatory diseases. *Theranostics*, 2015, **5**(2): 150.

15. Jacoby, C., *et al.* Probing different perfluorocarbons for in vivo inflammation imaging by ^{19}F MRI: image reconstruction, biological half-lives and sensitivity. *NMR Biomed*, 2014, **27**(3): 261–271.

16. Kadayakkara, D.K., *et al.* ^{19}F spin-lattice relaxation of perfluoropolyethers: dependence on temperature and magnetic field strength (7.0–14.1T). *J Magn Reson*, 2014, **242**: 18–22.

17. O'Hanlon, C.E., *et al.* NIR-labeled perfluoropolyether nanoemulsions for drug delivery and imaging. *J Fluor Chem*, 2012, **137**: 27–33.

18. Barnett, B.P., *et al.* Use of perfluorocarbon nanoparticles for non-invasive multimodal cell tracking of human pancreatic islets. *Contrast Media Mol Imaging*, 2011, **6**(4): 251–259.

19. Srinivas, M., *et al. In vivo* cytometry of antigen-specific t cells using ^{19}F MRI. *Magn Reson Med*, 2009, **62**(3): 747–753.

20. Janjic, J.M., *et al.* Self-delivering nanoemulsions for dual fluorine-19 MRI and fluorescence detection. *J Am Chem Soc*, 2008, **130**(9): 2832–2841.

21. Decato, S., *et al.* Synthesis and characterization of perfluoro--butyl semifluorinated amphiphilic polymers and their potential application in hydrophobic drug delivery. *Polym Chem*, 2014 **5**(22): 6461–6471.

22. Du, W., *et al.* Amphiphilic hyperbranched fluoropolymers as nanoscopic ^{19}F magnetic resonance imaging agent assemblies. *Biomacromolecules*, 2008, **9**(10): 2826–2833.

23. Powell, K.T., C. Cheng, and K.L. Wooley. Complex amphiphilic hyperbranched fluoropolymers by atom transfer radical self-condensing vinyl (co)polymerization. *Macromolecules*, 2007, **40**(13): 4509–4515.

24. Flaim, S.F. Pharmacokinetics and side effects of perfluorocarbon-based blood substitutes. *Artif Cells Blood Substit Immobil Biotechnol*, 1994, **22**(4): 1043–1054.

25. Patrick, M.J., *et al.* Intracellular pH measurements using perfluorocarbon nanoemulsions. *J Am Chem Soc*, 2013, **135**(49): 18445–18457.
26. Patel, S.K., *et al.* Two-color fluorescent (near-infrared and visible) triphasic perfluorocarbon nanoemuslions. *J Biomed Opt*, 2013, **18**(10): 101312.
27. Willmann, J.K., *et al.* Targeted microbubbles for imaging tumor angiogenesis: assessment of whole-body biodistribution with dynamic micro-PET in mice. *Radiology*, 2008, **249**(1): 212–219.
28. Perez, A.S., *et al.* Click-ready perfluorocarbon nanoemulsion for ^{19}F MRI and multimodal cellular detection. *ACS Nanosci Au*, 2022, **2**(2): 102–110.
29. Hingorani, D.V., *et al.* Cell penetrating peptide functionalized perfluorocarbon nanoemulsions for targeted cell labeling and enhanced fluorine-19 MRI detection. *Magn Reson Med*, 2020, **83**(3): 974–987.
30. Mountain, G.A., *et al.* Design and formulation of nanoemulsions using 2-(poly(hexafluoropropylene oxide)) perfluoropropyl benzene in combination with linear perfluoro(polyethylene glycol dimethyl ether). *J Fluor Chem*, 2014, **162**: 38–44.
31. Hitchens, T.K., *et al.* Combining perfluorocarbon and superparamagnetic iron-oxide cell labeling for improved and expanded applications of cellular MRI. *Magn Reson Med*, 2015, **73**(1): 367–375.
32. Kadayakkara, D.K., *et al.* In vivo observation of intracellular oximetry in perfluorocarbon-labeled glioma cells and chemotherapeutic response in the CNS using fluorine-19 MRI. *Magn Reson Med*, 2010, **64**(5): 1252–1259.
33. Vasudeva, K., *et al.* Imaging neuroinflammation *in vivo* in a neuropathic pain rat model with near-infrared fluorescence and ^{19}F magnetic resonance. *PLoS One*, 2014, **9**(2): e90589.
34. McClements, D.J. Nanoemulsions versus microemulsions: terminology, differences, and similarities. *Soft Matter*, 2012, **8**(6): 1719–1729.
35. Shakeel, F., *et al.* Nanoemulsions as potential vehicles for transdermal and dermal delivery of hydrophobic compounds: an overview. *Expert Opin Drug Deliv*, 2012, **9**(8): 953–974.
36. Kotta, S., *et al.* Exploring oral nanoemulsions for bioavailability enhancement of poorly water-soluble drugs. *Expert Opin Drug Deliv*, 2012, **9**(5): 585–598.
37. McClements, D.J., and J. Rao. Food-grade nanoemulsions: formulation, fabrication, properties, performance, biological fate, and potential toxicity. *Crit Rev Food Sci Nutr*, 2011, **51**(4): 285–330.

38. Rajpoot, P., K. Pathak, and V. Bali. Therapeutic applications of nanoemulsion based drug delivery systems: a review of patents in last two decades. *Recent Pat Drug Deliv Formul*, 2011, **5**(2): 163–172.

39. Sarker, D.K. Engineering of nanoemulsions for drug delivery. *Curr Drug Deliv*, 2005, **2**(4): 297–310.

40. Mitri, K., *et al.* Scale-up of nanoemulsion produced by emulsification and solvent diffusion. *J Pharm Sci*, 2012, **101**(11): 4240–4247.

41. Muller, R.H., D. Harden, and C.M. Keck. Development of industrially feasible concentrated 30% and 40% nanoemulsions for intravenous drug delivery. *Drug Dev Ind Pharm*, 2012, **38**(4): 420–430.

42. Windhab, E.J., *et al.* Emulsion processing—from single-drop deformation to design of complex processes and products. *Chem Eng Sci*, 2005, **60**(8–9): 2101–2113.

43. Tadros, T., *et al.* Formation and stability of nano-emulsions. *Adv Colloid Interface Sci*, 2004, **108–109**: 303–318.

44. Salager, J.L., *et al.* The emergence of formulation engineering in emulsion making-transferring know-how from research laboratory to plant. *Polym Int*, 2003, **52**(4): 471–478.

45. Hessien, M., *et al.* Stability and tunability of O/W nanoemulsions prepared by phase inversion composition. *Langmuir*, 2011, **27**(6): 2299–2307.

46. Sole, I., *et al.* Nano-emulsions prepared by the phase inversion composition method: preparation variables and scale up. *J Colloid Interface Sci*, 2010, **344**(2): 417–423.

47. Salager, J.L., *et al.* Using emulsion inversion in industrial processes. *Adv Colloid Interface Sci*, 2004, **108–109**: 259–272.

48. Scholz, P., and C.M. Keck. Nanoemulsions produced by rotor-stator high speed stirring. *Int J Pharm*, 2015, **482**(1–2): 110–117.

49. Sivakumar, M., S.Y. Tang, and K.W. Tan. Cavitation technology: a greener processing technique for the generation of pharmaceutical nanoemulsions. *Ultrason Sonochem*, 2014, **21**(6): 2069–2083.

50. Janjic, J.M., *et al.* Perfluorocarbon nanoemulsions with fluorescent, colloidal and magnetic properties. *Biomaterials*, 2014, **35**(18): 4958–4968.

51. Patel, S.K., *et al.* Cyclooxgenase-2 inhibiting perfluoropoly (ethylene glycol) ether theranostic nanoemulsions - in vitro study. *PLoS One*, 2013, **8**(2): e55802.

52. Janjic, J.M., *et al.* Suppressing inflammation from inside out with novel NIR visible perfluorocarbon nanotheranostics, in *Reporters, Markers, Dyes, Nanoparticles, and Molecular Probes for Biomedical Applications V*, San Fransisco, CA, USA, 2013.
53. Krafft, M.P., A. Chittofrati, and J.G. Riess. Emulsions and microemulsions with a fluorocarbon phase. *Curr Opin Colloid Interface Sci*, 2003, **8**(3): 251–258.
54. Rapoport, N. Phase-shift, stimuli-responsive perfluorocarbon nanodroplets for drug delivery to cancer. *Wiley Interdiscip Rev Nanomed Nanobiotechnol*, 2012, **4**(5): 492–510.
55. Kabalnov, A.S., and E.D. Shchukin. Ostwald ripening theory: applications to fluorocarbon emulsion stability. *Adv Colloid Interface Sci*, 1992, **38**: 69–97.
56. Craig, V.S.J., and M.Krafft. Editorial overview Hot Topic - Nanobubbles and Nanodroplets: from Basics to Applications. *Curr Opin Colloid Interface Sci*, 2021, **55**: 101516.
57. Rapoport, N., *et al.* Focused ultrasound-mediated drug delivery to pancreatic cancer in a mouse model. *J Ther Ultrasound*, 2013, **1**: 11.
58. Bartusik, D., and B. Tomanek. Application of 19F magnetic resonance to study the efficacy of fluorine labeled drugs in the three-dimensional cultured breast cancer cells. *Arch Biochem Biophys*, 2010, **493**(2): 234–241.
59. Fang, J.Y., *et al.* Acoustically active perfluorocarbon nanoemulsions as drug delivery carriers for camptothecin: drug release and cytotoxicity against cancer cells. *Ultrasonics*, 2009, **49**(1): 39–46.
60. Pan, D., *et al.* Anti-angiogenesis therapy in the Vx2 rabbit cancer model with a lipase-cleavable Sn 2 taxane phospholipid prodrug using alpha(v)beta(3)-targeted theranostic nanoparticles. *Theranostics*, 2014, **4**(6): 565–578.
61. Soman, N.R., *et al.* Synthesis and characterization of stable fluorocarbon nanostructures as drug delivery vehicles for cytolytic peptides. *Nano Lett*, 2008, **8**(4): 1131–1136.
62. Soman, N.R., *et al.* Molecularly targeted nanocarriers deliver the cytolytic peptide melittin specifically to tumor cells in mice, reducing tumor growth. *J Clin Invest*, 2009, **119**(9): 2830–2842.
63. Rapoport, N., *et al.* Ultrasound-mediated tumor imaging and nanotherapy using drug loaded, block copolymer stabilized perfluorocarbon nanoemulsions. *J Control Release*, 2011, **153**(1): 4–15.

64. Lanza, G.M., *et al.* Targeted antiproliferative drug delivery to vascular smooth muscle cells with a magnetic resonance imaging nanoparticle contrast agent: implications for rational therapy of restenosis. *Circulation*, 2002, **106**(22): 2842–2847.

65. Dewitte, H., *et al.* Design and evaluation of theranostic perfluorocarbon particles for simultaneous antigen-loading and ^{19}F-MRI tracking of dendritic cells. *J Control Release*, 2013, **169**(1–2): 141–149.

66. Lambert, E., V.S. Gorantla, and J.M. Janjic. Pharmaceutical design and development of perfluorocarbon nanocolloids for oxygen delivery in regenerative medicine. *Nanomedicine (Lond)*, 2019, **14**(20): 2697–2712.

67. Lambert, E., and J.M. Janjic. Quality by design approach identifies critical parameters driving oxygen delivery performance in vitro for perfluorocarbon based artificial oxygen carriers. *Sci Rep*, 2021, **11**(1): 5569.

68. Larue, L., *et al.* Fighting hypoxia to improve PDT. *Pharmaceuticals*, 2019, **12**(4): 163.

69. Krafft, M.P. Alleviating tumor hypoxia with perfluorocarbon-based oxygen carriers. *Curr Opin Pharmacol*, 2020, **53**: 117–125.

70. Zhang, Y., *et al.* Biomimetic nanoemulsion for synergistic photodynamic-immunotherapy against hypoxic breast tumor. *Angewandte Chemie*, 2021, **133**(19): 10742–10748.

71. Xavierselvan, M., *et al.* Photoacoustic nanodroplets for oxygen enhanced photodynamic therapy of cancer. *Photoacoustics*, 2022, **25**: 100306.

72. Lee, Y.-H., and Y.-C. Lin. Anti-EGFR indocyanine green-mitomycin C-loaded perfluorocarbon double nanoemulsion: a novel nanostructure for targeted photochemotherapy of bladder cancer cells. *Nanomaterials*, 2018, **8**(5): 283.

73. Wu, P., *et al.* Enhanced antitumor efficacy in colon cancer using EGF functionalized PLGA nanoparticles loaded with 5-Fluorouracil and perfluorocarbon. *BMC Cancer*, 2020, **20**(1): 1–10.

74. Lee, Y.-H., and Y.-T. Ma. Synthesis, characterization, and biological verification of anti-HER2 indocyanine green–doxorubicin-loaded polyethyleneimine-coated perfluorocarbon double nanoemulsions for targeted photochemotherapy of breast cancer cells. *J Nanobiotechnol*, 2017, **15**(1): 1–17.

75. Lee, Y.-H., *et al.* Indocyanine green-camptothecin co-loaded perfluorocarbon double-layer nanocomposite: a versatile

nanotheranostics for photochemotherapy and FDOT diagnosis of breast cancer. *Pharmaceutics*, 2021, **13**(9): 1499.

76. Liu, J., *et al.* Low-intensity focused ultrasound (LIFU)-activated nanodroplets as a theranostic agent for noninvasive cancer molecular imaging and drug delivery. *Biomaterials Sci*, 2018, **6**(11): 2838–2849.
77. Esser, A.K., *et al.* Nanotherapy delivery of c-myc inhibitor targets protumor macrophages and preserves antitumor macrophages in breast cancer. *Theranostics*, 2020, **10**(17): 7510.
78. Hingorani, D.V., *et al.* Cell penetrating peptide functionalized perfluorocarbon nanoemulsions for targeted cell labeling and enhanced fluorine-19 MRI detection. *Magn Reson Med*, 2020, **83**(3): 974–987.
79. Soman, N.R., *et al.* Synthesis and characterization of stable fluorocarbon nanostructures as drug delivery vehicles for cytolytic peptides. *Nano Letters*, 2008, **8**(4): 1131–1136.
80. Ding, L., *et al.* Nanoemulsion-assisted siRNA delivery to modulate the nervous tumor microenvironment in the treatment of pancreatic cancer. *ACS Appl Mater Interfaces*, 2022, **14**(8): 10015–10029.
81. Debela, D.T., *et al.* New approaches and procedures for cancer treatment: current perspectives. *SAGE Open Medicine*, 2021, **9**. DOI: 10.1177/20503121211034366.
82. Saxena, V., M. Sadoqi, and J. Shao. Degradation kinetics of indocyanine green in aqueous solution. *J Pharm Sci*, 2003, **92**(10): 2090–2097.
83. Akers, W.J., *et al.* Noninvasive photoacoustic and fluorescence sentinel lymph node identification using dye-loaded perfluorocarbon nanoparticles. *ACS Nano*, 2011, **5**(1): 173–182.
84. Janjic, J.M., *et al.* Perfluorocarbon nanoemulsions with fluorescent, colloidal and magnetic properties. *Biomaterials*, 2014, **35**(18): 4958–4968.
85. Lim, Y.T., *et al.* Multifunctional perfluorocarbon nanoemulsions for 19 F-based magnetic resonance and near-infrared optical imaging of dendritic cells. *Chem Commun*, 2009, (45): 6952–6954.
86. Soman, N.R., *et al.* Molecularly targeted nanocarriers deliver the cytolytic peptide melittin specifically to tumor cells in mice, reducing tumor growth. *J Clin Inv*, 2009, **119**(9): 2830–2842.
87. Patel, S.K., J. Williams, and J.M. Janjic. Cell labeling for 19F MRI: new and improved approach to perfluorocarbon nanoemulsion design. *Biosensors (Basel)*, 2013, **3**(3): 341–359.

88. Patel, S.K., *et al.* Two-color fluorescent (NIR and visible) triphasic perfluorocarbon nanoemuslions. *J Biomed Opt*, 2013, **18**(10): 101312.

89. Herneisey, M., *et al.* Development and characterization of resveratrol nanoemulsions carrying dual-imaging agents. *Ther Deliv*, 2016, **7**(12): 795–808.

90. Waiczies, H., *et al.* Perfluorocarbon particle size influences magnetic resonance signal and immunological properties of dendritic cells. *PLoS One*, 2011, **6**(7): e21981.

91. Hitchens, T.K., *et al.* 19F MRI detection of acute allograft rejection with in vivo perfluorocarbon labeling of immune cells. *Magn Reson Med*, 2011, **65**(4): 1144–1153.

92. Ruiz-Cabello, J., *et al.* In vivo "hot spot" MR imaging of neural stem cells using fluorinated nanoparticles. *Magn Reson Med*, 2008, **60**(6): 1506–1511.

93. Partlow, K.C., *et al.* 19F magnetic resonance imaging for stem/progenitor cell tracking with multiple unique perfluorocarbon nanobeacons. *FASEB J*, 2007, **21**(8): 1647–1654.

94. Zhang, H. Perfluoro-15-crown-5 ether-labeled dendritic cells, in *Molecular Imaging and Contrast Agent Database (MICAD)*, 2004, Bethesda, MD.

95. Thakkar, D., *et al.* Overcoming biological barriers with ultrasound. *AIP Conf Proc*, 2012, **1481**: 381–387.

96. Janjic, J.M., and E.T. Ahrens. Fluorine-containing nanoemulsions for MRI cell tracking. *Wiley Interdiscip Rev Nanomed Nanobiotechnol*, 2009, **1**(5): 492–501.

97. Patel, S.K., *et al.* Theranostic nanoemulsions for macrophage COX-2 inhibition in a murine inflammation model. *Clin Immunol*, 2015, **160**(1): 59–70.

98. de Oca-Ávalos, J.M.M., R.J. Candal, and M.L. Herrera. Nanoemulsions: stability and physical properties. *Curr Opin Food Sci*, 2017, **16**: 1–6.

99. Lee, S.J., *et al.* Interaction of melittin peptides with perfluorocarbon nanoemulsion particles. *J Phys Chem B*, 2011, **115**(51): 15271–15279.

100. Lim, Y.T., *et al.* Multifunctional perfluorocarbon nanoemulsions for (19)F-based magnetic resonance and near-infrared optical imaging of dendritic cells. *Chem Commun (Camb)*, 2009, (45): 6952–6954.

101. Patel, S.K., J. Williams, and J.M. Janjic. Cell labeling for 19F MRI: new and improved approach to perfluorocarbon nanoemulsion design. *Biosensors*, 2013, **3**: 341–359.

102. Ajazuddin, *et al.* Recent expansions in an emergent novel drug delivery technology: emulgel. *J Control Release*, 2013, **171**(2): 122–132.

103. Liu, L., C. Bagia, and J.M. Janjic. The first scale-up production of theranostic nanoemulsions. *BioResearch Open Access*, 2015, **4**(1): 218–228.

104. Fernandez Campos, F., *et al.* Development and characterization of a novel nystatin-loaded nanoemulsion for the buccal treatment of candidosis: ultrastructural effects and release studies. *J Pharm Sci*, 2012, **101**(10): 3739–3752.

105. Hung, C.F., *et al.* Development and evaluation of emulsion-liposome blends for resveratrol delivery. *J Nanosci Nanotechnol*, 2006, **6**(9–10): 2950–2958.

106. Neves, A.R., *et al.* Novel resveratrol nanodelivery systems based on lipid nanoparticles to enhance its oral bioavailability. *Int J Nanomedicine*, 2013, **8**: 177–187.

107. Sanna, V., *et al.* Resveratrol-loaded nanoparticles based on poly(epsilon-caprolactone) and poly(d,l-lactic-co-glycolic acid)-poly(ethylene glycol) blend for prostate cancer treatment. *Mol Pharm*, 2013, **10**(10): 3871–3881.

108. Zillich, O.V., *et al.* Release and in vitro skin permeation of polyphenols from cosmetic emulsions. *Int J Cosmet Sci*, 2013, **35**(5): 491–501.

109. Stoll, G., *et al.* Visualization of inflammation using (19) F-magnetic resonance imaging and perfluorocarbons. *Wiley Interdiscip Rev Nanomed Nanobiotechnol*, 2012, **4**(4): 438–447.

110. Balducci, A., *et al.* A novel probe for the non-invasive detection of tumor-associated inflammation. *Oncoimmunology*, 2013, **2**(2): e23034.

111. Kadayakkara, D.K., *et al.* Inflammation driven by overexpression of the hypoglycosylated abnormal mucin 1 (MUC1) links inflammatory bowel disease and pancreatitis. *Pancreas*, 2010, **39**(4): 510–515.

112. Weise, G., *et al.* In vivo imaging of inflammation in the peripheral nervous system by (19)F MRI. *Exp Neurol*, 2011, **229**(2): 494–501.

113. Hertlein, T., *et al.* Visualization of abscess formation in a murine thigh infection model of Staphylococcus aureus by 19F-magnetic resonance imaging (MRI). *PLoS One*, 2011, **6**(3): e18246.

114. Kadayakkara, D.K., *et al.* Assaying macrophage activity in a murine model of inflammatory bowel disease using fluorine-19 MRI. *Lab Invest*, 2012, **92**(4): 636–645.

115. Weibel, S., *et al.* Imaging of intratumoral inflammation during oncolytic virotherapy of tumors by 19F-magnetic resonance imaging (MRI). *PloS ONE*, 2013, **8**(2): e56317.

116. Balducci, A., *et al.* Visualizing arthritic inflammation and therapeutic response by fluorine-19 magnetic resonance imaging (19F MRI). *J Inflammation (London, UK)*, 2012, **9**(1): 24–34.

117. Ebner, B., *et al.* Early assessment of pulmonary inflammation by 19F MRI in vivo. *Circulation: Cardiovascular Imaging*, 2010 **3**(2): 202–210.

118. Hitchens, T.K., *et al.* 19F MRI detection of acute allograft rejection with in vivo perfluorocarbon labeling of immune cells. *Magn Reson Med*, 2011, **65**(4): 1144–1153.

119. Balducci, A., *et al.* A novel probe for the non-invasive detection of tumor-associated inflammation. *Oncoimmunology*, 2013, **2**(2): e23034.

120. Tang, S.Y., Shridharan, and M. Sivakumar, Impact of process parameters in the generation of novel aspirin nanoemulsions - Comparative studies between ultrasound cavitation and microfluidizer. *Ultrason Sonochem*, 2013, **20**(1): 485–497.

121. Qian, C., and D.J. McClements. Formation of nanoemulsions stabilized by model food-grade emulsifiers using high-pressure homogenization: factors affecting particle size. *Food Hydrocolloids*, 2011, **25**(5): 1000–1008.

122. Pinnamaneni, S., N.G. Das, and S.K. Das. Comparison of oil-in-water emulsions manufactured by microfluidization and homogenization. *Pharmazie*, 2003, **58**(8): 554–558.

123. Maa, Y.F., and C.C. Hsu. Performance of sonication and microfluidization for liquid-liquid emulsification. *Pharm Dev Technol*, 1999, **4**(2): 233–240.

124. Lee, L., and I.T. Norton. Comparing droplet breakup for a high-pressure valve homogeniser and a Microfluidizer for the potential production of food-grade nanoemulsions. *J Food Eng*, 2013, **114**(2): 158–163.

125. Tang, S.Y., P. Shridharan, and M. Sivakumar. Impact of process parameters in the generation of novel aspirin nanoemulsions – comparative studies between ultrasound cavitation and microfluidizer. *Ultrason Sonochem*, 2013, **20**(1): 485–497.

126. Mao, L., *et al.* Effects of homogenization models and emulsifiers on the physicochemical properties of β-carotene nanoemulsions. *J Disp Sci Technol*, 2010, **31**(7): 986–993.

127. Bonaccorso, A., *et al.* Quality by design tools reducing the gap from bench to bedside for nanomedicine. *Eur J Pharm Biopharm*, 2021, **169**: 144–155.

128. Herneisey, M., *et al.* Quality by design approach using multiple linear and logistic regression modeling enables microemulsion scale up. *Molecules*, 2019, **24**(11): 2066.

Chapter 6

Choosing and Optimizing a Pulse Sequence for Fluorine-19 MRI

Ruud B. van Heeswijk
Department of Radiology, Lausanne University Hospital (CHUV) and University of Lausanne (UNIL), Lausanne, Switzerland
ruud.van-heeswijk@chuv.ch

6.1 Introduction

Although already introduced in the 1970s [1], fluorine-19 magnetic resonance imaging (^{19}F MRI) in combination with perfluorocarbons (PFCS) has seen a renewed and increasingly widespread interest since its recent application to cell tracking [2] and inflammation imaging [3] in the 2000s, as well as several efforts to translate it to the clinical setting [4]. The advantages of ^{19}F MRI for tracking biological processes are numerous: it makes use of inert tracers that avoid harmful radiation, it is highly specific, it can be readily quantified, it can be used to monitor biological processes over time [5, 6], and it can easily be combined with anatomical MRI [3, 7]. However, opposed to all these advantages there is one large

Fluorine Magnetic Resonance Imaging: Methods and Applications in Biomedicine
Edited by Eric T. Ahrens and Ulrich Flögel
Copyright © 2025 Jenny Stanford Publishing Pte. Ltd.
ISBN 978-981-5129-37-3 (Hardcover), 978-1-003-53004-6 (eBook)
www.jennystanford.com

drawback: ^{19}F MRI has a relatively poor detection sensitivity compared to other molecular imaging modalities [8]. This is the case for all types of MRI (including anatomical ^{1}H imaging), since the low nuclear polarization is limited to only a few out of a million nuclei. In the case of ^{19}F MRI, this is drawn to the foreground, because we want to detect ^{19}F concentrations in the micro- and millimolar range instead of the 50 molar concentration of water that ^{1}H MRI targets. The low sensitivity of ^{19}F MRI thus directly leads to relatively high detection limits compared to nuclear methods or even ^{1}H MRI with indirectly detected (super-)paramagnetic contrast agents that affect the abundant water. Thus, we focus on methodological development to enhance sensitivity.

The ^{19}F MR detection sensitivity can be optimized in several ways. This chapter will focus on optimizing image acquisition and reconstruction, specifically, pulse sequences and their parameters for a combination of an organ, a compound, and a magnetic field strength. There are many other ways to increase the sensitivity of ^{19}F MRI, such as smart nanoparticle design [9, 10], the use of higher magnetic field strengths [11], and sensitive radiofrequency (RF) coils (including cryo-coils) [12], but these are discussed elsewhere in this volume.

This chapter is meant to serve as a primer for newcomers to the field, as a guide for those looking to start a new ^{19}F MRI project, and as an update for those that are well-established in the field. Several general definitions, considerations and caveats are first introduced. This is followed by a discussion of the three main pulse sequence families for efficient ^{19}F MRI: turbo spin echo, gradient echo, and ultrashort-echo-time sequences. Moreover, we will describe imaging of compounds with multiple resonances and briefly, image reconstruction with compressed sensing. Finally, a comparison between pulse sequences and reconstruction methods is presented to help provide guidance for a given situation.

6.2 General Considerations

6.2.1 Definition of Terms

We will start with a small refresher of methodological terms that will help us for the evaluation of the performance of a pulse sequence.

For ^{19}F MRI, the *detection limit* is the lowest concentration (or quantity) that can be reliably detected—we want this to be as low as possible. In our discussion here, we use the Rose criterion [13], which states that the lowest signal-to-noise ratio (SNR) that can be detected is SNR = 4. The measurement *sensitivity* is usually defined as the inverse of the detection limit, it should not be confused with sensitivity as the true positive rate that is often used in medicine. The sensitivity can also be expressed quantitatively as the *SNR efficiency*, which is the SNR per square root of the acquisition time of the pulse sequence.

Precision describes how variable the measurement is due to random errors, and directly influences the sensitivity and detection limit. It can be qualitatively expressed as the inverse of the standard deviation of a series of measurements; the lower the standard deviation of a repeated measurement or over a known homogeneous region of pixels, the higher the precision.

Accuracy in this context is how well our measurement reflects the ground truth if known, or alternatively, a gold-standard measurement. The term accuracy itself is mostly used in a qualitative sense; it can be quantified by a *bias, i.e.*, the systematic error in a measurement. High accuracy also requires linearity of the measurement; if there is a bias, it should be constant for the entire range of measurement.

A precise measurement is also *repeatable*, meaning that it should result in the same results when performed with the same setup by the same researcher. If the same measurement is performed by other researchers and leads to the same results, it is instead *reproducible*. It should also be noted that a repeatable measurement should be *robust*, meaning that it should be sensitive and accurate in the presence of small variations in experimental conditions, which in the case of MRI, often mean magnetic field inhomogeneities.

Overall, an ideal ^{19}F MRI pulse sequence should thus be precise, accurate, and robust, and should result in low detection limits and standard deviations.

6.2.2 Caveats for Pulse Sequence Comparisons

Optimization of a pulse sequence usually means focusing on how to achieve the highest possible *theoretical* sensitivity by varying the pulse sequence parameters as a function of a set of relaxation times.

However, we should be aware that there are always properties of a pulse sequence beyond this type of optimization that may strongly influence its suitability for the application. These "secondary" properties may render a direct comparison of the performance of multiple pulse sequences less straightforward, and thus, one should first compare the conditions where each pulse sequence performs best.

A secondary property that is often overlooked is the k-space trajectory. Most pulse sequences can be combined with linear Cartesian, centric Cartesian, radial, spiral, propeller, and other trajectories to fill up k-space. Each of these approaches has pros and cons, and the sensitivity of pulse sequences that are acquired with different k-space trajectories *cannot* easily be compared to one another on a theoretical level. A turbo spin echo pulse sequence with a centric readout on the one hand will for example theoretically result in high sensitivity, but the k-space pattern is highly sensitive to motion corruption and will also result in line broadening [14]. On the other hand, a gradient echo sequence with radial sampling has much lower nominal sensitivity due to both the sequence and the sampling pattern, but its performance will be robust in the presence of motion, and it also does not result in blurring of small signal foci; thus this sequence may be more appropriate in certain applications. Similar consideration should be given to the effects of the magnetic field strength. A pulse sequence that performs well on a 9.4T animal scanner may not work acceptably on a 3.0T clinical scanner, for example due to insufficient chemical shift dispersion at the lower magnetic field strength, and the opposite may occur for another pulse sequence due to the increased magnetic field inhomogeneity at 9.4T.

When comparing sequences, the choice of RF coil should also be taken into account. A small surface coil will lead to a much higher local sensitivity compared to a large volume coil, but the distance from a surface coil at which various pulse sequences still generate signal significantly differs. Consequently, it is often challenging to compare pulse sequence sensitivities among different studies.

The most appropriate method to directly compare pulse sequences is a well-characterized phantom experiment in which conditions can be reproduced for all pulse sequences. To this end, a quantitative measure of the sensitivity should be used. The SNR efficiency (SNR_{eff}) is often used as such a measure, and is defined as the time-normalized SNR of the pulse sequence. SNR_{eff} depends on several pulse sequence parameters [15, 16] and can be defined as:

$$SNR_{eff} = \frac{SNR}{\sqrt{T_{acq}}} \propto V\frac{S}{\sigma_N}\sqrt{\frac{NA \cdot N_{enc}}{BW \cdot TR}} \quad (6.1)$$

where V is the voxel volume, T_{acq} the acquisition time, S the signal amplitude, σ_N the standard deviation of the apparent noise, NA the number of averages, N_{enc} the total number of encoding steps (a multiplication of the encoding steps in all three dimensions), BW the readout bandwidth, and TR the repetition time. Several of these can be set directly as sequence parameters (V,NA, N_{enc}, BW, TR), but often come with consequences for other metrics. The signal S is the result of interactions of many hardware, acquisition, and reconstruction parameters, as is σ_N to a lesser degree. Even when comparing pulse sequences *in vitro* with the same experimental setup, the most relevant metric to assess the performance should be carefully chosen.

6.3 Optimizing Sensitivity of Pulse Sequences

There are several families of pulse sequences that that can be used for highly sensitive ^{19}F MRI (Fig. 6.1). Three pulse sequence families will be described. The Turbo Spin Echo (TSE), also known as Fast Spin Echo (FSE) or Rapid Acquisition with Relaxation Enhancement (RARE) will be treated first. We will then look at the gradient-recalled echo (GRE) or gradient echo (GE) and one of its most sensitive variants—the balanced Steady-State Free Precession (bSSFP) sequence. Lastly, the optimization of ultrashort echo-time (UTE) pulse sequences will be described.

For each pulse sequence we describe the advantages and disadvantages of its application and how the sensitivity should be optimized. This optimization can usually be achieved either directly through equations that analytically describe the signal, or through simulations of the Bloch equations. The latter requires a (usually small) script to be written in a programming language such a Matlab or Python, but once written these simulations are highly flexible and allow for the addition of specific assumptions, special conditions, and supposed imperfections.

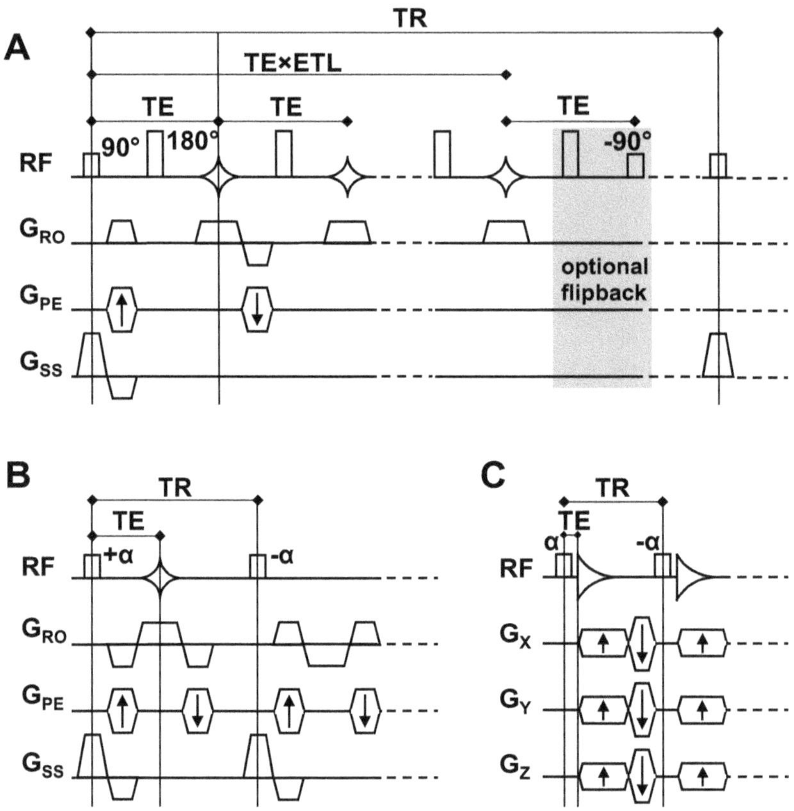

Figure 6.1 Pulse sequence diagram of three high-sensitivity pulse sequences. (A) A TSE pulse sequence with a centric Cartesian readout and optional flipback module. After a 90° excitation pulse, 180° refocusing pulses are played out each echo time (TE). (B) A linear Cartesian bSSFP pulse sequence with a readout with alternating polarity. (C) A 3D radial bSSFP-UTE pulse sequence. Note that the gradient directions are named x, y, and z, since they are all used for the readout.

6.3.1 Turbo Spin Echo

In a TSE pulse sequence, all magnetization is excited with a 90° RF pulse, refocused with a 180° RF pulse that cancels signal loss due to magnetic field inhomogeneities (T_2* decay), then forms an echo. The echo is recorded at the echo time (TE). The 180° pulse and readout are then repeated ETL times (Fig. 6.1A), where ETL stands for echo train length. Each subsequent echo is used to encode a different k-space line while its signal decreases with T_2 relaxation. Since all magnetization is used, sufficient T_1 recovery is needed to grow new signal, and the repetition time (TR) is always set long. Due to the need for both high-flip-angle (and slice-selective) RF pulses, the TE is also relatively long, and thus there is high RF power deposition (Table 6.1). This means that while the pulse sequence is SNR efficient and the refocusing renders it insensitive to main magnetic field (B_0) inhomogeneities and susceptibility (χ) effects, it is also relatively slow, and may be less suitable for PFCs with short relaxation times. Also, due to the long echo train, the sequence is sensitive to motion distortion.

The SNR efficiency of TSE pulse sequences can be increased by making use of stimulated echoes [14]. By moderating the strength of the crusher gradient pairs around the 180° pulses, imperfectly excited regions (such as at the edge of the slice profile) will see magnetization refocused as stimulated echoes on top of the spin echoes, thus increasing the total signal. The stimulated echoes also help with the limited penetration depth of the TSE sequence when used with transmit (Tx) surface coils. Because of the need for 180° refocusing pulses, the decreasing effective flip angle with distance from a Tx surface coil causes a disproportionally small echo to be formed away from the RF coil; stimulated echoes will partially compensate for this loss.

Table 6.1 Advantages and disadvantages of three sensitive MR pulse sequences

Pulse Sequence	Advantages	Disadvantages
TSE	SNR-efficient, insensitive to B_0 inhomogeneities and χ effects, flexible k-space filling	Sensitive to motion, slow, high power deposit, unsuited for short-T_2 species, limited penetration with Tx surface RF coils
bSSFP	SNR-efficient, fast, flexible k-space filling	Very sensitive to B_0 inhomogeneities, very high power deposit, requires ramp-up for steady state
UTE	SNR-efficient, very fast, great for low-T_2 species, very robust: insensitive to B_0/B_1 inhomogeneities and χ effects	Forced to use radial sampling, half-echo readout

The TSE sequence can be used with several k-space sampling patterns, including SNR-efficient centric Cartesian (which may lead to blurring in the case of short T_2 relaxation times), robust linear Cartesian sampling, or radial sampling when motion is a serious concern.

In the case of a longer T_2 value and a shorter echo time, or in the presence of motion that precludes the use of longer echo trains, the choice can also be made to use a flip-back module that restores the transverse magnetization as longitudinal magnetization (Fig. 6.1A, grey zone). This will occur at the cost of any T_1 recovery during the echo train, so should only be used if there is significant transverse signal left at the end of the echo train.

A final point of flexibility is that TSE can be efficiently used with both 3D and multi-slice 2D (MS2D) imaging. 3D TSE on the one hand has the advantage that the entire volume is excited at each readout, and that the SNR thus goes up by a factor of $\sqrt{2}$ with each doubling of the number of slices in the 3D stack. On the other hand, MS2D has the advantage that a number of slices $<TR/(TE \times ETL)$ can be acquired "for free"—without any extra scan time—and thus greatly increases the SNR efficiency. However, it should be noted that the thinly selective excitation of MS2D TSE has more restrictive requirements on the RF pulse and will often lead to longer echo times. Due to the imperfect slice profile, the spacing of the slices also cannot be contiguous. Nevertheless, in most cases with a limited number of slices, MS2D TSE will be more SNR efficient than its 3D counterpart.

Once the choice to use a TSE pulse sequence is made, the optimal pulse sequence parameters need to be set. Many parameters (such as voxel volume and the total acquisition time) will be set according to the requirements of the experiment. This leaves the readout bandwidth (BW), TE, ETL, and TR free to be optimized. In principle TE should be set to the shortest possible value to avoid T_2 decay, but since there is an interplay between BW and TE, their optimal choice depends on the T_2 value of the PFC used, as well as on hardware parameters such as the maximum gradient strength and slew rate. For longer T_2 values, BW should be minimized at the cost of a longer TE in order to maximize the SNR efficiency. There is a similar interplay between the ETL and TR, which is less dependent on hardware and was initially explored for ^{19}F MRI by Mastropietro and colleagues [17]. Because of the many interdependences, this optimization is best performed through numerical simulations of the Bloch equations [18, 19]. From such optimizations it can be readily observed that the optimum is quite broad, and both ETL and TR can be changed by >25% before there is a 10% drop in SNR efficiency (Fig. 6.2). This is reassuring because the relaxation times of perfluorocarbons tend to change significantly as a function of temperature and oxygenation of the target tissue [18].

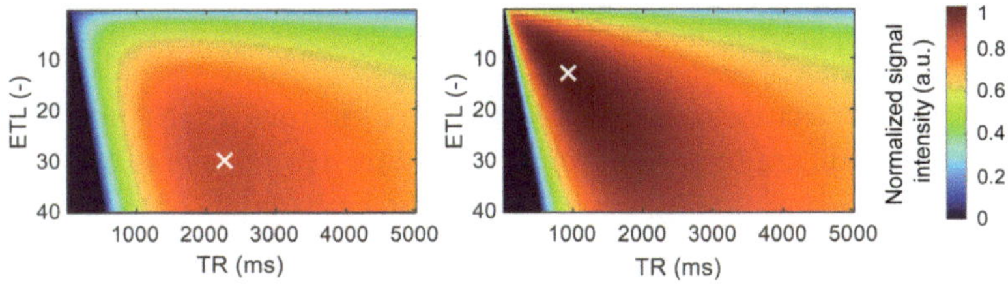

Figure 6.2 Optimization of TSE pulse sequence parameters. Bloch simulations of the TSE pulse sequence with the flip-back pulse turned off (left) and on (right) result in broad optimal regions, while the absolute optimum is indicated with a cross. Other parameters were chosen for a clinical 3T scanner and included BW = 130 Hz/px, TE = 12 ms, and PFC relaxation times T_1/T_2 = 1200/400 ms, as observed for PFTBH. Note that the optimum in the left plot is ~87% of that of the right plot and that the dark regions on the bottom left are impossible combinations where ETL × TE > TR.

6.3.2 Gradient-Recalled Echo and bSSFP

As its name implies, in gradient-recalled echo (GRE) imaging an RF pulse with a small flip angle is followed by a dephasing gradient and an opposed rephasing gradient, which results in an echo at a time TE [20]. Since macroscopic dephasing is not refocused as it is in TSE imaging, the signal at TE is determined by T_2* and the pulse sequence is sensitive to magnetic field inhomogeneities. Residual magnetization after the readout is often fully dephased (also called spoiled or crushed) with a strong variable gradient after the readout. The low flip angle means that most of the longitudinal magnetization remains after a readout, and a subsequent excitation can be played out quickly (on the order of 5–100 ms) after the spoiler, which makes the pulse sequence fast compared to TSE. GRE can be used with even greater flexibility in the choice of the k-space trajectory. Trajectories such as spiral and propellor types of k-space sampling can also be used with GRE. However, the low flip angle and spoiling make GRE a relatively SNR-inefficient pulse sequence. If GRE imaging is used due to the requirements of the experimental conditions (for example when imaging flow over a large area [21], RF excitation should be performed with the Ernst angle [22]:

$$\alpha_{\mathrm{opt}} = arccos\left(e^{-\mathrm{TR}/\mathrm{T}_1}\right) \tag{6.2}$$

Here a T_1/T_2 pair of 1200 ms/400 ms with TR = 10 ms would thus lead to a relatively low $\alpha_{\mathrm{opt}} = 7.3°$.

Several changes can be made to GRE to improve its SNR efficiency. First, the full gradient spoiling can be reduced to achieve steady-state free precession (SSFP), which retains several signal components to generate a larger signal. Nowadays, SSFP has been mostly replaced by balanced SSFP (bSSFP), in which all gradients are fully balanced (Fig. 6.1B), and which has alternating opposite-angle RF pulses [23]. bSSFP can be used with very high flip angles; in its steady state all magnetization bounces back and forth between two positions that have large transverse and longitudinal components, which means that it generates a large signal but also does not need to wait for T_1 recovery with a long TR (Table 6.1).

bSSFP requires a series of 5–30 ramp-up RF pulses to get into its steady state and thus loses some SNR efficiency when used for 2D imaging; when used in 3D imaging this small penalty becomes negligible. The sequence has a very high SNR efficiency for PFCs with a high T_2/T_1 ratio. Because of its balanced nature, SNR efficiency can furthermore be optimized as a function of the relaxation times and is independent of the other pulse sequence parameters:

$$\alpha_{\mathrm{opt}} = arccos\left(\frac{\mathrm{T}_1 - \mathrm{T}_2}{\mathrm{T}_1 + \mathrm{T}_2}\right) \tag{6.3}$$

As the T_2/T_1 ratio approaches 1, this optimal angle becomes higher (Fig. 6.3); our T_1/T_2 = 1200 ms/400 ms pair would lead to a relatively high $\alpha_{\mathrm{opt}} = 60°$.

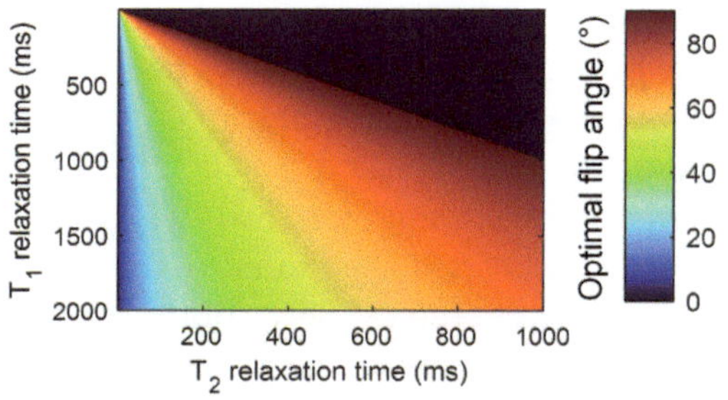

Figure 6.3 Optimization of the flip angle for a bSSFP pulse sequence. Independent of other pulse sequence parameters, the bSSFP flip angle should be chosen purely as a function of the T_1 and T_2 values of the PFC. The dark (zeroed) region on the top right contains impossible combinations with $T_2 > T_1$.

The main disadvantage of bSSFP also originates from its steady state; if there is too much intra-voxel dephasing, the transverse component of the steady state quickly falls apart and no signal remains. This readily occurs in the presence of relatively small main magnetic field (B_0) inhomogeneities and leads to dark bands in the image, although it can often be countered by high-performance shimming. Because of this B_0 sensitivity and because of the tendency of T_1/T_2 to increase with magnetic field strength for fluorine, the use of bSSFP is usually limited to lower-field scanners or magnetically homogeneous body regions such as the brain.

6.3.3 Ultrashort Echo Time

Finally, we consider the oldest and most straightforward (but perhaps slightly less known) pulse sequence family based on the free induction decay (FID). In this type of sequence the signal is recorded as soon as possible after an RF pulse is played out. The main imaging sequence derived from the FID is the UTE (ultrashort echo time) [24], although this is a slight misnomer since no echo is formed, and only half of k-space is traversed during a single readout (Fig. 6.1C). Since the time between the end of the RF pulse and the start of the recording should be kept to an absolute minimum to minimize T_2* relaxation, sampling usually starts a few microseconds after the end of the RF pulse (to avoid its ringing effects), including during the ramp-up of the encoding gradient. No phase encoding or compensation gradients are played out prior to the readout, which means that the pulse sequence will always be radial. The echo time is measured from the centroid (geometric center) of the RF pulse to the start of the readout [25]. This distinction between centroid and the center is especially relevant when a 2D UTE sequence is used; in this case, a half-pulse is often used to move the centroid to later in the pulse and thus to limit the T_2* decay during the RF pulse.

The microsecond TE make this pulse sequence very robust against perturbations that dephase and decrease transverse magnetization such as flow, susceptibility changes, B_0 inhomogeneities, and of course, short T_2 relaxation times. UTE is therefore the pulse sequence of choice for PFCs with very short T_2 relaxation times or nanoparticles that are mixed with paramagnetic compounds [26, 27]. If bound to a radial in-plane readout scheme, UTE can acquire single 2D slices [28], and can be extended to full 3D radial sampling with further SNR penalty [29], or can be used with a stack-of-stars trajectory [30] to retain SNR.

The UTE readout can be combined with most variations of the gradient echo pulse sequence. When the transverse magnetization is spoiled after the readout, the SNR efficiency will be relatively low, but the sequence is the most robust. If the relaxation times are much longer than the RF pulse length, the optimal flip angle can be calculated as the Ernst angle (Equation 6.2). When very short T_2* values are encountered, the optimal flip angle should be numerically simulated instead [28, 31]. The UTE readout can also be combined with a FISP sequence [32] or bSSFP sequence [29] to increase SNR efficiency. bSSFP-UTE (Fig. 6.1C) is the most SNR-efficient variant, but is more sensitive to B_0 inhomogeneities like its bSSFP parent. For longer relaxation times, the optimal flip angle follows that of regular bSSFP (Equation 6.3), while for very short T_2 values, numerical simulations should again be performed. There are also several pulse sequences that are similar to UTE and can be employed with even shorter echo times, such as the zero-echo-time (ZTE) sequence [32] and pointwise encoding time reduction with radial acquisition (PETRA) [33].

One of the practical disadvantages of UTE pulse sequences that hinders adoption is that they often require significant calibration of the image reconstruction, which is dependent on the scanner hardware characteristics. Since the readout already starts during the ramp-up of the gradient, and this ramp-up is usually performed at the maximum slew rate, the true timing and gradient strength dynamics need to be precisely characterized and calibrated in order to avoid image blurring.

6.4 Image Reconstruction and Compressed Sensing

Over the past decade it has become clear that advances in MRI are no longer limited to improvements in image acquisition alone and that reconstruction has become increasingly important; acquisition and reconstruction are inseparable and must be considered as a whole. Significant advances in image reconstruction have directly impacted image acquisition and data sampling strategies. While the old adage of "you are only as good as your acquired data" still holds, advances in reconstruction have enabled higher precision and can also improve detection sensitivity with the same data.

Several postprocessing steps have been around since the beginning of MRI. Zero-filling and zero-padding have been used to decrease the pixel size without affecting the SNR (since the zeros contain no signal or noise), with the only caveat that the transition from the data to the zeros should be relatively smooth to avoid sharp discontinuities that result in Gibbs rigging artifacts. Image [34] and k-space [35] filters have been used to suppress noise ("denoise"), usually at the cost of some spatial resolution. While such filtering techniques improve the precision of a ^{19}F MR measurement, it is important to note that they do not change the detection limits, since signal that is smaller than the noise level cannot be distinguished from that noise, and will be removed. Furthermore, and more importantly, because the noise characteristic of the images can be arbitrarily altered by changing filter parameters, the SNR is no longer a reliable measure of detection sensitivity.

Compressed sensing (CS) [36] goes beyond these techniques and, with a properly tailored acquisition scheme, can be used to accurately reconstruct highly undersampled data (Fig. 6.4). It is thus mainly a technique to accelerate the acquisition and not to increase the sensitivity. However, as always with MRI, the time gained by accelerating/undersampling the acquisition can be invested to increase the sensitivity, as discussed below. When reconstructing undersampled k-space data with a normal Fourier transform, the resulting image will contain undersampling artifacts proportional to the degree of undersampling and dependent on the manner in which the undersampling is performed. CS, on the other hand, can be used to reconstruct an artifact-free image based on three conditions:

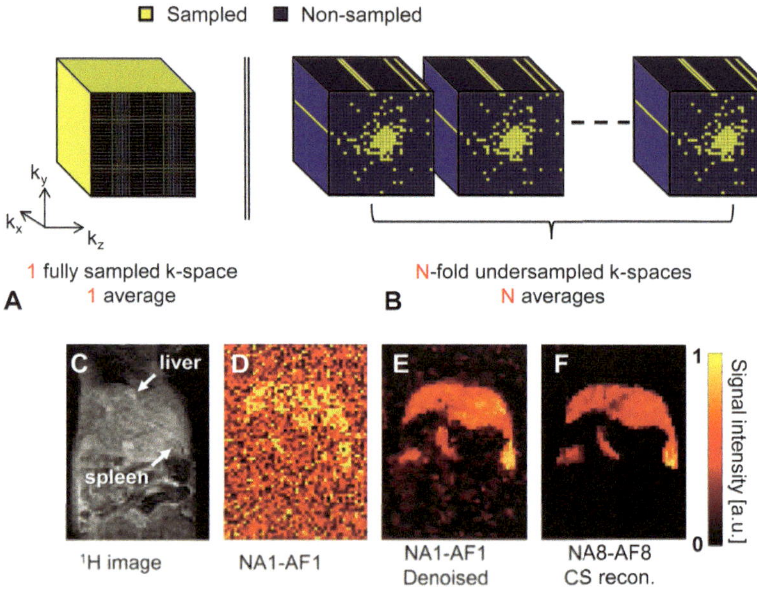

Figure 6.4 An example of compressed sensing in fluorine-19 MRI. (A) A fully sampled k-space with each line acquired once. (B) An N-fold-undersampled k-space, undersampled with a variable density function that fully samples its center and gradually undersamples its periphery, but with each line acquired N times. Sampling strategies A and B have the same total sample count and acquisition time. (C) *In vivo* coronal ^{1}H GRE image of the mouse abdomen. (D–F) *In vivo* images of the mouse abdomen with three strategies. NA1-AF1 ^{19}F image without denoising (D) and with denoising reconstruction (E); The liver and spleen of the mouse are more clearly visible in the latter. NA = number of averages, AF = acceleration factor. (F) NA8-AF8 ^{19}F image with CS reconstruction; the liver and spleen are more detailed and there is less residual background signal than in the denoised counterparts. Adapted from Darçot *et al.* [37] with permission.

1. The image of interest must have a sparse representation (*i.e.*, only a few non-zero coefficients), either in the image domain or another transform domain.

2. The applied undersampling pattern must generate incoherent (*i.e.*, noise-like) aliasing artifacts in that sparse domain.
3. A nonlinear reconstruction must be used to ensure both sparsity of the image representation and consistency of the acquired data.

The seminal paper of by Lustig *et al.* [36] gives an excellent overview of compressed sensing and the importance of these three conditions. Typically, ^{19}F MR images are already sparse in the image domain as they often contain only a few small regions of non-zero signal. However, for more conventional MR images, a sparsifying transform is generally needed to map the data from the image domain to another mathematical transform domain where their representation will be sparse. Many sparsifying transforms are available, and its choice depends on the imaging technique and the nature of the image. To meet the second condition, random undersampling should be used to avoid coherent aliasing artifacts. However, complete random undersampling is not practical for two reasons. First, k-space trajectories need to be relatively smooth to meet physiological and hardware constraints of the gradient system. Second, because most of the energy of the image is concentrated in the k-space center, this region should be more densely sampled than the periphery. Therefore, pseudo-random undersampling was developed to provide good compromise between aliasing incoherence and sampling feasibility and practicality.

To reconstruct the acquired undersampled data, the compressed sensing algorithm solves an optimization problem to find the right balance between data consistency and sparsity (condition 3, above). The unconstrained convex optimization problem of CS can thus be formulated as the following minimization problem:

$$\underset{m}{\operatorname{argmin}} \| \mathcal{F}_u m - y \|_2^2 + \lambda \| \Psi m \|_1, \tag{6.4}$$

where m is the reconstructed image, y is the acquired raw data, Ψ is a sparsifying transform operator such as the wavelet transform [38], $\|\cdot\|_1$ is the ℓ_1-norm, $\|\cdot\|_2^2$ is the ℓ_2-norm, λ is the regularization parameter, and $\mathcal{F}_u$ is the undersampled Fourier operator. The reconstruction is thus performed by minimizing the ℓ_1 norm of a transformed image, subject to data fidelity constraints (the ℓ_2 term). More regularization terms with different domains can be added if desired. In practice, these optimization problems are, for example, solved using the nonlinear conjugate gradient descent algorithm with backtracking line search as originally proposed [36], or alternatively, using the alternating direction method of multipliers (ADMM) algorithms [39].

The regularization weight λ is usually empirically chosen or optimized according to the desired result. As mentioned above for filters, because the regularization has a denoising effect on the image, our view is that image noise and the SNR can no longer be used to calculate precision and sensitivities. The regularization parameter λ can simply be increased during the reconstruction to increase image sparsity and thus the denoising effect, which artificially boosts the SNR. CS-based image reconstructions should therefore first be thoroughly characterized versus a non-CS ground truth through metrics such as the Dice similarity coefficient (DSC) [40], root mean square error (RMSE), or the structural similarity index measure (SSIM) [41]. Without a ground truth (for example a fully sampled high-SNR image), these metrics cannot be calculated and the fidelity or improvement in image quality of the CS-reconstructed image may be difficult to quantify.

In the case of fully sampled raw data, the fully sampled Fourier operator F can be used, and the algorithm of Equation 6.4 behaves as an iterative wavelet denoising filter [42]:

$$\underset{m}{\operatorname{argmin}} \| Fm - y \|_2^2 + \lambda \| \Psi m \|_1. \tag{6.5}$$

CS with ^{19}F MRI was first proposed for fast catheter tracking [43], followed by its application for accelerated cell tracking [44]. Liang *et al.* [45] explored the efficiency of several CS algorithms at low SNR and concluded that the CS algorithm developed by Lustig *et al.* [36] remained the most efficient in terms of preserving the features of the signal of interest. The possibility to flexibly adapt the scan duration [46] was explored for ^{19}F chemical shift imaging. Recently, several groups [37, 47] combined CS and ^{19}F MRI with signal averaging (Fig. 6.4). Here, individual k-space lines are averaged multiple times, which elevates small signals above the noise floor. Darçot *et al.* [37] demonstrated that while wavelet denoising can be effectively used to improve the sensitivity of a fully sampled image, a CS-averaged acquisition further improves this sensitivity and also improves the robustness of the images against motion artifacts.

CS is thus a powerful technique that can be used to accelerate acquisitions, improve sensitivity, and make the acquisitions more robust against motion artifacts; these attributes make CS an attractive option for potential clinical use [48].

6.5 Imaging of Multiple Resonances

^{19}F MRI of PFCs comes with unique challenges and opportunities due to spectral properties of PFCs and the nature of the fluorine atom. A fluorine nucleus in a PFC molecule is on average surrounded by 9 electrons instead of the single electron of hydrogen, making fluorine extremely electronegative and causing the range of chemical shifts of fluorinated groups to be much larger than those of hydrogen. On the one hand, this means that compounds with multiple resonances such perfluorooctyl bromide (PFOB) will see these resonances spread over a frequency range of several kilohertz, which will cause large chemical shift displacement (CSD) artifacts in images. On the other hand, the resonances of multiple compounds will be well-separated, and pulse sequences can be designed to efficiently exploit this difference to generate discrete images of multiple compounds. In this section, several techniques to remove CSD artifacts from images of compounds with multi-resonance spectra are described. Additionally, we describe techniques that can be used both for CSD artifact correction and for the separation of multiple compounds, which can be used for 'multi-color' imaging.

6.5.1 Imaging Single Compounds with Multiple Resonances

Several of the PFCs with the most promising biocompatibility profile, *i.e.*, they are retained a short time in the liver and spleen, have complex MR spectra with multiple resonances [49]. These resonances are spread over a broad frequency range and create CSD artifacts that dilute the signal over multiple voxels, making the interpretation of *in vivo* images highly ambiguous. Furthermore, fluorine groups on the same molecule can J-couple to one another [29, 50], which both further splits the resonances and drastically shortens the T_2 relaxation times—a double hit on the signal obtained with most pulse sequences.

The most straightforward solution to avoid CSD artifacts is to selectively excite only one resonance [50]. For molecules such as PFTBH (perfluoro-t-butylcyclohexane, also known as Oxycyte and ABL-101 [19]) this is straightforward because even at 3T there is a separation of several kilohertz between the large CF_3 singlet that can be used for imaging and the CF-CF_2 multiplet that could cause CSD artifacts. However, in the case of PFOB, the CF_3 resonance is close to the CF_2Br resonance and requires a narrow-band excitation pulse to exclusively excite it [50]. At lower magnetic field strengths, the excitation bandwidth should be narrower, and the RF pulse length needs to be longer. However, all intra-molecular coupling will be removed from the resonance, which should yield a long T_2 relaxation time that can be exploited for high SNR efficiency with several types of pulse sequences.

A variation on this selective excitation is to target a large group of J-coupled atoms and to image them very fast with a UTE sequence, before dephasing occurs. This was demonstrated to be effective for the CF_2 multiplet of PFOB with a bSSFP-UTE acquisition by Goette and colleagues [29]; in a direct comparison, the sequence outperformed the selective TSE acquisition of the CF_3 resonance. This approach requires that no other resonances are close to the multiplet.

CSD artifacts can also be removed through deconvolution if the ^{19}F spectrum is well-known. Deconvolution is the inverse of applying a filter; knowledge of the readout bandwidth and the spectrum can be used to collapse the CSD artifacts back onto the pixel where the signal originated. Several iterations of this technique were proposed soon after the introduction of ^{19}F MRI [51, 52], which highlights the strong need for CSD corrections of biologically interesting PFCs with complex spectra. These techniques mostly lead to a decrease in SNR or equivalence to single-peak acquisition due to the high uncertainty and noise amplification in the deconvolution algorithms. More recently, Meissner *et al.* [53] improved upon these techniques by iteratively optimizing the point spread function of the deconvolution filter, and doing so with complex images. This regularized deconvolution resulted in increased sensitivity in both high- and low-SNR regimes. A disadvantage of all these techniques is that the CSD artifact is also present in the through-plane direction when a slice selection gradient is used. For 2D imaging in particular, one should be certain that the compound is not present outside the imaged slice, since the off-resonance peaks may displace back into the slice. Conversely, non-selective 3D imaging or 3D imaging with a very weak slab selection gradient should avoid this complication.

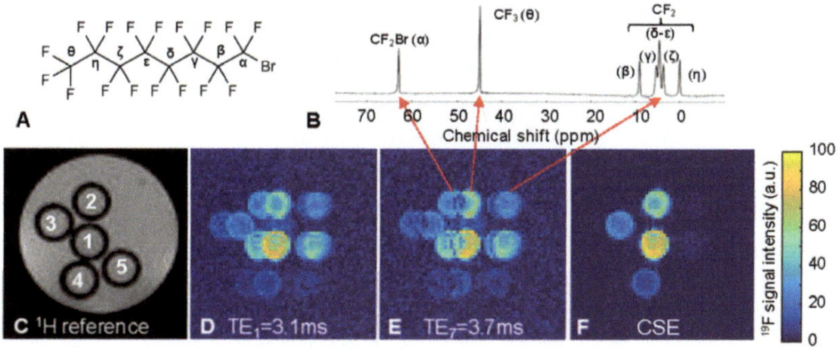

Figure 6.5 CSE-bSSFP imaging of PFOB. (A) The molecular structure of PFOB. (B) A ^{19}F MR spectrum of PFOB, with three separate resonance structures. (C) An axial ^{1}H reference image of a phantom with numbered tubes with different PFC concentrations. Tube 5 does not contain PFOB. (D–E) Direct-FFT ^{19}F images at the shortest and longest TE. The CSD artifacts can be directly traced back to the structures of the spectrum (red arrows). Note the negligible signal decrease between the two images. (F) The CSE-reconstructed image has high SNR and only very faint CSD artifacts (4% of the original intensity). Adapted from van Heeswijk *et al.* [54] with permission.

The chemical shift information can also be directly encoded in the image reconstruction [54]. To avoid the high uncertainty of chemical shift encoding (CSE) from a single k-space and echo time (similar to deconvolution in image space), multiple images with different echo times are acquired. This encoding reinforces the restoration of the signals to the correct pixels, similar to a ^{1}H multi-echo water-fat separation sequence. Pulse-sequence-specific modulations of the signal at off-resonance frequencies need to be encoded in this reconstruction for it to work correctly. This is can be achieved analytically for the GRE sequence since the signal intensity only depends on the T_2* decay at TE and the RF pulse profile, but for bSSFP, one needs to perform simulations with the Bloch equations. The resulting CSE-bSSFP pulse sequence (Fig. 6.5) makes use of all resonances and profits from the intrinsically high sensitivity of the bSSFP method, and its SNR efficiency compares well to that of other multi-resonance sequences [54].

6.5.2 Simultaneously Imaging Multiple Compounds

The techniques described above for single compounds and thus known ratios between the resonances can be extended to situations with unknown ratios between the resonances. This enables the separation of unwanted signals from those of the PFC of interest, as well as the separation of multiple PFCs that target different biological mechanisms.

A common confounder in preclinical ^{19}F MRI is isoflurane, an inhaled anesthetic that accumulates in adipose tissue and has two J-coupled fluorinated groups that can lead to image artifacts overlapping with tissues of interest [55]. Isoflurane signals can be avoided by several minor modifications to standard pulse sequences: (1) by shifting the isoflurane signals outside the region of interest using a narrow acquisition bandwidth, (2) by suppression of isoflurane signals with a frequency-selective suppression pulse before signal acquisition, or (3) by preventing isoflurane excitation with a 3D pulse sequence with a narrow excitation bandwidth [56].

If multiple compounds need to be imaged, chemical shift imaging [57] can be used for a very precise and model-free separation of the resonances. Chemical shift imaging comes at a significant cost in scan time in order to encode the spectral dimension, although this can be reduced with compressed sensing [46] and by limiting the spatial resolution.

A similarly model-free multi-resonance pulse sequence can be made by simply using a single-resonance-selective pulse sequence multiple times in a row at different frequencies, if the time is available [58, 59]. Recently, Flögel and colleagues [60] demonstrated that a single-resonance-selective TSE sequence can also be extended in a manner similar to multi-slice TSE in order to acquire the extra frequencies without an increase in scan time. Here, instead of acquiring different slices during the waiting time between the end of the echo train and the next excitation, different selective frequencies are acquired (Fig. 6.6). This multi chemical shift selective imaging (mCSSI) has the robustness and sensitivity of regular TSE, can be easily combined with 3D imaging, and can readily be used for the separation of several resonances and compounds without an acquisition time penalty. As mentioned above, to selectively excite a single resonance, the RF excitation pulse needs to be somewhat longer than usual, which will lead to a prolonged TE, although this is offset by the reduction in J-coupled T_2 relaxation. However, the chemical shift difference between the resonances at lower (clinical) magnetic field strengths may not be sufficient large to be selectively excited.

Other CSD artifact removal techniques have also been used for multi-compound imaging. Schoormans *et al.* [61] extended deconvolution-based CSD removal in order separately quantify multiple compounds. They used two GRE acquisitions of the same location but with perpendicular readout directions, such that the CSD artifacts would show in different directions. The two acquisitions were then combined into an iterative image reconstruction that prevents noise amplification and results in a single separate image for each compound.

Finally, much like Dixon-type fat-water separation, the chemical shift encoding (CSE) approach can also be extended to include multiple compounds [62]. Here, the compounds are modelled as separate entries in a non-linear least squares estimation reconstruction directly from k-space, which is combined with multiple GRE acquisitions with different TEs. The main limitations of this technique are similar to those of others that acquire all resonances; a 3D acquisition should be used to prevent through-plane CSD artifacts, and the reconstruction is dependent on the data behaving like the signal model.

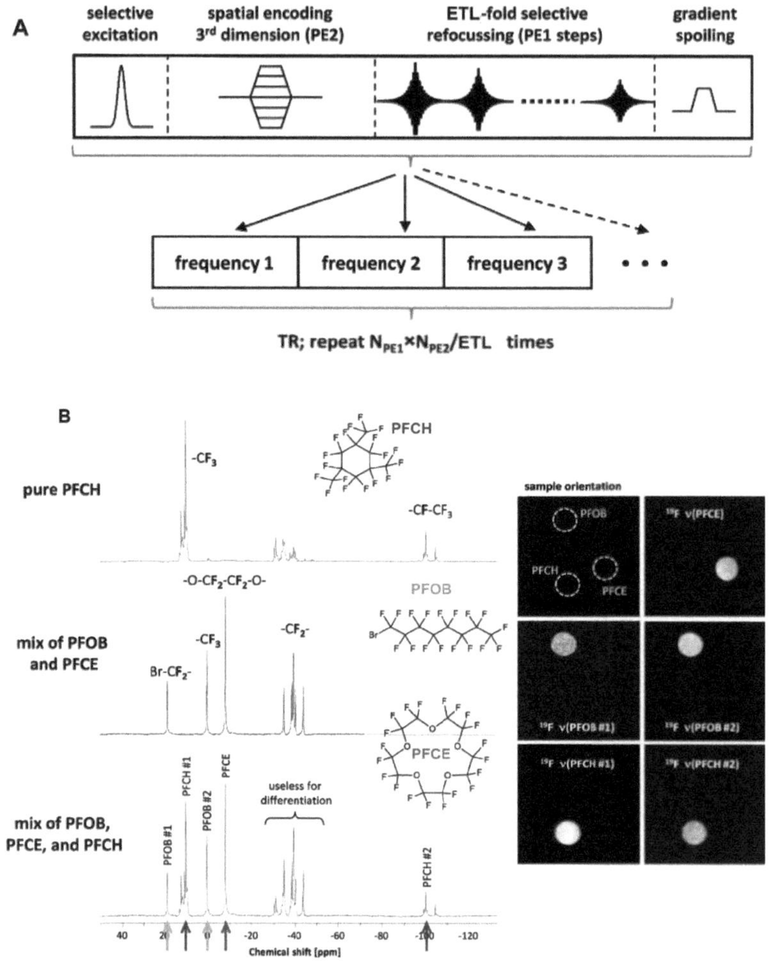

Figure 6.6 Overview of a multi chemical shift selective imaging (mCSSI) pulse sequence. (A) Narrow bandwidth excitation is followed by phase encoding of the third dimension and acquisition of the echo train with selective refocusing pulses. Gradient spoiling is required to destroy remaining transverse magnetization at the end of each cycle. This procedure is repeated for the number of excitation frequencies within one TR. The method provides a complete 3D dataset for each of the excitation frequencies and allows the simultaneous acquisition of several frequencies within one scan. PE_n = phase encoding. (B) Left: ^{19}F MR spectra of pure PFCH (top), a mixture of PFOB + PFCE (middle) and a combination of all three PFCs (bottom). The arrows in the bottom spectrum indicate which signals were used for mCSSI. Right: ^{19}F mCSSI images acquired from the individual resonance frequencies indicated on the left demonstrate artifact-free imaging of the respective five signals. Adapted from Flögel *et al.* [60] with permission.

6.6 Picking the Best Pulse Sequence

From the above discussion, it should be clear that the best choice of pulse sequence strongly depends on the properties and needs of the chosen ^{19}F MR application. While theoretical sensitivities (including k-space trajectories) can be calculated for a head-to-head comparison, most practical applications will see many pulse sequence types fall away because of artifacts or instabilities as laid out in Table 6.1.

A point that is not often raised is that not all pulse sequences are available on all systems—while UTE-type pulse sequences may appear ideal for many applications, they often require significant setup time and calibration to work near their theoretical sensitivity. It should be noted that with the advent of nanoparticles that include a paramagnetic agent, and thus have very short relaxation times, implementing sequences such as UTE becomes worthwhile.

Compressed sensing has seen its first few applications in the field of ^{19}F MRI, and most pulse sequences can be adapted for use with CS image reconstruction. This will lead to a decrease in scan time or an increase in precision and sensitivity when combined with averaging, without sacrificing much else. Implementing CS thus appears to be a worthwhile effort for most applications.

Iterative reconstruction and CS have also enabled the advancement of several multi-resonance imaging techniques to the point where they have sensitivities similar to those of regular pulse sequences. Conversely, several of these techniques depend on multiple assumptions on the behavior of the signal and might not handle magnetic field inhomogeneities well, while others require chemical shift differences that might not be attainable at all magnetic field strengths.

In conclusion, as we have seen ^{19}F MRI mature and diversify over the last decade, the size and diversity of our pulse sequences arsenal have grown with it. While the large choice of pulse sequence options might initially seem overwhelming, there is now a tailored solution for nearly every application.

Acknowledgements

The author would like to thank Matthias Stuber, PhD for insightful discussions on pulse sequences and Jérôme Yerly, PhD for guidance through the world of compressed sensing. This work was financially supported by the Swiss National Science Foundation (SNSF grants 32003B_182615 and CRSII5_202276).

References

1. G. N. Holland, P. A. Bottomley, and W. S. Hinshaw. ^{19}F magnetic resonance imaging. *J. Magn. Reson.*, **28**, 133–136, 1977.
2. E. T. Ahrens, R. Flores, H. Xu, and P. A. Morel. In vivo imaging platform for tracking immunotherapeutic cells. *Nat. Biotechnol.*, **23**, 983–987, 2005, doi: 10.1038/nbt1121.
3. U. Flogel, *et al.* In vivo monitoring of inflammation after cardiac and cerebral ischemia by fluorine magnetic resonance imaging. *Circulation*, **118**, 140–148, 2008, doi: CIRCULATIONAHA.107.737890 [pii] 10.1161/CIRCULATIONAHA.107.737890.
4. E. T. Ahrens, B. M. Helfer, C. F. O'Hanlon, and C. Schirda. Clinical cell therapy imaging using a perfluorocarbon tracer and fluorine-19 MRI. *Magn. Reson. Med.*, **72**, 1696–1701, 2014, doi: 10.1002/mrm.25454.
5. S. Temme, *et al.* Technical advance: monitoring the trafficking of neutrophil granulocytes and monocytes during the course of tissue inflammation by noninvasive 19F MRI. *J. Leukoc. Biol.*, **95**, 689–697, 2013, doi: 10.1189/jlb.0113032.
6. R. B. van Heeswijk, *et al.* Fluorine MR imaging of inflammation in atherosclerotic plaque in vivo. *Radiology*, **275**, 421–429, 2015, doi: 10.1148/radiol.14141371.
7. R. B. van Heeswijk, *et al.* Selective in vivo visualization of immune-cell infiltration in a mouse model of autoimmune myocarditis by fluorine-19 cardiac magnetic resonance. *Circ. Cardiovasc. Imaging*, **6**, 277–284, 2013, doi: 10.1161/CIRCIMAGING.112.000125.
8. J. Ruiz-Cabello, B. P. Barnett, P. A. Bottomley, and J. W. Bulte. Fluorine (19F) MRS and MRI in biomedicine. *NMR Biomed.*, **24**, 114–129, 2011, doi: 10.1002/nbm.1570.
9. I. Tirotta, *et al.* A superfluorinated molecular probe for highly sensitive in vivo(19)F-MRI. *J. Am. Chem. Soc.*, **136**, 8524–8527, 2014, doi: 10.1021/ja503270n.

10. M. Bouchoucha, R. B. van Heeswijk, Y. Gossuin, F. Kleitz, and M.-A. Fortin. Fluorinated mesoporous silica nanoparticles for binuclear probes in ^{1} H and ^{19} F magnetic resonance imaging. *Langmuir*, **33**(40), 10531–10542, 2017, doi: 10.1021/acs.langmuir.7b01792.
11. S. Waiczies, *et al.* Fluorine-19 MRI at 21.1 T: enhanced spin-lattice relaxation of perfluoro-15-crown-5-ether and sensitivity as demonstrated in ex vivo murine neuroinflammation. *Magma N. Y. N.*, **32**(1), 37–49, 2019, doi: 10.1007/s10334-018-0710-z.
12. S. Waiczies, *et al.* Enhanced fluorine-19 MRI sensitivity using a cryogenic radiofrequency probe: technical developments and ex vivo demonstration in a mouse model of neuroinflammation. *Sci. Rep.*, **7**, 9808, 2017, doi: 10.1038/s41598-017-09622-2.
13. R. Watts, and Y. Wang. k-space interpretation of the Rose Model: noise limitation on the detectable resolution in MRI. *Magn. Reson. Med.*, **48**, 550–554, 2002, doi: 10.1002/mrm.10220.
14. M. A. Bernstein, K. F. King, and Z. J. Zhou. *Handbook of MRI Pulse Sequences*, Academic Press, Amsterdam ; Boston, 2004.
15. R. W. Brown, Y.-C. N. Cheng, E. M. Haacke, M. R. Thompson, and R. Venkatesan. *Magnetic Resonance Imaging: Physical Principles and Sequence Design*, 2nd edn, Hoboken, New Jersey, John Wiley & Sons, Inc., 2014.
16. C. Faber, and F. Schmid. Pulse sequence considerations and schemes, in *Fluorine Magnetic Resonance Imaging* (U. Flögel, and E. T. Ahrens, eds.), 1st edn, Jenny Stanford Publishing, Singapore, 2017, pp. 3–27.
17. A. Mastropietro, *et al.* Optimization of rapid acquisition with relaxation enhancement (RARE) pulse sequence parameters for ^{19}F-MRI studies. *J. Magn. Reson. Imaging JMRI*, **40**(1), 162–170, 2014, doi: 10.1002/jmri.24347.
18. R. Colotti, *et al.* Characterization of perfluorocarbon relaxation times and their influence on the optimization of fluorine-19 MRI at 3 tesla. *Magn. Reson. Med.*, **77**, 2263–2271, 2017, doi: 10.1002/mrm.26317.
19. E. Darçot, R. Colotti, D. Brennan, G. A. Deuchar, C. Santosh, and R. B. Heeswijk. A characterization of ABL-101 as a potential tracer for clinical fluorine-19 MRI. *NMR Biomed.*, **33**(1), 2020, doi: 10.1002/nbm.4212.
20. B. Hargreaves. Rapid gradient-echo imaging. *J. Magn. Reson. Imaging*, **36**(6), 1300–1313, 2012, doi: 10.1002/jmri.23742.
21. R. B. van Heeswijk, Y. Pilloud, U. Flogel, J. Schwitter, and M. Stuber. Fluorine-19 magnetic resonance angiography of the mouse. *PloS One*, **7**, e42236, 2012, doi: 10.1371/journal.pone.0042236.

22. R. R. Ernst, and W. A. Anderson. Application of fourier transform spectroscopy to magnetic resonance. *Rev. Sci. Instrum.*, **37**(1), 93–102, 1966, doi: 10.1063/1.1719961.
23. O. Bieri, and K. Scheffler. Fundamentals of balanced steady state free precession MRI: fundamentals of balanced SSFP MRI. *J. Magn. Reson. Imaging*, **38**(1), 2–11, 2013, doi: 10.1002/jmri.24163.
24. M. Weiger, and K. P. Pruessmann. Short-T2 MRI: principles and recent advances. *Prog. Nucl. Magn. Reson. Spectrosc.*, **114–115**, 237–270, 2019, doi: 10.1016/j.pnmrs.2019.07.001.
25. M. D. Robson, and P. D. Gatehouse. Consequences of T_2 relaxation during half-pulse slice selection for ultrashort TE imaging. *Magn. Reson. Med.*, **64**(2), 610–615, 2010, doi: 10.1002/mrm.22323.
26. A. M. Neubauer, *et al.* Gadolinium-modulated 19F signals from perfluorocarbon nanoparticles as a new strategy for molecular imaging. *Magn. Reson. Med.*, **60**(5), 1066–1072, 2008, doi: 10.1002/mrm.21750.
27. A. A. Kislukhin, H. Xu, S. R. Adams, K. H. Narsinh, R. Y. Tsien, and E. T. Ahrens. Paramagnetic fluorinated nanoemulsions for sensitive cellular fluorine-19 magnetic resonance imaging. *Nat. Mater.*, **15**(6), 662–668, 2016, doi: 10.1038/nmat4585.
28. M. Carl, J.-T. A. Chiang, and J. Du. Maximizing MR signal for 2D UTE slice selection in the presence of rapid transverse relaxation. *Magn. Reson. Imaging*, **32**(8), 1006–1011, 2014, doi: 10.1016/j.mri.2014.05.001.
29. M. J. Goette, J. Keupp, J. Rahmer, G. M. Lanza, S. A. Wickline, and S. D. Caruthers. Balanced UTE-SSFP for 19F MR imaging of complex spectra. *Magn. Reson. Med.*, **74**, 537–543, 2015, doi: 10.1002/mrm.25437.
30. A. Balasch, *et al.* Tiny golden angle stack-of-stars (tygaSoS) free-breathing functional lung imaging. *Magn. Reson. Imaging*, **82**, 24–30, 2021, doi: 10.1016/j.mri.2021.06.016.
31. F. Springer, G. Steidle, P. Martirosian, C. D. Claussen, and F. Schick. Effects of in-pulse transverse relaxation in 3D ultrashort echo time sequences: analytical derivation, comparison to numerical simulation and experimental application at 3T. *J. Magn. Reson.*, **206**(1), 88–96, 2010, doi: 10.1016/j.jmr.2010.06.010.
32. F. Schmid, C. Höltke, D. Parker, and C. Faber. Boosting ^{19}F MRI-SNR efficient detection of paramagnetic contrast agents using ultrafast sequences: optimizing SNR efficiency in ^{19}F MRI. *Magn. Reson. Med.*, **69**(4), 1056–1062, 2013, doi: 10.1002/mrm.24341.

33. D. M. Grodzki, P. M. Jakob, and B. Heismann. Ultrashort echo time imaging using pointwise encoding time reduction with radial acquisition (PETRA). *Magn. Reson. Med.*, **67**(2), 510–518, 2012, doi: 10.1002/mrm.23017.

34. C. Jacoby, *et al.* Probing different perfluorocarbons for in vivo inflammation imaging by 19F MRI: image reconstruction, biological half-lives and sensitivity. *NMR Biomed.*, **27**, 261–271, 2014, doi: 10.1002/nbm.3059.

35. M. Moreno López, J. M. Frederick, and J. Ventura. Evaluation of MRI denoising methods using unsupervised learning. *Front. Artif. Intell.*, **4**, 75, 2021, doi: 10.3389/frai.2021.642731.

36. M. Lustig, D. Donoho, and J. M. Pauly. Sparse MRI: the application of compressed sensing for rapid MR imaging. *Magn. Reson. Med.*, **58**, 1182–1195, 2007, doi: 10.1002/mrm.21391.

37. E. Darçot, *et al.* Compressed sensing with signal averaging reduces motion artifacts in fluorine-19 MRI. *Proc. Int. Soc. Magn. Reson. Med.*, 3072, 2017.

38. S. Sardy, P. Tseng, and A. Bruce. Robust wavelet denoising. *IEEE Trans. Signal Process.*, **49**(6), 1146–1152, 2001, doi: 10.1109/78.923297.

39. S. Boyd, N. Parikh, E. Chu, B. Peleato, and J. Eckstein. Distributed optimization and statistical learning via the alternating direction method of multipliers. *Found. Trends® Mach. Learn.*, **3**(1), 1–122, 2010, doi: 10.1561/2200000016.

40. L. R. Dice. Measures of the amount of ecologic association between species. *Ecology*, **26**(3), 297–302, 1945, doi: 10.2307/1932409.

41. Z. Wang, A. C. Bovik, H. R. Sheikh, and E. P. Simoncelli. Image quality assessment: from error visibility to structural similarity. *IEEE Trans. Image Process. Publ. IEEE Signal Process. Soc.*, **13**(4), 600–612, 2004, doi: 10.1109/tip.2003.819861.

42. D. L. Donoho, and I. M. Johnstone. Adapting to unknown smoothness via wavelet shrinkage. *J. Am. Stat. Assoc.*, **90**(432), 1200–1224, 1995, doi: 10.1080/01621459.1995.10476626.

43. C. O. Schirra, D. Brunner, J. Keupp, R. Razavi, T. Schaeffter, and S. Kozerke. Compressed sensing for highly accelerated 3D visualization of 19F-catheters. *Proc. Int. Soc. Magn. Reson. Med.*, **17**, 4405, 2009.

44. J. Zhong, P. H. Mills, T. K. Hitchens, and E. T. Ahrens. Accelerated fluorine-19 MRI cell tracking using compressed sensing. *Magn. Reson. Med.*, **69**, 1683–1690, 2013, doi: 10.1002/mrm.24414.

45. S. Liang, T. Dresselaers, K. Louchami, C. Zhu, Y. Liu, and U. Himmelreich. Comparison of different compressed sensing algorithms for low SNR ^{19}F MRI applications-Imaging of transplanted pancreatic islets and cells labeled with perfluorocarbons. *NMR Biomed.*, **30**(11), 2017, doi: 10.1002/nbm.3776.

46. T. Kampf, *et al.* Application of compressed sensing to in vivo 3D ^{19}F CSI. *J. Magn. Reson. San Diego Calif 1997*, **207**(2), 262–273, 2010, doi: 10.1016/j.jmr.2010.09.006.

47. L. Starke, A. Pohlmann, C. Prinz, T. Niendorf, and S. Waiczies. Performance of compressed sensing for fluorine-19 magnetic resonance imaging at low signal-to-noise ratio conditions. *Magn. Reson. Med.*, **84**(2), 592–608, 2020, doi: 10.1002/mrm.28135.

48. E. Darçot, *et al.* Towards quantification of inflammation in atherosclerotic plaque in the clinic – characterization and optimization of fluorine-19 MRI in mice at 3 T. *Sci. Rep.*, **9**(1), 17488, 2019, doi: 10.1038/s41598-019-53905-9.

49. J. G. Riess. Perfluorocarbon-based oxygen delivery. *Artif. Cells Blood Substit. Immobil. Biotechnol.*, **34**, 567–580, 2006, doi: P17005M83N567254 [pii] 10.1080/10731190600973824.

50. C. Giraudeau, *et al.* A new paradigm for high-sensitivity 19F magnetic resonance imaging of perfluorooctylbromide. *Magn. Reson. Med.*, **63**, 1119–1124, 2010, doi: 10.1002/mrm.22269.

51. L. J. Busse, *et al.* Deconvolution techniques for removing the effects of chemical shift in 19F nuclear magnetic resonance imaging of perfluorocarbon compounds. *Med. Phys.*, **13**(4), 518–524, 1986, doi: 10.1118/1.595960.

52. H. K. Lee, and O. Nalcioglu. Reblurred deconvolution method for chemical shift removal in F-19 (PFOB) MR imaging. *J. Magn. Reson. Imaging*, **2**, 53–61, 1992.

53. M. Meissner, M. Reisert, T. Hugger, J. Hennig, D. von Elverfeldt, and J. Leupold. Revealing signal from noisy 19F MR images by chemical shift artifact correction. *Magn. Reson. Med.*, **73**(6), 2225–2233, 2015, doi: 10.1002/mrm.25370.

54. R. B. van Heeswijk, *et al.* Chemical shift encoding (CSE) for sensitive fluorine-19 MRI of perfluorocarbons with complex spectra. *Magn. Reson. Med.*, **79**, 2724–2730, 2018, doi: 10.1002/mrm.26895.

55. C. Constantinides, *et al.* Temporal accumulation and localization of isoflurane in the C57BL/6 mouse and assessment of its potential contamination in ^{19}F MRI with perfluoro-crown-ether-labeled cardiac

progenitor cells at 9.4 Tesla. *J. Magn. Reson. Imaging*, 2016, doi: 10.1002/jmri.25564.

56. A. H. J. Staal, A. Veltien, M. Srinivas, and T. W. J. Scheenen. ^{19}F MRI imaging strategies to reduce isoflurane artifacts in in vivo images. *Mol. Imaging Biol.*, 2021, doi: 10.1007/s11307-021-01653-6.
57. J. S. van Gorp, *et al.* (19)F MRSI of capecitabine in the liver at 7 T using broadband transmit-receive antennas and dual-band RF pulses. *NMR Biomed.*, **28**, 1433–1442, 2015, doi: 10.1002/nbm.3390.
58. C. Chirizzi, *et al.* Multispectral MRI with dual fluorinated probes to track mononuclear cell activity in mice. *Radiology*, **291**(2), 351–357, 2019, doi: 10.1148/radiol.2019181073.
59. K. C. Partlow, *et al.* ^{19}F magnetic resonance imaging for stem/progenitor cell tracking with multiple unique perfluorocarbon nanobeacons. *FASEB J.*, **21**(8), 1647–1654, 2007, doi: 10.1096/fj.06-6505com.
60. U. Flögel, *et al.* Multi-targeted 1H/19F MRI unmasks specific danger patterns for emerging cardiovascular disorders. *Nat. Commun.*, **12**(1), 5847, 2021, doi: 10.1038/s41467-021-26146-6.
61. J. Schoormans, *et al.* An iterative sparse deconvolution method for simultaneous multicolor 19F-MRI of multiple contrast agents. *Magn. Reson. Med.*, **83**(1), 228–239, 2020, doi: 10.1002/mrm.27926.
62. K. D. Ludwig, D. Hernando, N. T. Roberts, R. B. van Heeswijk, and S. B. Fain. A chemical shift encoding (CSE) approach for spectral selection in fluorine-19 MRI. *Magn. Reson. Med.*, **79**, 2183–2189, 2018, doi: 10.1002/mrm.26874.

Chapter 7

^{19}F for Quantitative Tracking of Immune Cells Involved in Cancer Progression and Therapy

Olivia C. Sehl,[a] Ashley V. Makela[b] and Paula J. Foster[a,c]
[a]*Western University, Ontario, Canada*
[b]*Michigan State University, USA*
[c]*Robarts Research Institute, Ontario, Canada*
pfoster4@uwo.ca

7.1 The Tumor Microenvironment

The tumor microenvironment (TME) consists of cancer cells, stromal cells such as fibroblasts, pericytes, endothelial cells and immune cells, as well as noncellular components such as collagen and laminin. The cancer cells interact dynamically with the TME controlling the function of the other cells to promote cancer development and metastasis.

Fluorine Magnetic Resonance Imaging: Methods and Applications in Biomedicine
Edited by Eric T. Ahrens and Ulrich Flögel
Copyright © 2025 Jenny Stanford Publishing Pte. Ltd.
ISBN 978-981-5129-37-3 (Hardcover), 978-1-003-53004-6 (eBook)
www.jennystanford.com

All the immune cell types can have both pro- and anti-tumorigenic effects. The main players include T lymphocytes, B lymphocytes, natural killer (NK) cells, dendritic cells (DC), myeloid derived suppressor cells (MDSC), neutrophils and tumor-associated macrophages (TAMs).

Tumors have significant heterogeneity in their immune cell composition. The proportions and abundance of specific immune cell types and their spatial distribution in the TME are crucial prognostic and predictive factors. Three major tumor phenotypes have been proposed: inflamed (immune cells in the parenchyma), immune-excluded (immune cells in the surrounding stroma) and immune-desert. These categorizations have been used to predict response to different cancer treatments.

TAMs are the most abundant immune cell type within the TME. The activity of TAMs in cancers is usually pro-tumorigenic (M2). Pre-clinical and clinical studies have shown that an abundance of TAMs in the TME is associated with a poor prognosis in several types of cancer. TAMs represent a possible target for improving cancer therapy. Anti-cancer therapies have been investigated that deplete the M2 TAMs, activate anti-tumorigenic TAMs (M1), or shift the M2 phenotype to M1 [1, 2]. Depletion of TAMs has been shown to improve the efficacy of chemotherapy.

7.2 Cancer Immunotherapy

Cancer immunotherapy is a type of treatment that boosts the body's natural defenses to fight cancer. There are several types of immunotherapies including monoclonal antibodies and tumor-agnostic treatments, such as immune checkpoint inhibitors, oncolytic virus therapy, and adoptive cell immunotherapy. Adoptive cell immunotherapy involves reprogramming patients' own immune cells to find and attack cancer cells throughout the body [3]. This can be done using DC, peripheral blood mononuclear cells (PBMC), T cells, or NK cells. Typically, the immune cells are removed from a patient's blood, exposed to tumor antigens, activated, or genetically manipulated *ex vivo* and re-infused back into the patient.

DC have long been a focus of cancer immunotherapy. DC are known as professional antigen presenting cells [4]. Since DC are relatively scarce, they are often derived *ex vivo* by differentiating monocytes isolated from a subject's own peripheral blood or bone marrow. An important role of DC is to cross-present antigens to naïve T cells to initiate T cell activation. DC activation and maturation can be induced in a laboratory setting with tumor-associated antigens or tumor neoantigens. Once administered *in vivo*, therapeutic DC cross-present tumor antigens to activate T cells to promote an anti-tumor response. Importantly, transferred DC must migrate from the site of administration to secondary lymphoid organs, such as the lymph nodes or spleen, to interact with $CD4^+$ helper and $CD8^+$ cytotoxic T cells. PBMC, which are a heterogeneous cell mixture, comprised of DC, T cells, NK cells, B cells, and monocytes, are an attractive alternative to DC therapies, because the extensive *ex vivo* culture required to produce monocyte-derived DC is not required [5]. PBMC are the basis of Sipuleucel-T (Provenge®) FDA approved immunotherapy for prostate cancer which was approved in 2010 [6, 7].

T lymphocytes (T cells) originate from the bone marrow and develop in the thymus. Mature T cells express a T cell receptor (TCR), which is necessary to recognize specific antigens, and co-receptors CD4 or CD8. Naïve T cells are mature T cells that have left the thymus but have not yet been exposed to their specific antigen. As described, antigen presenting cells activate T cells by presenting the specific antigen to the T cell in the lymph nodes. These effector T cells undergo clonal expansion and recognise antigens in the periphery, leading to helper immune responses (CD4+ T cells) or cytotoxic functions (CD8+ T cells). T cell immunotherapy for cancer involves expanding effector T cell populations that recognize cancer antigens to increase the adaptive immune responses towards the tumor [8]. T cells can be collected from peripheral blood by leukapheresis, isolated, transduced to express a tumor antigen specific TCR, then activated and expanded *in vitro*. The TCR-transduced cell product is then reinfused to the patient.

A novel and effective T cell therapeutic strategy involves the use of a chimeric antigen receptor (CAR) [9] introduced *via* vector transduction. CARs recombine the antigen recognition portion of a monoclonal antibody with the signal transduction domain of a TCR. This allows for non-MHC (major histocompatibility complex) tumor antigen recognition by T cells, which is beneficial since cancer cells downregulate MHC 1 expression to evade immune recognition. There are currently three approved CAR-T cell immunotherapies, all for hematological malignancies, which are cancers affecting the blood, bone marrow and lymph nodes.

NK cells are lymphocytes that are part of the innate immune system. Like T cells, NK cells have shown promising results against hematological malignancies. NK cells can be derived from peripheral blood and induced pluripotent stem cells. A major benefit of NK cells is that they can be of allogenic source, meaning the cells can come from a donor instead of the patient, and each donor could supply >100 treatments. This would greatly reduce the cost to prepare cell therapies, compared to T cell products, which are autologous, *i.e.*, from the patient [10].

7.3 Imaging Cells

Despite the immense potential of immunotherapy, clinical results have been variable and discordant because of dissimilarities and disparities in cell source, preparation, and route of administration. It remains difficult, if not impossible to spatially and quantitatively track the migration and accumulation of cells in tumors and lymphoid organs and the immune response that follows. Many critical questions about the presence, numbers, persistence, and delivery of cells remain unanswered. For example, in DC immunotherapy, the magnitude of an anti-tumor response is proportional to the quantity of antigen presenting cells that reach a target lymph node [11], and therefore, it is crucial to know whether the injected cells have migrated to the target, and how many. Similarly, while there is strong evidence for a link between TAM content and outcome there are few strategies for measuring the accumulation and density of TAMs in tumors. *In vivo* cellular imaging has the potential to answer these questions. The ideal approach for monitoring immunotherapies or immune cell traffic in tumors is non-invasive, non-ionizing, sensitive, specific, and quantitative—providing an accurate measure of cell number.

Cellular magnetic resonance imaging (MRI) combines the ability to obtain anatomical MR images with the use of contrast agents or tracers for labeling specific cells. Fluorine-19 (^{19}F) based cellular MRI uses perfluorocarbons (PFC) to label cells, either *in vitro* prior to their injection or implantation, or administered intravenously (iv) leading to uptake of particles *in situ* by macrophages. *In vitro* cell labeling is performed by simple co-culture. The PFCs are not degraded in cells and there is no evidence of exocytosis thus labeling is stable and live cells can be tracked over time. PFCs released from dead cells are cleared *via* the liver and lungs. PFCs commonly administered iv and used for MRI circulate for 12–48 hours and passively target macrophages; there is no evidence that the PFCs are taken up by nonphagocytic cells *in situ*. PFC-labeled cells can be imaged by ^{19}F MRI with high specificity since there is very little endogenous fluorine in biological tissues. To localize the ^{19}F signal, anatomical images are also collected using proton MRI. ^{19}F images are typically presented with a color scale to represent numerical values and overlaid onto the corresponding proton images. ^{19}F MRI is quantitative since the signal intensity is linearly related to the number of ^{19}F atoms. A sample of PFC labeled cells can be analyzed by nuclear magnetic resonance (NMR) spectroscopy to determine the average ^{19}F per cell and this can be used to estimate the number of cells which contribute to the signal in the ^{19}F image. The ability to determine cell number *in vivo* from ^{19}F images is the most compelling reason for using ^{19}F MRI over other cellular imaging techniques. A shortcoming of ^{19}F MRI, especially concerning the imaging of cells in the tumor microenvironment or in cancer immunotherapy, is the cellular detection limit. The current detection limit is thousands to tens of thousands of cells. ^{19}F MRI has been used to track all the above-described immune cells in preclinical models of cancer immunotherapy, and for experimental studies which investigate TAM recruitment and response to cancer therapies.

7.4 Fluorine-19 MRI of Therapeutic Immune Cells for Cancer

7.4.1 Dendritic Cells

Dendritic cells (DC) were the first cell type to be labeled with PFC and imaged with ^{19}F MRI [12]. DC are naturally phagocytic and incorporate high levels of PFC, on the order of 10^{12}–10^{13} ^{19}F atoms/cell [13–17]. Importantly, PFC labeling of DC does not alter the phenotype, viability, migration ability, or the antigen-specific response of these cells (Fig. 7.1a–c) [12–15].

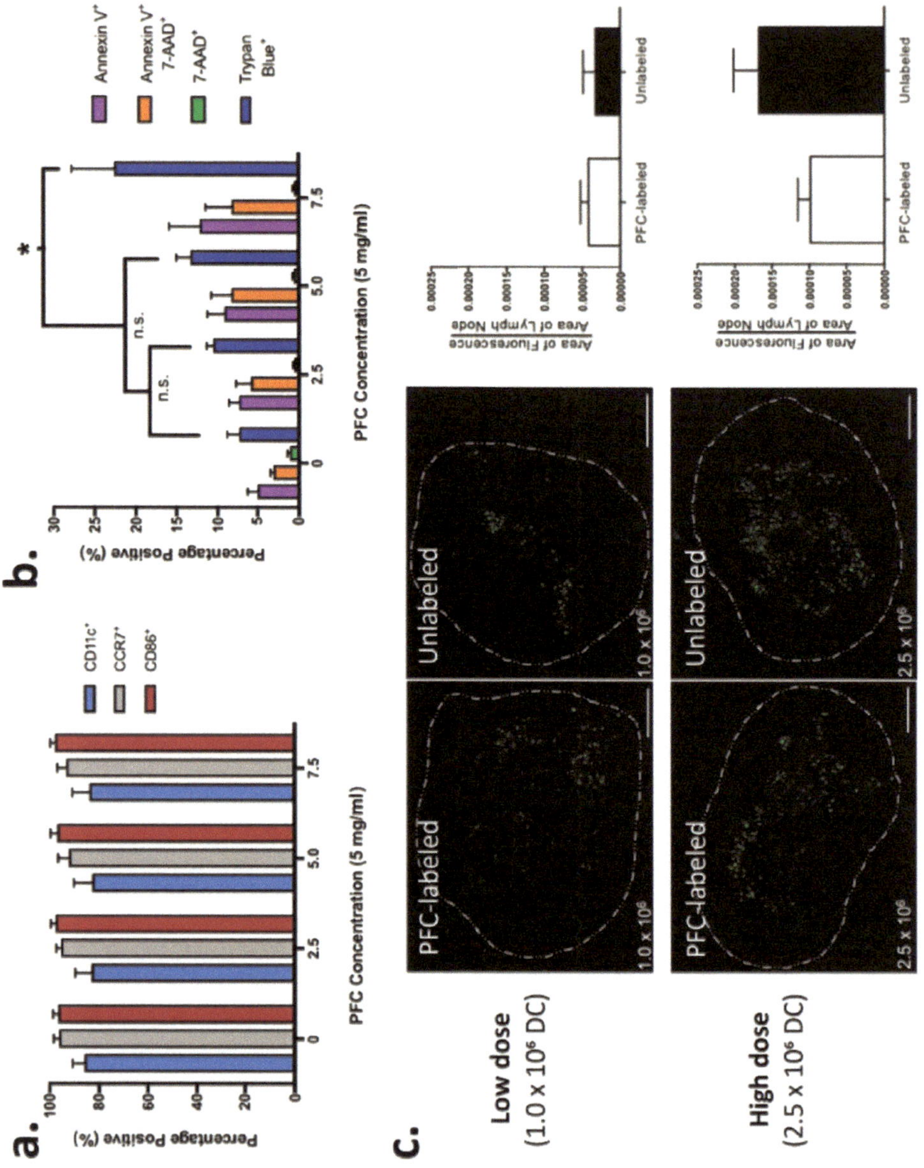
a.
Percentage Positive (%)
PFC Concentration (5 mg/ml)
CD11c⁺
CCR7⁺
CD86⁺
b.
Percentage Positive (%)
PFC Concentration (5 mg/ml)
n.s.
n.s.
*
Annexin V⁺
Annexin V⁺ 7-AAD⁺
7-AAD⁺
Trypan Blue⁺
c.
Low dose (1.0 x 10⁶ DC)
High dose (2.5 x 10⁶ DC)
PFC-labeled
Unlabeled
Area of Fluorescence / Area of Lymph Node

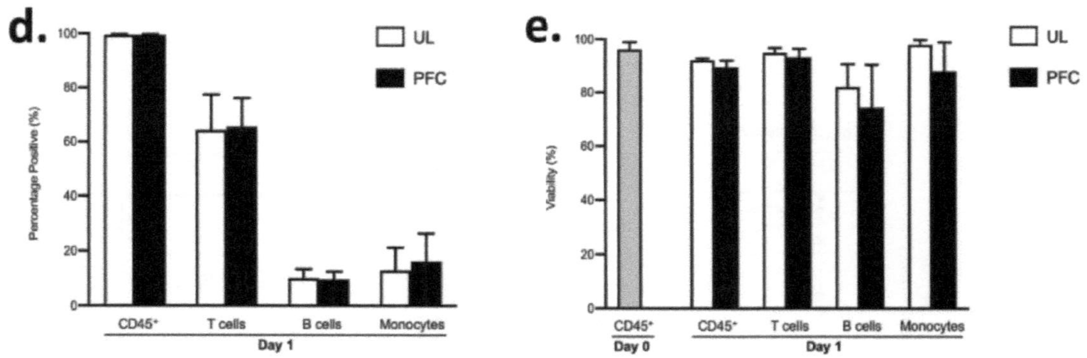

Figure 7.1 PFC labeling does not alter DC or PBMC phenotype, viability, or migration to popliteal lymph nodes. (a) Flow cytometry showed no differences in expression of CDllc$^+$ DC cell surface markers CD11c (blue), CCR7 (grey), and CD86 (red) with PFC labeling concentrations 0, 2.5, 5, and 7.5 mg/mL. (b) DC were stained with Annexin V/7-Aminoactinomycin D (7-AAD) and trypan blue exclusion assay to determine viability. PFC labeling did not alter DC viability at concentrations of 2.5 mg/mL and 5.0 mg/mL, however a significant increase in trypan blue$^+$ DC (decrease in viability) was measured at 7.5 mg/mL. (c) PFC labeling does not alter DC migration *in vivo*. PFC$^+$ and unlabeled CFSE-fluorescent DC were administered to the mouse footpad in low dose (1 million DC) or high dose (2.5 million DC) (n = 7). Two days later, cryosections of popliteal lymph nodes revealed no significant differences in CFSE fluorescence quantification in PFC$^+$ and unlabeled DC at either administration dose. (d) PFC labeling did not alter PBMC lineage composition (UL = unlabeled). (e) The viability of cryopreserved

(*Caption Continued*)

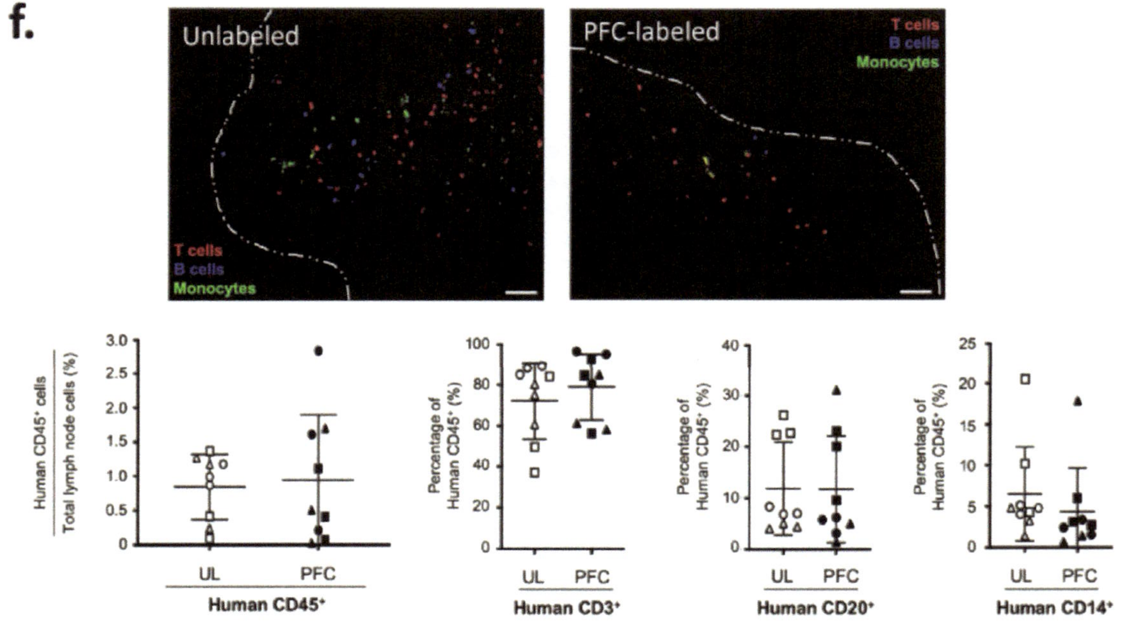

PBMC was measured with flow cytometry and LIVE/DEAD^{TM} fixable aqua dead cell stain post-thaw (day 0) and after PFC labeling (day 1). No significant viability differences were measured between PFC-labeled and unlabeled PBMC as a whole (CD45$^+$), or in enriched cell lineages (T cells, B cells, monocytes). (f) PFC labeling does not alter the migration in any of the three main PBMC cell lineages. Each cell lineage was stained with a fluorescent marker: T cells Far Red$^+$, B cells CTV$^+$, and monocytes CFSE$^+$, then recombined. PFC$^+$ and unlabeled PBMC were administered to hind mouse footpads two days later, and cryosections of popliteal lymph nodes were prepared (top) or digested for assessment by flow cytometry. There were no significant differences in the number of migrated CD45$^+$ bulk PBMC, CD3$^+$ T cells, CD20$^+$ B cells, and CD14$^+$ monocytes for PFC-labeled and unlabeled (UL) PBMC (n = 3). Figure reprinted from Ref. [5] under CC BY-NC-ND and Ref. [14] under CC-BY.

The most common mouse model for studying DC immunotherapy has been adoptive transfer *via* intradermal administration of DC to the mouse footpad(s) [18]. In this model activated DC will upregulate expression of chemokine receptor 7 (CCR7), promoting migration of DC to the draining popliteal lymph node *via* lymphatic vessels, with only 3–5% of the cells reach the draining lymph node after 24–48 hours [19, 20]. As the number of DC reaching the lymph nodes is directly related to the ensuing immune response, efforts around improving migration can be complimented by the *in vivo* cell quantitation provided by ^{19}F MRI.

Fink *et al.* [14] pre-treated the popliteal lymph nodes with a pro-inflammatory cytokine IL-1β to investigate if DC migration would improve. In this study, the number of PFC$^+$ DC was reported by measuring ^{19}F signal in the lymph nodes and calculating how many DC this signal represents, with knowledge of average ^{19}F DC loading by ^{19}F NMR spectroscopy. ^{19}F MRI quantification revealed that after 3 million DC were injected into the footpad there were between $4 - 20 \times 10^3$ DC in IL-1β pre-treated lymph nodes and between 4.3–17×10^3 DC in non-treated lymph nodes. Therefore, pre-treatment did not result in significantly enhanced DC migration. The lowest ^{19}F signal detected in the popliteal lymph nodes at 9.4 Tesla was $6.8 \pm 1.4 \times 10^{15}$ ^{19}F spins ($4.0 \pm 0.8 \times 10^3$ DC with labeling of 1.7×10^{12} ^{19}F spins/cell).

In the same study, ^{19}F-labeled DC were imaged at 3 Tesla. To achieve ^{19}F-cell detection, a small (4.3 cm) dual-tuned ^{1}H/^{19}F surface coil and a balanced steady state free precession (bSSFP) imaging sequence were used to enhance sensitivity. In mice bearing B16-F10 melanoma tumors, 3 million DC were administered to each footpad and migrating DC (range $3.63 - 19.1 \times 10^3$ cells) were detected in 9/10 popliteal lymph nodes (n = 5) (Fig. 7.2a). The lowest ^{19}F signal detected in the popliteal lymph nodes at 3 Tesla was 6.17×10^{15} ^{19}F spins (3.63×10^3 DC with labeling of 1.7×10^{12} ^{19}F spins/cell). This DC treatment led to a mild therapeutic effect towards the tumors, as monitored by bioluminescence imaging and measurements of tumor volume.

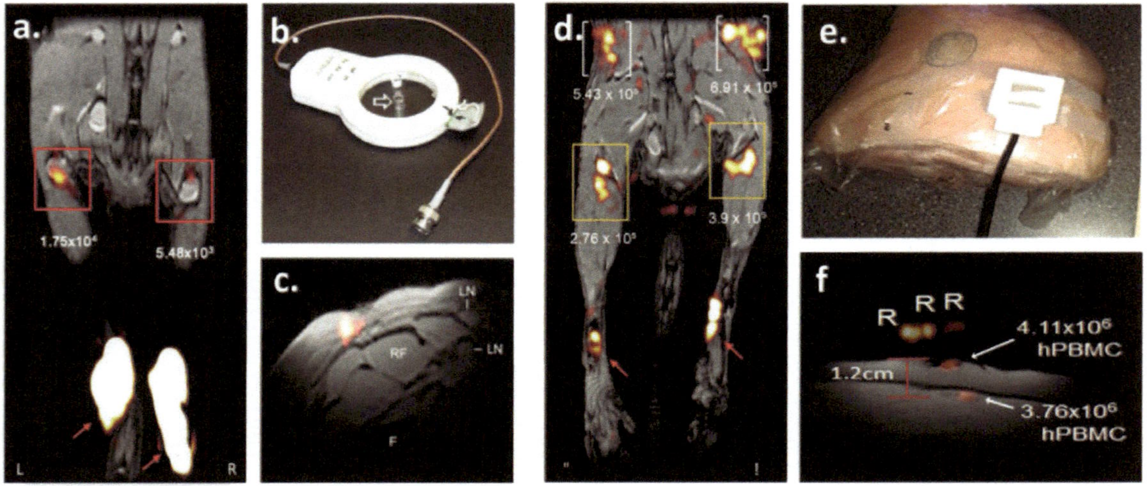

Figure 7.2 Preclinical and clinical ^{19}F detection of DC and PBMC at 3 Tesla. (a) The migration of PFC$^+$ DC in mice from the site of injection (red arrows) to the draining popliteal lymph nodes (red boxes) could be detected after 2 days. The number of PFC$^+$ DC measured at the lymph nodes is indicated. (b) A 3 Tesla compatible dual-tuned (^{1}H/^{19}F) surface coil for ^{19}F detection was used to obtain image (c). (c) Clinical ^{19}F detection of 10 million PFC$^+$ DC in the right quadricep of a 53-year-old female patient. Intradermally administered DC were not detected at the draining inguinal lymph node (LN) after 4 or 24 hours. (d) The migration of PFC$^+$ PBMC in mice from the site of injection (red arrows) to the draining popliteal (orange boxes) and inguinal (white brackets) lymph nodes could be detected after 24 hours. The number of PFC$^+$ PBMC measured at the lymph nodes is indicated. (e) 4.5 million PFC$^+$ PBMC were injected intradermally and intramuscularly in a pig leg specimen and imaged with a 3 Tesla dual-tuned surface coil. (f) The resulting ^{1}H/^{19}F image shows detection of PFC$^+$ PBMC at these two positions. External reference phantoms were used for ^{19}F quantification and the corresponding PBMC cell number is indicated. Figure reprinted from Ref. [5] under CC BY-NC-ND and Refs. [14, 15, 21] under CC-BY.

The first ^{19}F MRI cell tracking in clinical trials involved DC immunotherapy in 5 patients with metastatic (stage 4) colorectal cancer [15]. Three patients received 10 million PFC-labeled DC intradermally in the quadriceps near the inguinal region. The clinical imaging protocol used a 3 Tesla MRI scanner, a 7 cm diameter $^{1}H/^{19}F$ surface coil (shown in Fig. 7.2b) and a standard fast low angle shot (FLASH) imaging sequence with a 9.5 min scan time. ^{19}F signal was visualized at the injection site in images obtained 4 and 24 hours after administration (one patient image from 4 h timepoint shown in Fig. 7.2c). Quantification revealed a 50% reduction in ^{19}F signal over this period which can be attributed to the death of DC (and clearance of PFC) and/or cell migration from the injection site. However, no ^{19}F signal was detected at the nearby draining lymph node. In two patients that received lower doses (1 million DC) ^{19}F signal was not reliably detected. When translating ^{19}F MRI of DC from mouse to humans, several strategies can be explored for improving cellular sensitivity including enhanced cell labeling, improved pulse sequences, image post-processing and optimal radiofrequency coils.

7.4.2 Peripheral Blood Mononuclear Cells

The first report of PBMC labeling with PFCs was in 2018 [21]. Human PBMC were acquired from healthy volunteers, or prostate cancer patients, and labeled with a clinical grade PFC. Greater than 99% of PBMC were labeled with agent without affecting functionality or viability (Fig. 7.1d–f). The functionality of APCs was assessed with a mixed lymphocyte reaction, in which PFC-labeled or unlabeled APCs from one donor were incubated with T lymphocytes of another donor for 4 days. There were no differences in the amount of T cell proliferation and activation between these groups, indicating that PFC labeling does not influence antigen presentation by PBMC. The PFC labeling efficiency for PBMC was significantly lower than that reported for other cell types (10^{10} – 10^{11} vs 10^{12}) [13, 22]. The labeling of PBMC also varied between donors. These two observations are related to the heterogeneity of the cell population and differences in the cell lineage percentages with each donor. Cells with a smaller cytoplasmic volume (*i.e.,* lymphocytes) incorporate less PFC than larger leukocytes (*i.e.,* monocytes and DC). Additionally, cells with greater endocytic and phagocytic ability can label with more PFC per cell. Therefore, donor PBMC which consist of a higher percentage of monocytes and DC (and less lymphocytes), and cell preparations which contain only DC, would have overall higher ^{19}F per cell.

In a subsequent study [23], it was determined that pre-treatment of PBMC with granulocyte macrophage colony-stimulating factor (GM-CSF) in culture improved the PBMC migration to lymph nodes. Three million GM-CSF treated PFC-labeled PBMC were administered intradermally to the mouse footpad and migration of these cells to the draining popliteal lymph node was detected with 9.4 T ^{19}F MRI in 6 of 8 mice. In comparison, only 1 of 8 mice that received untreated PFC-labeled PBMC had ^{19}F signal in popliteal lymph nodes. The PFC-labeled and GM-CSF treated PBMC were imaged at 3 Tesla. Following footpad injection (3 million PBMC right, and 6 million PBMC left), the migration of these PBMC could be detected in popliteal and inguinal lymph nodes after 24 and 48 hours (Fig. 7.2d). Quantification of cell number revealed that greater than 5% of PBMC could be detected in each of these lymph nodes, owing to improved migration ability of these GM-CSF treated PBMC.

The translational potential of ^{19}F MRI for tracking human PBMC has been studied on a clinical system, where cells were injected intradermally in a pig leg specimen to mimic the quadriceps in patients [21] (Fig. 7.2e). The clinical imaging protocol used a 3 T MRI scanner, a dual-tuned 4.3 cm diameter $^{1}H/^{19}F$ surface coil and a bSSFP imaging sequence with a scan time of 15 minutes. 4.5 million PBMC were injected both at the surface and at a depth of 1.2 cm, equivalent to the depth of a human lymph node (Fig. 7.2f). ^{19}F signal was observed in both locations.

7.4.3 T Cells

PFC-labeling of T cells has been demonstrated for small cultures of T cells for over a decade [24–26], and more recently, large clinical scale manufacturing [27]. ^{19}F cell tracking of T cells allows researchers to study fundamental questions about the fate of these cells, including their trafficking patterns and persistence, and assess methods which have the potential to improve T cell targeting to tumors. T cells are expected to migrate to secondary lymphoid tissues, including the spleen, lymph nodes, and mucosal-associated lymphoid tissues, due to the expression of chemokine receptors. Activated T cells are also expected to extravasate from the circulation and home to sites of inflammation and to tumors where specific antigens are expressed. Dead T cells release the PFC which is taken up predominantly by Kupffer cells of the liver [28]. The route of T cell administration will also play a role in determining trafficking patterns and timing of T cell arrival to organs [29]. For example, cells administered intravenously by the mouse tail vein are expected to trap within the lung capillaries, by size exclusion, before other patterns of migration are observed [30].

The first T cell tracking with ^{19}F in mice was conducted in 2008 [24]. Ten million PFC^{+} naïve T cells were administered to the intraperitoneal cavity and ^{19}F MRI was conducted 2 days later. ^{19}F images showed multiple punctate regions, where the trafficking of naïve T cells to abdominal lymph nodes occurred. In a subsequent study, longitudinal ^{19}F cell tracking of activated ovalbumin-specific T cells labeled with PFC was demonstrated following intraperitoneal administration to BALB/c mice [25]. These mice did not have tumors, but an inflammation model was established by administering the TCR-specific antigen, ovalbumin, subcutaneously. ^{19}F images over 21 days showed ^{19}F signal (associated with PFC^{+} activated T cells) in a single location, the lymph node draining the antigen. By day 7, about 40% of the T cells had accumulated in the draining lymph node (Fig. 7.3a), and over time, gradual clearance of PFC^{+} T cells was detected. It was speculated that the transferred activated T cells initially traffic to all lymph nodes at early timepoints and antigen-specific accumulation occurs preferentially in the draining lymph node over time. Importantly, in control mice which received antigen-negative injections, T cells were detected only in the mesenteric region, and in control mice that received heat-killed ovalbumin-specific T cells, ^{19}F signal was only observed near the bladder (Fig. 7.3a). Therefore, only the viable, antigen-specific T cells showed purposeful migration to the lymph node draining the antigen.

The first PFC labeling of CAR T cells was conducted by Chapelin *et al.* [28]. CAR T cells typically incorporate $\sim10^{11}$ ^{19}F/atoms per cells with nearly 100% labeling efficiency [28, 31, 32] and PFC is localized intracellularly in lysosomal compartments surrounding the nucleus in confocal and electron microscopy [28, 31]. With ^{19}F NMR, Chapelin *et al.* [28] confirmed that there was no residual (extracellular) PFC when labeling CAR T cells with PFC, indicating any initial ^{19}F signal observed was due to PFC-labeled CAR T cells. To investigate biodistribution of CAR T cells, Chapelin *et al.* [28] administered 20 million PFC^{+} CAR T cells or PFC^{+} control T cells iv to mice bearing subcutaneous glioblastoma tumors. Mice were euthanized and organs were extracted after 2, 7, or 14 days and the ^{19}F content was measured from specimens using NMR (Fig. 7.3b). On day 2, CAR T cells were located largely in the liver (15% of cells), lungs, and spleen (2.5%), and to a lesser extent, lymph nodes and tumors. Not all cells were accounted for, and there were several organs with undetectable ^{19}F levels, where the limit of detection is between $7 - 40 \times 10^3$ cells. There was less ^{19}F measured in the lymph nodes and thymus at later timepoints, concomitant with an increase in liver ^{19}F signal, indicating the clearance of PFC from dead T cells. Importantly, the trafficking of cells to tumors was significantly higher for CAR T cells compared to untransduced T cells on day 2, although the quantification of ^{19}F indicated that only approximately 85,000 CAR T cells (0.425% of injected dose) arrived at the tumor, they appeared to have invoked a significant reduction in tumor burden after 7 days, compared to mice receiving untransduced T cells and untreated mice.

To improve ^{19}F detection for T cells, there has been recent efforts to increase uptake of PFC by CAR T cells by conjugating a cell penetrating peptide to the PFC nanoemulsion [32]. This approach increased the ^{19}F uptake in CAR T cells by 8 times and this group demonstrated improved CAR T cell detection *in vivo*.

Intratumoral administration of CAR T cells may greatly improve the delivery and therapeutic outcomes for solid tumors [33–36]. Two publications [31, 33] have demonstrated PFC+ CAR T cell detection in mouse models, for multiple days following intratumoral injection (Fig. 7.3c).

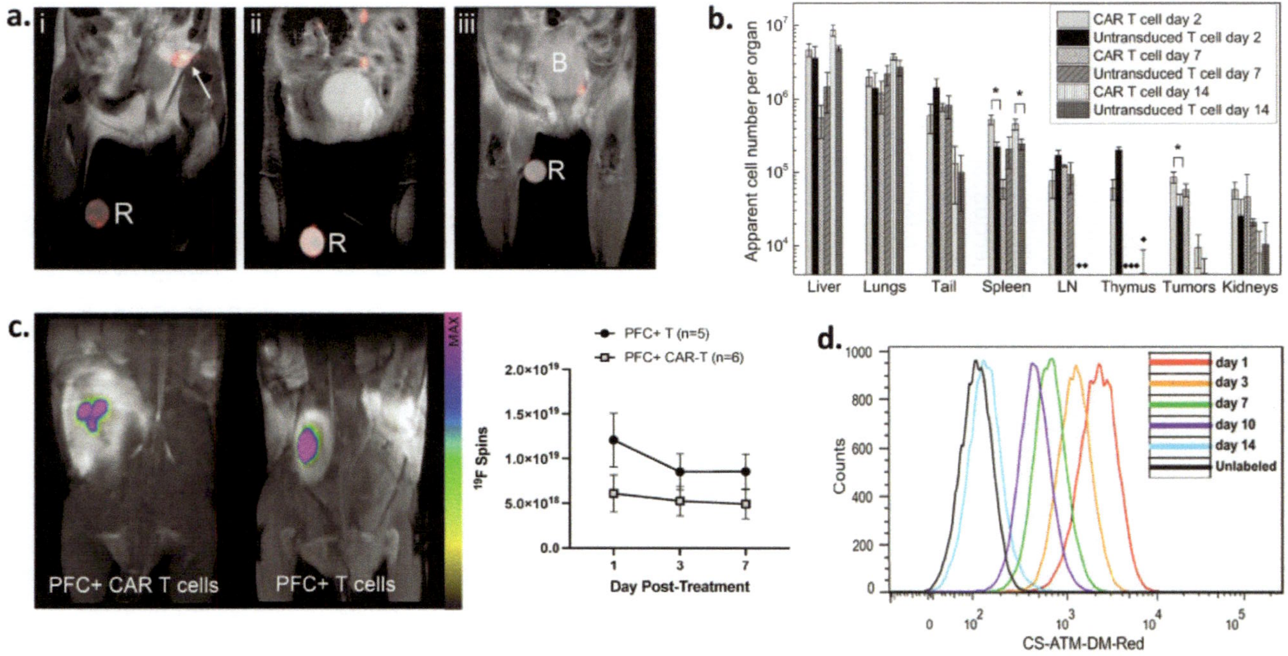

Figure 7.3 (a) ^{19}F MRI detection of antigen-specific T cell trafficking to the lymph node draining the antigen (i). Control mice with untransduced T cells show ^{19}F signal in the abdominal lymph nodes (ii), and mice that received heat-killed antigen-specific T cells show ^{19}F signal in the bladder (B) region (iii). (b) Biodistribution of PFC^+ CAR T cells and PFC^+ T cells in flank glioblastoma-bearing mice as measured by *ex vivo* ^{19}F NMR, with organs collected 2, 7, and 14 days after intravenous infusion of the cells. (c) ^{19}F MRI detection of PFC^+ CAR T cells and PFC^+ untransduced T cells, on the day after direct intratumoral administration. ^{19}F imaging was repeated on day 3 and 7, where quantification of ^{19}F atoms remained constant over this time and ^{19}F atoms detected from PFC^+ CAR T cells and PFC^+ untransduced T cells did not significantly differ. (d) A red fluorescent version of PFC was used to label CAR T cells and the fluorescence was measured longitudinally with flow cytometry. The fluorescence intensity persisted but decreased over 14 days (to baseline) because of CAR T cell division *in vitro*. Figures reprinted from Refs. [24, 28, 31] under CC-BY.

7.4.4 NK Cells

NK cell immunotherapies in humans have been performed using both iv infusions and direct intratumoral injections [37, 38]. A protocol for labeling NK cells with PFC was reported in 2016 [39, 40]. ^{19}F MRI paired with anatomical imaging can help to recognize if transferred cells have been administered to the correct location. Kennis *et al.* [41] have demonstrated the successful, partially successful, and failed injection of PFC-labeled NK cells to medulloblastoma tumors in mice, which requires a complex intracranial injection. Successful injections were identified as ^{19}F signal, originating from PFC-labeled NK cells, within the tumor. Mice with successful injections of NK cells showed significant suppression of tumor growth as measured by bioluminescence imaging, compared to NK cells administered to the contralateral side of the brain and mice which did not receive NK cells. Bouchlaka *et al.* [39] demonstrate that ^{19}F-labeled NK cells administered intratumorally to neuroblastoma and lymphoma tumors in mice could be detected for multiple days. The number of cells present within the tumor were quantified from ^{19}F images and diminished over time.

NK cells administered to sites distant to tumors or by iv infusion must traffic to tumors to exert cytotoxic effects against the cancer cells. In melanoma-bearing mice, Bouchlaka *et al.* [39] administered 10 million PFC^{+} NK cells subcutaneously on the contralateral side (distant from the tumor). The ^{19}F signal at the site of administration diminished over 15 days (to 65%) which is indicative of NK cell migration, however migrated NK cells were not detected at the tumor or other organs. The inability to detect ^{19}F-labeled NK cells in the tumor may be due to their inability to migrate to the tumor or may be related to a low number of NK cells successfully homing to the tumor which is below the ^{19}F detection threshold. Furthermore, the authors report the inability to detect NK cells after iv systemic administration. Limited ^{19}F sensitivity is challenging for this application. Clinically, NK cells are administered by infusion, and in the case of hematologic malignancies, it is crucial for NK cells to efficiently migrate to the bone marrow [42].

7.5 Fluorine-19 MRI of Tumor-Associated Macrophages

As mentioned, the systemic administration of PFCs results in uptake by phagocytic cells within the body. In healthy subjects, PFCs given iv are taken up by cells of the reticuloendothelial system (RES) including Kupffer liver macrophages, spleen, lymph nodes and bone marrow (Fig. 7.4).

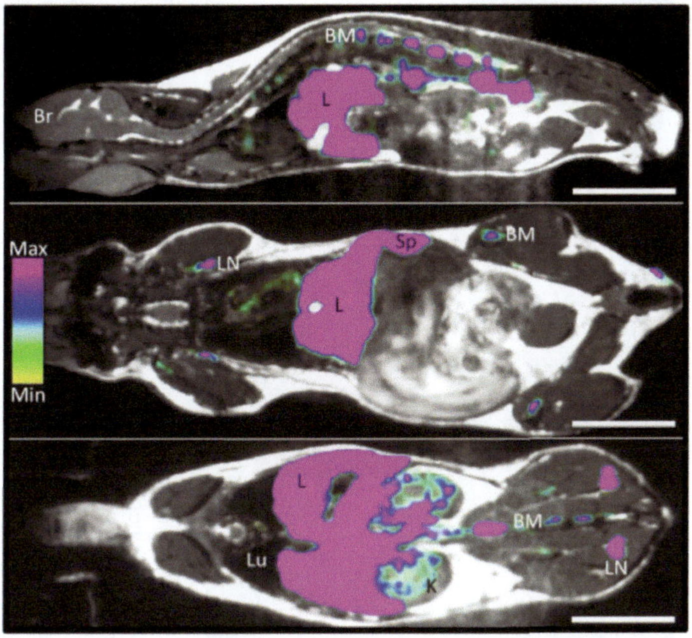

Figure 7.4 Whole body $^{1}H/^{19}F$ MRI after systemic PFC administration into a healthy mouse. Sagittal (top) and coronal views (middle, bottom) of a mouse 24 hours after iv administration of a PFC agent. PFC accumulation is noted in the liver (L), bone marrow (BM), lymph nodes (LN), spleen (S) and kidneys (K). Color scale bar represents numerical minimum/maximum values (9597/24583) to demonstrate differences in ^{19}F signal. Scale bars: 10 mm. Figure reprinted from Ref. [43] under CC BY.

In tumor-bearing subjects, PFCs are also taken up by tumor-associated macrophages (TAMs) and ^{19}F MRI can be used to provide information on the location and density of TAMs within

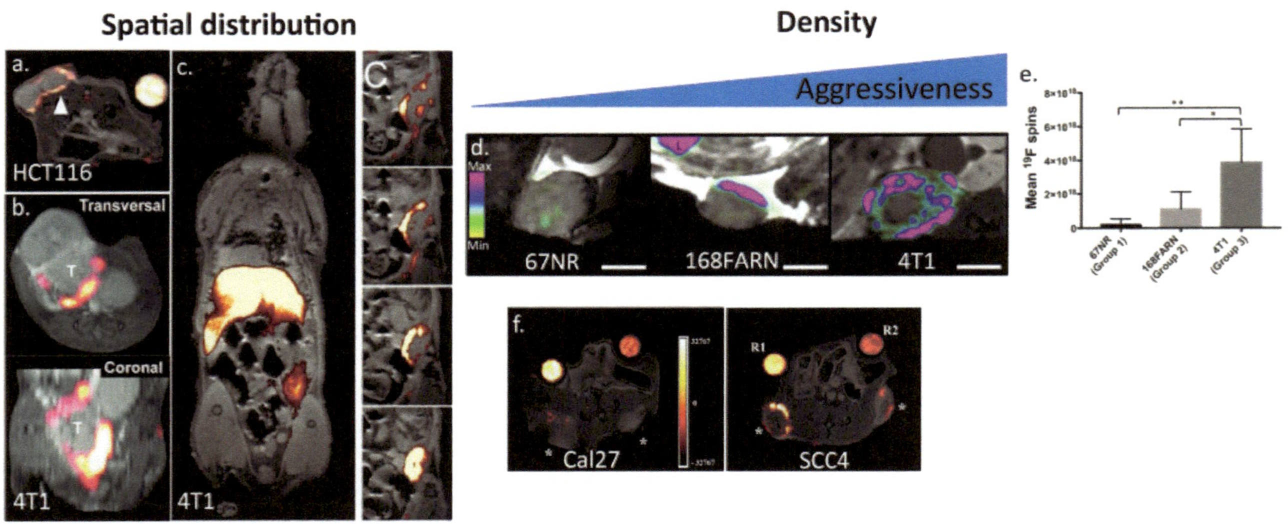

Figure 7.5 Fluorine-19 MRI identifies tumor associated macrophage distribution and density *in vivo*. Tumor associated macrophage (TAM) distribution is identified around the tumor periphery in (a). HCT116 subcutaneous tumors and (b) & (c) 4T1 tumors. TAM content, related to ^{19}F signal, was identified in (d). 67NR, 168FARN and 4T1 tumors (tumors differ in aggressiveness; low to high, respectively), all at 21 days post cancer cell implantation. (e) More ^{19}F signal was quantified in 4T1 versus 67NR and 168FARN tumors. (f) Human-derived tumor models also display this trend, with less ^{19}F signal within Cal27 versus SCC4 tumors (less to more aggressive; tumors noted adjacent to *). $**p < 0.01$, $*p < 0.05$. Scale bars = 5 mm. Reprinted from Refs. [46, 47] under CC BY-NC, Ref. [48] (this research was originally published in JNM ©SNMMI) and Refs. [43, 45] under CC BY.

the tumor. The spatial distribution of TAMs within the tumor may have clinical significance; macrophages surrounding the rim of the tumor are considered within the "invasive edge" and play a role in tumor progression and aid in the escape of tumor cells [44]. Several studies have shown ^{19}F signal related to TAMs localized to the tumor periphery in preclinical cancer models (Fig. 7.5a–c). Makela *et al.* demonstrated a correlation between TAM content measured by ^{19}F MRI and known tumor aggressiveness [43]. Figure 7.5d shows ^{19}F images of three types of murine tumors derived from the same parental cell line; 67NR is a non-aggressive, non-metastatic cell line; 168FARN is mildly aggressive with little propensity to metastasize; while 4T1 cells are highly aggressive and metastatic, resembling a late-stage human breast carcinoma. Qualitative differences in ^{19}F signal were obvious and quantitation confirmed that TAM content was significantly higher in the 4T1 tumors versus either 67NR or 168FARN tumors (Fig. 7.5e) [43]. Similar findings have been reported for human-derived tumors in mice; twice the amount of ^{19}F signal was measured within the more aggressive CAL27 (double hit) versus SCC4 (single hit) human tongue squamous cancer cells (Fig. 7.5f) [45]. The ability to detect these differences highlights the potential of ^{19}F MRI as a biomarker for potential treatability and malignancy.

Most studies which have used ^{19}F to image TAMs perform a single iv injection of PFCs and images are typically acquired one or two days later to allow time for uptake by cells. The persistence and stability of ^{19}F signal in TAMs has been demonstrated in a 4T1 breast cancer model where PFCs were injected iv once and images obtained every 12 hours between 12 and 60 hours post injection (Fig. 7.6a) [43]. No differences in ^{19}F signal within the tumor were noted among any of the time points studied. In other studies, macrophage infiltration in a model of spinal cord injury compared ^{19}F images acquired over time after a single PFC administration, as well as images acquired over time after multiple PFC administrations, starting at 24 hours post injection [49]. After a single injection of PFC, the ^{19}F signal in the inflamed spinal cord was highest at 1 day post PFC injection (2 days post injury, DPI) and subsequently decreased after 2 days post PFC injection (3 DPI), plateauing at around half of the peak signal intensity from 5 DPI through 14 DPI (Fig. 7.6b). When multiple PFC injections were administered every 3 days the ^{19}F signal remained

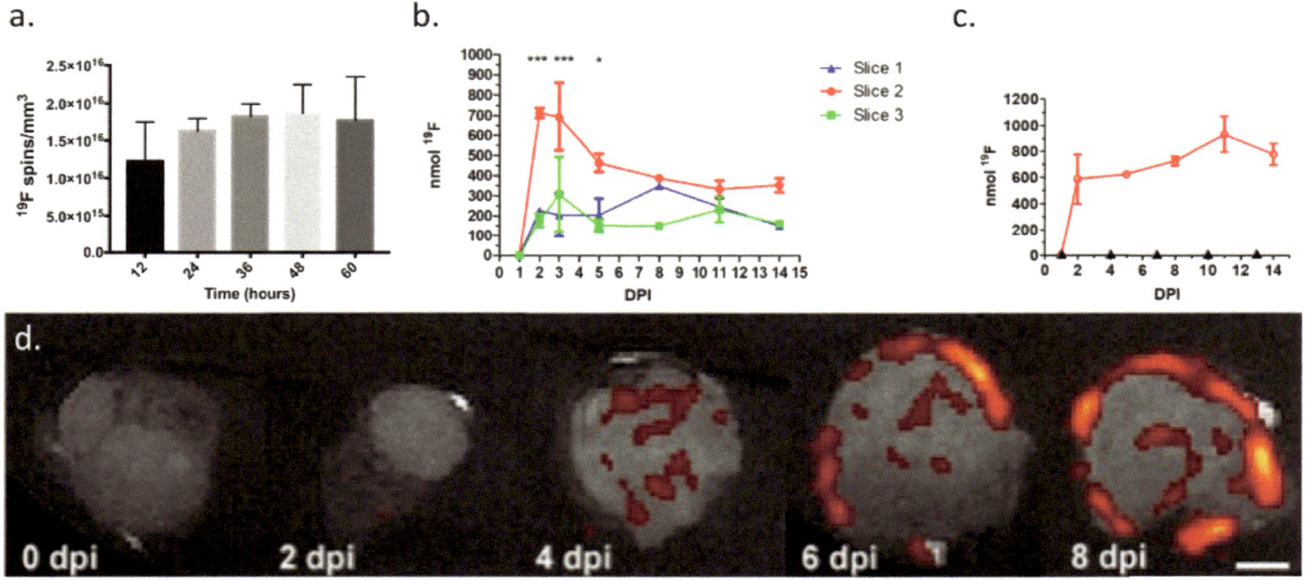

Figure 7.6 Timing of PFC administration and ^{19}F imaging. (a) Serial imaging of mice bearing 4T1 tumors at 21 days post cancer cell implantation. ^{19}F imaging was performed at 12, 24, 36, 48 and 60 hours after a single iv PFC injection. ^{19}F spins/mm^3 were quantified for all mice at each time point and no significant difference was noted among these time points, confirming persistence and stability of PFC-labeled TAMs. (b) In a model of spinal cord injury, PFC accumulation was monitored in the inflamed spinal cord after a single PFC iv injection. There was a significant increase in ^{19}F accumulation (nmol ^{19}F) within the spinal cord (red circle) at 1, 2 and 3 days post PFC administration (2, 3 and 4 DPI), but not in adjacent slices (slice 1 and 3, blue triangle and green square). ^{19}F content appeared to plateau at 5–14 DPI. (c) In the same model, after multiple PFC administration (days identified by black triangles), followed by ^{19}F MRI 1 day after each injection, there was no difference in ^{19}F content over a 14 day period. (d) Different ^{19}F imaging time-points after oncolytic virotherapy identifies differences in myeloid cell recruitment and distribution within the tumor. *$p < 0.05$, ***$p < 0.001$. DPI, days post injury (b–c) or days post infection (d). Reprinted from Refs. [43, 49, 52] under CC.

constant. ^{19}F MRI was performed 1 day after each injection and ^{19}F signal remained high over the 14 days (Fig. 7.6c), suggesting that multiple PFC administrations leads to labeling of additional recruited macrophages. We note that many other inflammation models, such as arthritis and inflammatory bowel disease [50, 51], are effectively imaged using similar methods.

Weibel *et al.* used ^{19}F MRI to examine macrophage accumulation during the development of melanoma tumors treated with oncolytic virotherapy. Oncolytic virotherapy uses tumor specific viruses (in this case GLV-1h68) which selectively target and kill cancer cells without harming normal tissue [53]. The cancer cell death triggers the recruitment of macrophages in response to viral presence, and the products due to lytic destruction of the tumor cells can shift macrophage polarization into an M1-like anti-tumoral state [54]. In this model ^{19}F signal was first apparent in tumors imaged 4 days post infection and appeared scattered within the tumor. In tumors imaged at 6 or 8 days post infection the intratumoral ^{19}F signal was greater and, at this point, appeared along the tumor periphery (Fig. 7.6d). These differences in the spatial distribution of ^{19}F signal at different times in tumor development have also been reported by Makela *et al.* using a mouse model of breast cancer [55].

Hasan *et al.* examined the effect of temozolomide (TMZ) [56] and Na/H exchanger 1 (NHE1) inhibitor, each a common therapeutic used for glioblastoma, in mice bearing orthotopic gliomas. TMZ alone creates an increase in immunosuppressive Tregs in glioblastoma multiforme patients [57] and stimulates Th2 polarization of $CD11b^+$ glioma-associated microglia and myeloid cells (GAM) [58]. However, when combined with an additional therapeutic, anti–IL-2Rα monoclonal antibody or NHE1 inhibitor, respectively, results in positive therapeutic outcomes relating to a proinflammatory immune response. When using pharmacological inhibition of NHE1 with TMZ, there was an increase in immune cell infiltration, identified by increased ^{19}F signal (Fig. 7.7). Although each of these studies monitored the influx, or increase in macrophages within these tumors, the opposite affect could potentially be monitored if a therapeutic were to deplete TAMs in TME.

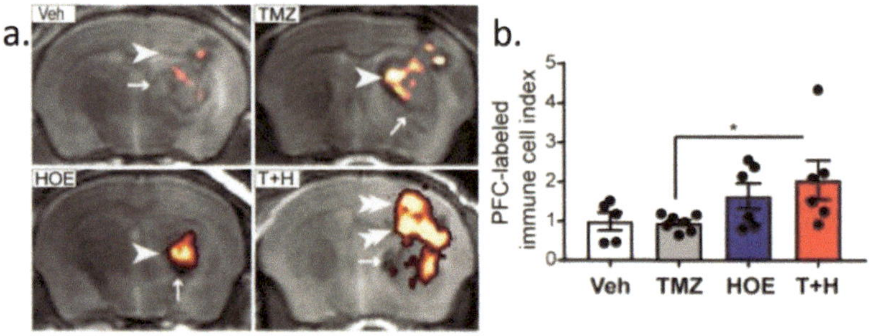

Figure 7.7 ^{19}F MRI monitors immune cell presence during cancer therapeutics. (a) ^{1}H/^{19}F MRI overlays identify presence of GAMs after various treatments in an orthotopic glioma model. Immune cell infiltration index was calculated by ^{19}F signal/tumor volume and data normalized to control. (b) An increase in PFC-labeled immune cell index was identified when TMZ was combined with NHE1 inhibitor (T+H), versus TMZ alone, suggesting that this combination therapy can recruit proinflammatory immune cells. *$p < 0.05$. GAM, glioma associated microglia and myeloid cells; Veh, vehicle control; TMZ, temozolomide, HOE, HOE642 NHE1 specific inhibitor; T+H, TMZ+HOE642. Figure reprinted from Ref. [58] under CC-BY.

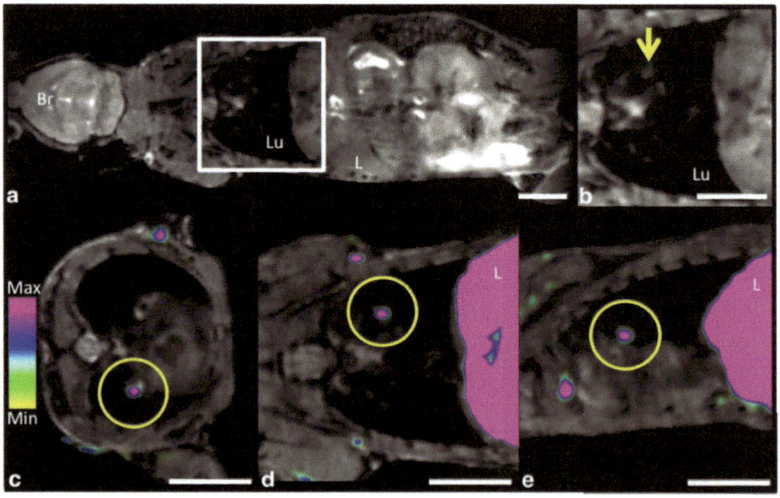

Figure 7.8 Metastasis associated macrophages identified using fluorine-19 MRI. Whole mouse body MRI identifies lung metastases in a mouse bearing a 4-week 4T1 primary tumor. (a) ^{1}H MRI provides anatomical context to associate ^{19}F signal with a lung metastasis. (b) A zoomed in region identifies a hyperintense area within the black lung corresponding to a lung metastasis (yellow arrow). (c)

Axial, (d) coronal and (e) Sagittal overlays identify ^{19}F signal in the same location as the hyperintense region in all 3 dimensions (circled in yellow). Color scale bar represents numerical minimum/maximum values (c, d, and e 15515/27438) to demonstrate differences in ^{19}F signal. Scale bars are 5 mm. Br, brain; L, liver; Lu, lung. Figure reprinted from Ref. [43] under CC-BY.

Macrophages present within metastases are referred to as metastasis-associated macrophages or MAMs. Qian *et al.* have shown that these macrophages have a distinct phenotype and are recruited to lungs bearing metastatic cells to enhance seeding and growth of metastases [59]. Figure 7.8 shows the detection of ^{19}F signal co-localized to lung metastases identified by proton MRI, in a mouse model of metastatic breast cancer [43]. The use of ^{19}F MRI to identify and quantify MAMs in metastatic sites provides us with the opportunity to further study the role of MAMs and quantitatively monitor any therapeutic effects. Table 7.1 summarizes key features of the ^{19}F MRI studies on TAMs covered in this chapter.

Table 7.1 ^{19}F studies of TAMs using various PFC nanoemulsions

Tumor Type	Cell Line	Field Strength (T)	Application	Ref.
Murine breast	4T1	7	Multi-modal imaging can identify TAMs using a PFC–NIR probe.	[47]
	4T1	9.4	Comparing iron- and ^{19}F-based MRI to identify TAMs in early and late-stage tumors.	[55]
	4T1	7	Multi-modal imaging can be used to better understand the relationship between TAMs (^{19}F MRI) and tumor metabolism (PET).	[60]
	67NR, 168FARN, 4T1	9.4	^{19}F MRI used to quantify TAMs can be used as a biomarker of tumor aggressiveness and can identify MAMs.	[43]

(*Continued*)

Table 7.1 (*Continued*)

Tumor Type	Cell Line	Field Strength (T)	Application	Ref.
	4T1	7	Multi-modal imaging can identify HIFU lesions (ADC) and can quantify macrophage infiltration (^{19}F MRI).	[61]
	4T1	3, 9.4	^{19}F MRI studies at 3 T.	[62]
	4T1	11.7	A ^{19}F/PET nanoemulsion can be used as a multi-modal imaging tool.	[48]
Murine glioma	SB28	7	^{19}F MRI can identify immune cells in the TME during therapeutic response in a glioma model.	[58]
Human	1936-MEL (melanoma) A549 (lung) GI-101A (breast)	7, 11.7	^{19}F MRI can identify host immune response in tumors during oncolytic virotherapy.	[52]
	HCT116 (colorectal)	7	^{19}F MRI can guide PFCNE-enhanced HIFU ablation of tumors.	[46]
	Cal27, SCC4 (tongue squamous)	11.7	^{19}F MRI can quantify TAMs and distinguish between single- and double-hit tumors.	[45]

7.6 Conclusions

There is a growing need for methods enabling *in vivo* cell imaging to report on the fate of immunotherapeutic cells after their administration without the use of invasive biopsy. In this chapter, we described several studies which used *ex vivo* labeling of therapeutic cells with PFC prior to their transplantation. ^{19}F MRI on the day of transplantation provides important information about cell delivery, and subsequent imaging sessions can help determine whether the labeled therapeutic cells survive or have reached their target (*i.e.*, the tumor or the lymph nodes) and in what quantity. Importantly, these imaging techniques provide a platform that allows for rapid pre-clinical optimization to study and maximize therapeutic benefit. For example, optimal cell delivery doses, administration routes, and immunoadjuvant strategies, are being investigated and suitable for ^{19}F imaging studies.

Moreover, the presence of TAMs in the TME correlates with tumor aggressiveness, risk of metastasis, and worse clinical outcomes for several types of cancer. ^{19}F labeling of TAMs can be achieved through iv administration of PFC and subsequent ^{19}F imaging allows for assessment of the TAM and MAM density and distribution. This route of PFC administration for imaging studies has only been used preclinically to date. However, ^{19}F visualization and quantification of TAMs burden is amenable to future clinical studies.

References

1. L. Cassetta, and J. W. Pollard. Targeting macrophages: therapeutic approaches in cancer. *Nat. Rev. Drug Discov.*, **17**(12), 887–904, 2018.
2. J. L. Guerriero. Macrophages: the road less traveled, changing anticancer therapy. *Trends Mol. Med.*, **24**(5), 472–489, 2018.
3. C. Hayes. Cellular immunotherapies for cancer. *Ir. J. Med. Sci.*, **190**(1), 41–57, 2021.
4. M. Sadeghzadeh, *et al.* Dendritic cell therapy in cancer treatment: the state-of-the-art. *Life Sci.*, 254(March), 117580, 2020.
5. C. Fink, *et al.* Quantification and characterization of granulocyte macrophage colony-stimulating factor activated human peripheral blood mononuclear cells by fluorine-19 cellular MRI in an immunocompromised mouse model. *Diagn. Interv. Imaging*, **101**(9), 577–588, 2020.
6. P. W. Kantoff, *et al.* Sipuleucel-T immunotherapy for castration-resistant prostate cancer. *N. Engl. J. Med.*, **363**(5), 411–422, 2010.
7. C. E. Handy, and E. S. Antonarakis. Sipuleucel-T for the treatment of prostate cancer: novel insights and future directions. *Futur. Oncol.*, **14**(10), 907–917, 2018.

8. Ö. Met, K. M. Jensen, C. A. Chamberlain, M. Donia, and I. M. Svane. Principles of adoptive T cell therapy in cancer. *Semin. Immunopathol.*, **41**(1), 49–58, 2019.
9. S. A. Holstein, and M. A. Lunning. CAR T-cell therapy in hematologic malignancies: a voyage in progress. *Clin. Pharmacol. Ther.*, **107**(1), 112–122, 2020.
10. J. A. Myers, and J. S. Miller. Exploring the NK cell platform for cancer immunotherapy. *Nat. Rev. Clin. Oncol.*, **18**(2), 85–100, 2021.
11. A. Martín-Fontecha, *et al.* Regulation of dendritic cell migration to the draining lymph node: impact on T lymphocyte traffic and priming. *J. Exp. Med.*, **198**(4), 615–621, 2003.
12. E. T. Ahrens, R. Flores, H. Xu, and P. A. Morel. In vivo imaging platform for tracking immunotherapeutic cells. *Nat. Biotechnol.*, **23**(8), 983–987, 2005.
13. B. M. Helfer, *et al.* Functional assessment of human dendritic cells labeled for in vivo 19 F magnetic resonance imaging cell tracking. *Cytotherapy*, **12**(2), 238–250, 2010.
14. C. Fink, M. Smith, J. M. Gaudet, A. Makela, P. J. Foster, and G. A. Dekaban. Fluorine-19 cellular MRI detection of in vivo dendritic cell migration and subsequent induction of tumor antigen-specific immunotherapeutic response. *Mol. Imaging Biol.*, **19**, 2019.
15. E. T. Ahrens, B. M. Helfer, C. F. O'Hanlon, and C. Schirda. Clinical cell therapy imaging using a perfluorocarbon tracer and fluorine-19 MRI. *Magn. Reson. Med.*, **72**(6), 1696–1701, 2014.
16. F. Bonetto, *et al.* A novel 19F agent for detection and quantification of human dendritic cells using magnetic resonance imaging. *Int. J. Cancer*, **129**(2), 365–373, 2011.
17. M. Srinivas, L. J. Cruz, F. Bonetto, A. Heerschap, C. G. Figdor, and I. J. M. de Vries. Customizable, multi-functional fluorocarbon nanoparticles for quantitative in vivo imaging using 19F MRI and optical imaging. *Biomaterials*, **31**(27), 7070–7077, 2010.
18. D. Baumjohann, A. Hess, L. Budinsky, K. Brune, G. Schuler, and M. B. Lutz. In vivo magnetic resonance imaging of dendritic cell migration into the draining lymph nodes of mice. *Eur. J. Immunol.*, **36**(9), 2544–2555, 2006.
19. L. H. Butterfield. Dendritic cells in cancer immunotherapy clinical trials: are we making progress?. *Front. Immunol.*, **4**(December), 3–9, 2013.

20. G. A. Dekaban, *et al.* Tracking and evaluation of dendritic cell migration by cellular magnetic resonance imaging. *Wiley Interdiscip. Rev. Nanomedicine Nanobiotechnol.*, **5**(5), 469–483, 2013.

21. C. Fink, *et al.* 19F-perfluorocarbon-labeled human peripheral blood mononuclear cells can be detected in vivo using clinical MRI parameters in a therapeutic cell setting. *Sci. Rep.*, **8**(1), 1–13, 2018.

22. L. C. Rose, *et al.* Fluorine-19 labeling of stromal vascular fraction cells for clinical imaging applications. *Stem Cells Transl. Med.*, **4**, 1472–1481, 2015.

23. C. Fink, *et al.* Quantification and characterization of granulocyte macrophage colony-stimulating factor activated human peripheral blood mononuclear cells by fluorine-19 cellular MRI in an immunocompromised mouse model. *Diagn. Interv. Imaging*, **101**, 577–588, 2020.

24. J. M. Janjic, M. Srinivas, D. K. K. Kadayakkara, and E. T. Ahrens. Self-delivering nanoemulsions for dual fluorine-19 MRI and fluorescence detection. *J. Am. Chem. Soc.*, **130**(9), 2832–2841, 2008.

25. M. Srinivas, M. S. Turner, J. M. Janjic, P. A. Morel, D. H. Laidlaw, and E. T. Ahrens. In vivo cytometry of antigen-specific T cells using 19F MRI. *Magn. Reson. Med.*, **62**(3), 747–753, 2009.

26. M. Srinivas, P. A. Morel, L. A. Ernst, D. H. Laidlaw, and E. T. Ahrens. Fluorine-19 MRI for visualization and quantification of cell migration in a diabetes model. *Magn. Reson. Med.*, **58**(4), 725–734, 2007.

27. C. F. O'Hanlon, T. Fedczyna, S. Eaker, W. D. Shingleton, and B. M. Helfer. Integrating a 19F MRI tracer agent into the clinical scale manufacturing of a T-cell immunotherapy. *Contrast Media Mol. Imaging*, **2017**, 9548478, 2017.

28. F. Chapelin, S. Gao, H. Okada, T. G. Weber, K. Messer, and E. T. Ahrens. Fluorine-19 nuclear magnetic resonance of chimeric antigen receptor T cell biodistribution in murine cancer model. *Sci. Rep.*, **7**(1), 1–12, 2017.

29. N. Sato, H. Wu, K. O. Asiedu, L. P. Szajek, G. L. Griffiths, and P. L. Choyke. 89Zr-oxine complex PET cell imaging in monitoring cell-based therapies. *Radiology*, **275**(2), 490–500, 2015.

30. M. Srinivas, *et al.* Imaging of cellular therapies. *Adv. Drug Deliv. Rev.*, **62**(11), 1080–1093, 2010.

31. V. P. Dubois, O. C. Sehl, P. J. Foster, and J. A. Ronald. Visualizing CAR-T cell immunotherapy using 3 tesla fluorine-19 MRI. *Mol. Imaging Biol.*, 0123456789, 2021.

32. D. V. Hingorani, F. Chapelin, E. Stares, S. R. Adams, H. Okada, and E. T. Ahrens. Cell penetrating peptide functionalized perfluorocarbon nanoemulsions for targeted cell labeling and enhanced fluorine-19 MRI detection. *Magn. Reson. Med.*, **83**(3), 974–987, 2019.
33. F. Chapelin, and H. Okada. Assessing oximetry response to chimeric antigen receptor T-cell therapy against glioma with 19 F MRI in a murine model. **19**, 1–9, 2021.
34. J. Tchou, *et al.* Safety and efficacy of intratumoral injections of chimeric antigen receptor (CAR) T cells in metastatic breast cancer. *Cancer Immunol. Res.*, **5**(12), 1152–1161, 2017.
35. F. Guo, and J. Cui. CAR-T in solid tumors: blazing a new trail through the brambles. *Life Sci.*, **260**(June), 2020.
36. P. Sridhar, and F. Petrocca. Regional delivery of chimeric antigen receptor (CAR) T-cells for cancer therapy. *Cancers (Basel).*, **9**(7), 1–10, 2017.
37. S. Liu, *et al.* NK cell-based cancer immunotherapy : from basic biology to clinical development. *J. Hematol. Oncol.*, **14**(1), 7, 2021.
38. S. Lorenzo-Herrero, A. López-Soto, C. Sordo-Bahamonde, A. P. Gonzalez-Rodriguez, M. Vitale, and S. Gonzalez. NK cell-based immunotherapy in cancer metastasis. *Cancers (Basel).*, **11**(1), 1–22, 2019.
39. M. N. Bouchlaka, *et al.* 19F-MRI for monitoring human NK cells in vivo. *Oncoimmunology*, **5**(5), 1–12, 2016.
40. S. S. Somanchi, B. A. Kennis, V. Gopalakrishnan, D. A. Lee, and J. A. Bankson. In vivo 19F-magnetic resonance imaging of adoptively transferred NK cells. *Methods Mol. Biol.*, **1441**, 317–332, 2016.
41. B. A. Kennis, *et al.* Monitoring of intracerebellarly-administered natural killer cells with fluorine-19 MRI. *J. Neurooncol.*, **142**(3), 395–407, 2019.
42. N. Sato, *et al.* In vivo tracking of adoptively transferred natural killer cells in rhesus macaques using 89 zirconium-oxine cell labeling and PET imaging. *Clin. Cancer Res.*, **26**(11), 2573–2581, 2020.
43. A. V. Makela, and P. J. Foster. Imaging macrophage distribution and density in mammary tumors and lung metastases using fluorine-19 MRI cell tracking. *Magn. Reson. Med.*, **80**(3), 1138–1147, 2018.
44. S. Aras, and M. R. Zaidi. TAMeless traitors: macrophages in cancer progression and metastasis. *Br. J. Cancer*, **117**, 1583–1591, 2017.
45. A. Khurana, *et al.* Visualization of macrophage recruitment in head and neck carcinoma model using fluorine-19 magnetic resonance imaging. *Magn. Reson. Med.*, **79**(4), 1972–1980, 2018.

46. S. H. Shin, *et al.* Tracking perfluorocarbon nanoemulsion delivery by 19FMRI for precise high intensity focused ultrasound tumor ablation. *Theranostics*, **7**(3), 562–572, 2017.
47. A. Balducci, *et al.* A novel probe for the non-invasive detection of tumor-associated inflammation. *Oncoimmunology*, **2**(2), e23034, 2013.
48. C. Wang, *et al.* Metallofluorocarbon nanoemulsion for inflammatory macrophage detection via PET and MRI. *J. Nucl. Med.*, **62**(8), 1146–1153, 2021.
49. F. Garello, *et al.* Imaging of inflammation in spinal cord injury : novel insights on the usage of PFC-based contrast agents. 1–27, 2021.
50. U. Flogel, *et al.* In vivo monitoring of inflammation after cardiac and cerebral ischemia by fluorine magnetic resonance imaging. *Circulation*, **118**(2), 140–148, 2008.
51. A. Balducci, B. M. Helfer, E. T. Ahrens, C. F. O'Hanlon 3rd, and A. K. Wesa. Visualizing arthritic inflammation and therapeutic response by fluorine-19 magnetic resonance imaging (19F MRI). *J. Inflamm. (Lond).*, **9**(1), 24, 2012.
52. S. Weibel, *et al.* Imaging of intratumoral inflammation during oncolytic virotherapy of tumors by 19F-magnetic resonance imaging (MRI). *PLoS One*, **8**(2), e56317, 2013.
53. H. Fukuhara, Y. Ino, and T. Todo. Oncolytic virus therapy: a new era of cancer treatment at dawn. *Cancer Sci.*, **107**(10), 1373–1379, 2016.
54. L. Hofman, S. E. Lawler, and M. L. M. Lamfers. The multifaceted role of macrophages in oncolytic virotherapy. *Viruses*, **13**(8), 1570, 2021.
55. A. V. Makela, J. M. Gaudet, and P. J. Foster. Quantifying tumor associated macrophages in breast cancer: a comparison of iron and fluorine-based MRI cell tracking. *Sci. Rep.*, 742109, 2017.
56. M. M. Mrugala, and M. C. Chamberlain. Mechanisms of disease: temozolomide and glioblastoma—look to the future. *Nat. Clin. Pract. Oncol.*, **5**(8), 476–486, 2008.
57. D. A. Mitchell, *et al.* Monoclonal antibody blockade of IL-2 receptor α during lymphopenia selectively depletes regulatory T cells in mice and humans. *Blood*, **118**(11), 3003–3012, 2011.
58. M. N. Hasan, *et al.* Blocking NHE1 stimulates glioma tumor immunity by restoring OXPHOS function of myeloid cells. *Theranostics*, **11**(3), 1295–1309, 2021.
59. B. Z. Qian, *et al.* CCL2 recruits inflammatory monocytes to facilitate breast-tumour metastasis. *Nature*, **475**(7355), 222–225, 2011.

60. S. H. Shin, S. H. Park, S. H. Kang, S. W. Kim, M. Kim, and D. Kim. Fluorine-19 magnetic resonance imaging and positron emission tomography of tumor-associated macrophages and tumor metabolism. *Contrast Media Mol. Imaging*, **2017**, 4896310, 2017.
61. S. H. Shin, S. H. Park, S. W. Kim, M. Kim, and D. Kim. Fluorine MR imaging monitoring of tumor inflammation after high-intensity focused ultrasound ablation. *Radiology*, **287**(2), 476–484, 2018.
62. A. V Makela, and P. J. Foster. Preclinical 19F MRI cell tracking at 3 Tesla. *Magn. Reson. Mater. Phys., Biol. Med.*, **32**(1), 123–132, 2019.

Chapter 8

Noninvasive *In Vivo* Imaging of Neutrophil Trafficking by ^{19}F MRI

Pascal Bouvain, Sebastian Temme, Maria Grandoch and Ulrich Flögel
Experimental Cardiovascular Imaging, Heinrich Heine University, Universitätsstraße 1, 40225 Düsseldorf, Germany
bouvain@hhu.de

8.1 The Role of Neutrophil Granulocytes during Inflammatory Processes

Daily exposure to pathogens is a constant battle and depends heavily on the innate immune system as the first response to invaders. Neutrophils are an important part of the innate immune system and armed with a variety of effective "weapons" to fight invading pathogenic microorganisms. In the recent years, however, it has been discovered that neutrophils not only eliminate pathogens, but also play sophisticated roles in the modulation of adaptive immune responses and promote angiogenesis and wound healing [1].

Fluorine Magnetic Resonance Imaging: Methods and Applications in Biomedicine
Edited by Eric T. Ahrens and Ulrich Flögel
Copyright © 2025 Jenny Stanford Publishing Pte. Ltd.
ISBN 978-981-5129-37-3 (Hardcover), 978-1-003-53004-6 (eBook)
www.jennystanford.com

Of note, neutrophils were first named by Paul Ehrlich when he recognized that most cells on a blood slide are neither stained with acid nor basic dyes, nor neutral dyes, and therefore he termed these cells neutrophil granulocytes [2].

8.1.1 Generation and Release of Neutrophil Granulocytes from the Bone Marrow

Neutrophils are mainly generated in the bone marrow by hematopoietic stem cells. The daily production of neutrophils in humans is approximately 10^{11} neutrophils [3] and about 10^7 in mice [4]. Under noninflammatory conditions, most neutrophils are located within the bone marrow from where they are constantly released into the blood stream. This process is controlled by the CXCL12 (CXC-motif chemokine 12)—CXCR4 (CXC chemokine receptor type 4) axis. Stromal cells in the bone marrow continuously secrete high levels of the chemokine CXCL12 [5], which binds to CXCR4 to retain neutrophils within the bone marrow [6, 7]. G-CSF (granulocyte colony stimulating factor) counteracts the CXCL12/CXCR4 axis by downregulation of CXCR4, resulting in an egress of neutrophils from the bone marrow into circulation [8]. Importantly, acute inflammation is often associated with high levels of G-CSF that leads to rapid and massive neutrophil release, where mature (segmented) as well as non-mature (banded) neutrophils are flushed into circulation [9].

8.1.2 Recruitment into Inflamed Tissue and Effector Functions of Neutrophil Granulocytes

Under healthy conditions, neutrophils are kept in an inactivated state patrolling through the vessels, but in case of infection or tissue damage, neutrophils become activated by cytokines (*e.g.*, IL-8, TNF-alpha, IL-6), pathogen-associated molecular patterns (LPS, fMLP) or danger/damage-associated molecular patterns (HMGB5, ATP). Subsequently, neutrophils are recruited into the affected tissue in a multistep process called leukodiapedesis that involves (I) tethering, (II) rolling, (III) adhesion, (IV) crawling and finally (V) transmigration (Fig. 8.1, upper).

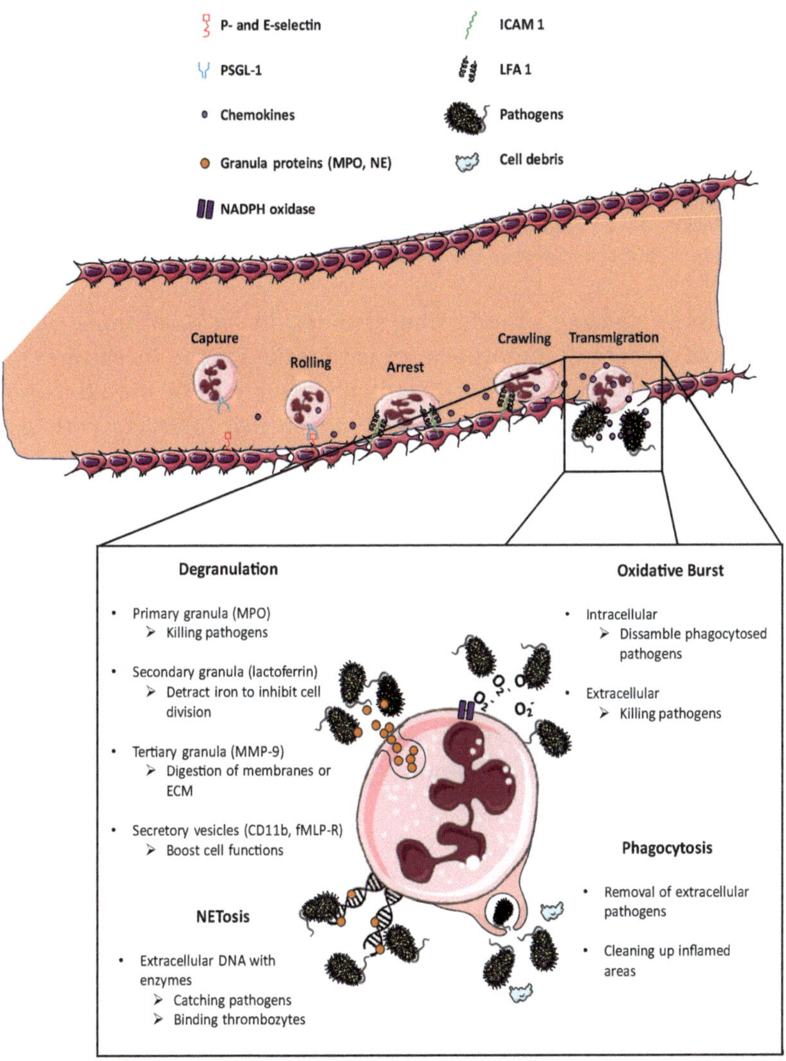

Figure 8.1 Leukodiapedesis and effector functions of neutrophils. The upper panel displays the multistep process for neutrophil infiltration into affected tissue that begins with the binding (capture) and rolling on activated endothelial cells [1]. Binding and rolling are mediated for example by PSGl-1 which interacts with P-selectin and E-selectin on endothelial cells. During rolling, signals from adhesion molecules and chemokines induce a conformational change in integrin molecules that results in a strong interaction between LFA-1/MAC-1 and ICAM-1 and leads to neutrophil "arrest". This strong interaction enables the "crawling" of the neutrophil granulocytes which is guided by chemokines

(Caption Continued)

decorated on the apical surface of the endothelium. Finally, neutrophils leave the blood stream and transmigrate into the inflamed tissue. To this end, neutrophils follow a chemokine gradient to find suitable places for migration into the affected tissue, which are often tricellular corners that display lower abundance of junction proteins [10]. Within the affected tissue, neutrophils release bioactive molecules, generate reactive oxygen species (ROS), secrete neutrophil extracellular traps (NETosis) and phagocytose pathogens and cell debris. Taken together, these effector functions help to fight invading pathogens, but also clean up the affected tissue to foster wound healing.

Tissue damage or infection also results in alterations of the luminal surface of endothelial cells to facilitate the recruitment of immune cells. For example, this can manifest itself by the exposition of P- and E-selectin on the apical side which interacts with that P-selectin glycoprotein1 (PSGL1) enabling the tethering and rolling of neutrophils on the endothelium.

Tight binding to the endothelial cells is mediated *via* integrins, like ICAM-1 (intercellular adhesion molecule) on endothelial cells and LFA-1 (lymphocyte function-associated antigen) found on the surface of neutrophils. Integrins are normally retained in a low-affinity conformation, but selectin binding or chemokines can induce a conformation switch to a high-affinity state resulting in strong adhesion to the endothelium. Sequential chemokine gradients guide the crawling of the cells to find suitable sites for transmigration into the inflamed tissue.

Neutrophils often infiltrate at tricellular corners of the endothelium, where the endothelial cell alignment is less ordered and the density of junction proteins is reduced [10]. The last step is the migration into the inflammatory foci, a process which can take place para- or trans-cellular.

After entering the site of inflammation, neutrophils execute multiple effector functions. Neutrophils are potent phagocytes and can rapidly engulf pathogens (non-sterile) to avoid spreading of the infection, or they remove cell debris (sterile) to clean up the damaged tissue. Phagocytosis is enhanced for example by complement factors or IgG antibodies, which decorate bacteria or infected/damaged cells or cell debris.

Neutrophils have different granules, categorized as primary, secondary, and tertiary, as well as secretory vesicles. Secretory vesicles are quickly released upon activation and contain primarily membrane proteins such as CD11b (integrin for adhesion to the endothelium) or the formyl peptide receptor (recognize foreign proteins) to increase their surface expression [11]. Tertiary granules contain matrix metalloproteinases, gelatinase or cathepsins that degrade basement membranes or extracellular matrix, thus paving the way for neutrophils to enter the inflamed area [12]. Secondary granules are characterized by high levels of lactoferrin which limits the available iron for bacteria, thereby inhibiting cell division. Finally, primary granules are filled with neutrophil elastase or myeloperoxidase to disintegrate the cell membrane of bacteria and induce cell death.

Reactive oxygen species (ROS, H_2O_2, HOCl) are also produced by the NADPH oxidase complex (which generates O_2^-) in combination with superoxide dismutase (*i.e.*, converts O_2^- to H_2O_2) and myeloperoxidase (utilizes H_2O_2 to produce HOCl), either within phagolysosomes or on the cell surface. ROS are highly reactive chemicals that can oxidize and modify cellular components like proteins, lipids, DNA and have a strong microbicidal activity. O_2^- anions have only a very short half-life and are quickly converted to H_2O_2. Myeloperoxidase then utilizes H_2O_2 and Cl^- to generate HOCl with strong antimicrobial activity [13]. However, excessive ROS can also induce severe damage to host cells and tissue, which has for example been observed in the lungs of patients suffering from severe SARS-CoV-2 infections [14]. Additionally, neutrophils can secrete DNA filaments (neutrophil extracellular traps, NETs) decorated with cytosolic, granular or nuclear proteins [15] (*e.g.*, lysozyme, myeloperoxidase) which help to catch and destroy extracellular pathogens [16]. However, dysregulated release of NETs—in particular in sterile inflammation—may cause exacerbated inflammation and host tissue damage [17–20]. Interestingly, NETs are also able to degrade cytokines and chemokines which can improve the resolution of inflammation [21] (Fig. 8.1, lower).

8.1.3 Neutrophils and Cardiovascular Diseases

It is well known that neutrophil granulocytes are essential for fighting pathogenic microorganisms such as bacteria, viruses, and fungi [16]. However, as mentioned above, their potent effector mechanisms, such as the release of ROS, proteolytic enzymes or NETosis, can also lead to massive tissue destruction and organ damage. This has for example been observed in severe cases of SARS-CoV-2 infection where dysregulated neutrophil activity fostered exaggerated thrombo-inflammatory responses [22]. There is evidence that this is also relevant to stroke or myocardial infarction, where high levels of circulating neutrophils are associated with poor prognosis [23], and more generally, to atherosclerotic-type cardiovascular events.

Myocardial infarction (MI) is caused by occlusion of a coronary artery, due to rupture or fissure of an atherosclerotic plaque. To re-establish blood flow, MI patients are treated by thrombolytic therapy or percutaneous coronary intervention (PCI). Both, ischemia and subsequent reperfusion, lead to tremendous tissue injury by induction of cardiomyocyte death, which initiates the release of damage-associated molecular patterns (DAMPs) and cytokines/chemokines finally resulting in an intense inflammatory response [24]. The subsequent infiltration of immune cells not only removes dead cells and debris but also stimulates reparative pathways. However, excessive or prolonged tissue inflammation can lead to adverse left ventricular (LV) remodeling and heart failure [25]. Within the first few hours post-MI, neutrophils are recruited to the injured myocardium. Neutrophil counts peak on day one post MI; thereafter neutrophils become apoptotic and are removed by local macrophages [26], a process which is called efferocytosis. Interestingly, efferocytosis polarizes macrophages to an anti-inflammatory state that fosters resolution of inflammation and wound healing.

Within the myocardium, neutrophils phagocyte cellular debris, degrade extracellular matrix and secrete chemotactic factors to promote wound healing. However, secretion of ROS and NETosis can damage surrounding tissue and have detrimental impact on the viable myocardium [27–29]. This is the reason why excessive infiltration of neutrophils and/or their delayed removal may exacerbate myocardial injury by prolonging the inflammatory phase. Massive neutrophil infiltration is therefore linked to worse outcome after MI [30, 31]. On the other hand, an absence of neutrophils, as seen for example in neutropenia, can lead to poorer outcomes after myocardial infarction [32].

8.2 Technologies for Imaging of Neutrophils

As neutrophils are one of the immune system's first responders, there is increasing interest in elucidating their properties under normal conditions and during periods of inflammation. Ideally, a suitable *in vivo* imaging technology would: (I) be specific for neutrophils, (II) have a high tissue penetration, (III) be compatible with whole-body imaging, (IV) be useful for long-term investigations and (V) be non-invasive.

Optical techniques like bioluminescence or intravital microscopy have been widely used for visualization of neutrophils [33, 34]. Intravital microscopy enables imaging of neutrophils in their natural environment for example to investigate migration over time. Commonly, fluorochrome-conjugated antibodies against Ly6G are used for optical imaging of neutrophils in mice. Although low doses of injected antibodies (1–40 μg) are sufficient for intravital microscopy, at least for several hours, higher doses can cause depletion of neutrophils which prevent long-term studies [33]. To overcome this problem, Hasenberg *et al.* developed a knock-in mouse model with stable tdTomato expression under the control of the Ly6G promoter resulting in fluorescently labelled neutrophils [35]. Nonetheless, intravital microscopy always relies on a small observation window which prevents whole-body investigation of trafficking.

An alternative to intravital microscopy is bioluminescence. Luminol is a small molecule which is oxidized by HOCl, leading to chemiluminescence. Gross *et al.* made use of systemic administered luminol to monitor myeloperoxidase (MPO) activity, an enzyme that is highly expressed and secreted by neutrophils and that generates HOCl [34]. However, myeloperoxidase is also expressed by macrophages and therefore luminol conversion by MPO is not neutrophil specific.

The clinical gold standard for imaging of neutrophils is white blood cell scintigraphy. Technetium-99 (^{99m}Tc) and indium-111 (^{111}In) are the most common radiopharmaceuticals for labeling cells, and although all white blood cells are labeled, the majority of them are neutrophils [36]. Furthermore, gamma radiation at low doses less than 175 Gy seem to have no influence on neutrophil migration, ROS secretion or phagocytosis [37]. While ^{99m}Tc has a short half-life of approximately 6 hrs and is used for imaging acute inflammation,

^{111}In is suited for visualization of chronic inflammation due to its half-life of 2.8 days [38].

Alternatively, gallium-67 (^{67}Ga) labeled citrate has been used for over 30 years for SPECT visualization of infections, especially neutrophils [37]. Ga acts as an iron analogue that binds to circulating transferrin and extravasates within inflammatory lesions. The reason neutrophils take up Ga-citrate is not completely understood, but it is postulated that it binds to iron-binding proteins like lactoferrin or accumulates with phagocytic uptake of bacteria containing Ga [39, 40]. Interestingly, exchange of ^{67}Ga with ^{68}Ga not only minimizes the half-life from 78 hours to 68 minutes but also improves sensitivity for PET [41]. The primary disadvantage of Ga-citrate imaging is nonspecific accumulation within inflammatory hot spots *via* the enhanced permeability and retention effect [42].

A more specific approach for targeting neutrophils is by using ^{68}Ga-pentixafor for identification of CXCR4 expression patterns in myocardium and other organs *via* PET [43]: This agent has been exploited for image-based theranostics [44] by blocking CXCR4 to increase the survival rate after acute myocardial infarction. Pentixafor is a cyclic pentapeptide that has very high affinity for CXCR4 [45]. However, CXCR4 is not only strongly expressed by neutrophils, but also by monocytes [46]. Furthermore, radioactive nuclides with a short half-life are required for PET and SPECT, thereby preventing long-term investigations of labeled cells and also exposes the patient to significant ionizing radiation (see also Table 8.1).

Table 8.1 Comparison of different imaging modalities for neutrophils

Imaging Approach	Speci-ficity	Penetration Depth	Whole Body	Long Term	Non Invasive
Intravital microscopy →Ly6G[33]	√	X	X	X	X
Bioluminescence →MPO[34]	X	X	√	√	√
Scintigraphy → ^{99m}Tc, ^{111}In[36]	X	√	√	√	X
SPECT → gallium-67[37]	X	√	√	√	X
PET →CXCR4[43]	X	√	√	X	X

8.3 Visualization of Neutrophils by $^1H/^{19}F$ MRI

Another clinically important imaging modality that can be used for neutrophil imaging is magnetic resonance imaging (MRI). MRI does not involve harmful radiation and provides excellent contrast between soft tissues. Moreover, conventional anatomical hydrogen (^1H) images can be combined with data obtained from fluorine-19 (^{19}F) MRI, localized magnetic resonance spectroscopy (MRS), chemical exchange saturation transfer (CEST), or mapping techniques to add additional information about inflammation, metabolism, or alterations in tissue architecture.

One of the MRI technologies that has gained increasing attention in the last decade is the visualization of inflammatory processes by ^{19}F MRI. Most often lipid-emulsified perfluorocarbon (PFC) nanoemulsions or PFC encapsulated in PLGA nanoparticles, both with hydrodynamic diameters ranging from 100–200 nm, are used as tracer materials for *in vivo* cell-tracking [47, 48]. After intravenous injection, phagocytic cells such as monocytes or macrophages avidly internalize PFC tracers. The ^{19}F-labeled cells accumulate in inflammatory hot-spots, and their anatomical localization as well as their relative quantity are elucidated by $^1H/^{19}F$ MRI.

8.3.1 Passive Targeting of Neutrophils

In pioneering studies, Flögel *et al.* in 2008 [49] described the first use of anatomical ^1H and ^{19}F MRI to visualize the accumulation of monocytes and macrophages within the infarcted heart and the ischemic brain. Subsequently, in murine models of pulmonary inflammation [50] and collagen-induced rheumatoid arthritis [51], it was found that in addition to monocytes and macrophages, neutrophils are labelled with PFC. Generally, neutrophils take up lower amounts of PFC and therefore, in most inflammatory disease models, the majority of the ^{19}F signal is related to monocytes and macrophages. However, under specific conditions, the ^{19}F signal can be nearly exclusively derived from neutrophils, as for example, in the early phase of a Matrigel/lipopolysaccharide (LPS) induced subcutaneous inflammation model [52]. Nevertheless, all approaches that utilize passive uptake of PFC by phagocytic cells lack a distinct specificity for neutrophils. Thus, under real-life conditions, a complex mixture of cell types will be present in an inflammatory foci, with neutrophils only weakly labeled compared to highly phagocytic monocytes and macrophages.

8.3.2 Specific Imaging of Neutrophils by Active Targeting of the Cell Surface Receptor CD177

To enable a selective visualization of neutrophils by ^{19}F MRI, PFCs have to be equipped with specific ligands to target neutrophil-specific cell-surface receptors, and in parallel, the phagocytic uptake by monocytes and macrophages has to be blocked or at least strongly impaired.

8.3.2.1 Preparation and characterization of neutrophil-specific ^{mNP}PFCs

In 1999 Mazzucchelli *et al.* used phage display screening to identify peptides that specifically bind to human neutrophils [53]. Later on, our group revealed that one of these peptides (hNP) binds to human CD177, a GPI-anchored cell surface protein that is expressed in most humans in a large fraction of the blood neutrophils (50–60%) [54]. Unfortunately, we also found that hNP does not bind to murine CD177 and thus was not useful for rodent model studies. However, nearly 20 years after identification of hNP, Miettinen *et al.* described a peptide that targets murine CD177 [55]. Interestingly, in contrast to humans, 100% of the blood neutrophils from C57BL/6J, Balb/c and NMRI mice express CD177 on their surface [54].

We recently used these two neutrophil-specific peptides (NPs) to generate PFCs that are specifically targeted to human and murine neutrophils, respectively [54] (^{mNP}PFCs, ^{hNP}PFCs, see Fig. 8.2A). Both peptides are additionally equipped with a triple glycine linker followed by a terminal cysteine for conjugation to maleimide groups on the surface of PFCs. PFCs were additionally equipped with polyethylene glycol (PEG) to avoid nonspecific uptake by monocytes and macrophages [56–58] (Fig. 8.2A). Incubation of blood-derived immune cells with ^{NP}PFCs enabled a highly specific labeling of neutrophils, with only minimal uptake by monocytes and lymphocytes (Fig. 8.2B). Furthermore, enhanced uptake of ^{NP}PFCs by neutrophils in comparison to nonspecific control PFCs (^{Con}PFCs) was also verified by ^{19}F MRI (Fig. 8.2C).

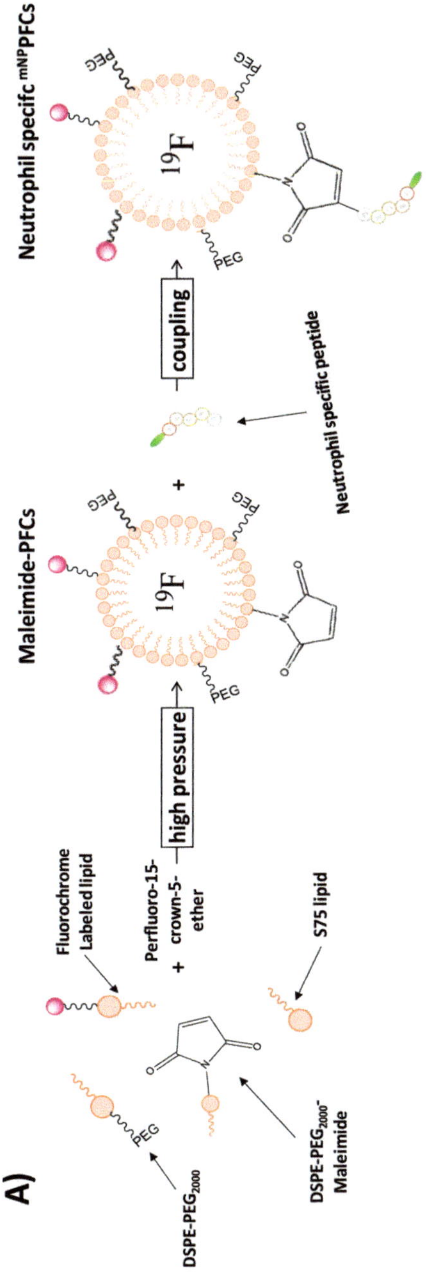
A)
DSPE-PEG2000
DSPE-PEG2000-Maleimide
Fluorochrome Labeled lipid
S75 lipid
+ Perfluoro-15-crown-5-ether
high pressure
Maleimide-PFCs
19F
PEG
+
Neutrophil specific peptide
coupling
Neutrophil specifc mNPPFCs

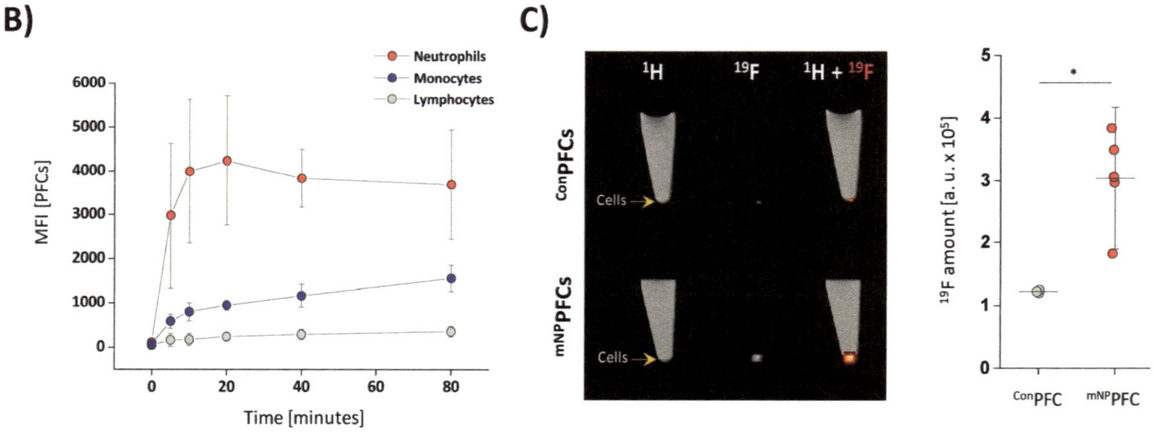

Figure 8.2 Preparation and characterization of neutrophil-specific ^{mNP}PFC. (A) For preparation of maleimide PFCs, lipids derived from soybeans (S75 lipids, Lipoid GmbH) are mixed with fluorochrome-labeled lipids (DSPE-PEG$_{2000}$-rhodamine; DSPE = 1,2-distearoyl-sn-glycero-3-phospho-ethanolamine), PEGylated lipids (DSPE-PEG$_{2000}$; PEG = polyethylene glycol) and PEGylated lipids that have a maleimide group on the distal end of the PEG-chain. The maleimide residues facilitate the site-directed coupling of the neutrophil-specific peptide (NP) to the surface of the PFC. PEG chains impair the phagocytic uptake of PFC, and the fluorescence-labeled lipids enable visualization *via* fluorescence-based imaging modalities. After generation of a crude pre-emulsion, the perfluorocarbon (*e.g.*, perfluoro-15-crown-5 ether) is added to the mixture and processed *via* high pressure homogenization to generate stable nanoemulsions with a diameter between 100–200 nm. Afterwards, NP is added to the emulsion and the C-terminal free cysteine of NP forms a stable thioether-bond with the maleimide group on the PFCs to form neutrophil-specific ^{NP}PFCs. (B) Incubation of ^{mNP}PFC with murine blood immune cells followed by flow cytometric analysis at distinct time points indicate a strong uptake by neutrophils and essentially no uptake by monocytes and lymphocytes. (C) Isolated bone marrow neutrophils were incubated with ^{mNP}PFC or ^{con}PFC, washed several times and centrifuged into an Eppendorf tube. Afterwards, the amount of fluorine within the cell pellets were determined *via* MRI. While the ^{mNP}PFC show strong signals, the ^{con}PFC displays only background signals. Source: Figures (B) and (C), reprinted with permission from Bouvain *et al.* [54].

8.3.2.2 Impact of inflammatory stimuli on CD177-targeting

The functional relevance of CD177 is only poorly understood, but some studies provide evidence that it is involved in enhanced migration [59], and its expression is upregulated in inflammatory bowel and Kawasaki disease [60, 61]. Interestingly, we found that CD177 expression on murine neutrophils is also upregulated after experimental myocardial infarction or subcutaneous implantation of a matrigel plug containing LPS for induction of subcutaneous inflammation [54]. Of note, CD177 was substantially upregulated on circulating neutrophils in the blood, but the highest expression levels were found on neutrophils directly isolated from the inflammatory hot-spot (Fig. 8.3A). Importantly, upregulation of the targeting receptor also strongly enhanced ^{mNP}PFC labelling of neutrophils within the affected tissue (Fig. 8.3B). We also demonstrated that a local stimulus from an implanted Matrigel/LPS plug resulted in increased labeling of neutrophils within their hematopoietic bone marrow niche (Fig. 8.3C). Taken together, this inflammation dependent upregulation of CD177 can strongly enhance the detection limit of neutrophil imaging and may also be exploited to assess the inflammatory state of neutrophils *in vivo*.

8.3.2.3 Imaging of neutrophil trafficking from the bone marrow to the infarcted heart

Inflammatory events trigger the quick and rapid release of neutrophils from the bone marrow into the blood stream from where they enter the affected hot spot. To visualize neutrophil trafficking from the bone marrow into an inflammatory lesion by combined $^1H/^{19}F$ MRI, we pre-labeled neutrophils in the bone marrow by repetitive injections of ^{mNP}PFCs prior to myocardial infarction. Subsequently, whole-body 3D ^1H and ^{19}F MRI before and one day after induction of myocardial infarction revealed neutrophil egress from the bone marrow and infiltration into the infarcted heart (Fig. 8.4A; A left = before MI, right after MI). ^{19}F signal in the femur strongly decreased after MI, while concomitantly ^{19}F signal appeared in the infarcted heart. A more detailed analysis of femur and heart is shown in Fig. 8.4B. Furthermore, we tracked neutrophil infiltration into the infarcted heart in a time-dependent manner, starting at 3 hrs, and up to 3 days after infarction (Fig. 8.4C), thus demonstrating the feasibility of imaging very early migration of neutrophils into the heart, as well as long-term monitoring.

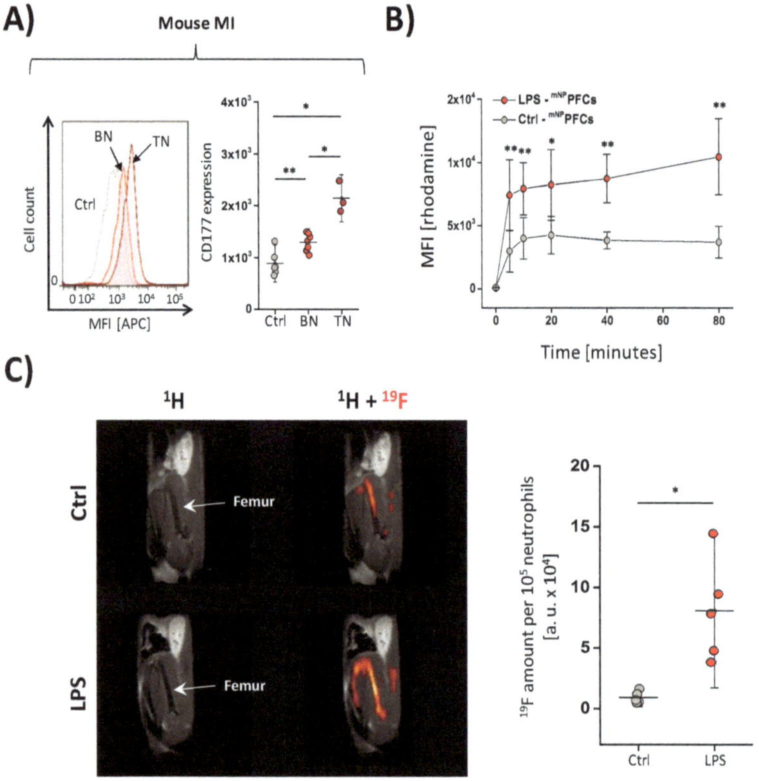

Figure 8.3 Impact of inflammatory stimuli on CD177 targeting. (A) Cell surface expression of CD177 on neutrophil granulocytes after myocardial infarction (MI). Neutrophils were isolated from healthy control mice (Ctrl), as well as from blood and hearts from mice with MI. CD177 expression on blood neutrophils (BN) is increased after MI; the highest expression levels are found on neutrophils obtained from the infarcted tissue (TN). (B) Time course of ^{NP}PFC-uptake by blood-derived neutrophils from untreated control mice (Ctrl) and mice with subcutaneously implanted Matrigel doped with LPS. (C) *In vivo* imaging of bone marrow neutrophils in mice with matrigel/LPS (LPS, lower panel) or control mice (Ctrl, upper panel) after intravenous delivery of ^{mNP}PFCs. Anatomical ^{1}H images (grey) are merged with ^{19}F datasets (hot-iron). Quantification of the ^{19}F labelling of the bone marrow neutrophils is shown on the right. Source: Reprinted with permission from Bouvain *et al.* [54].

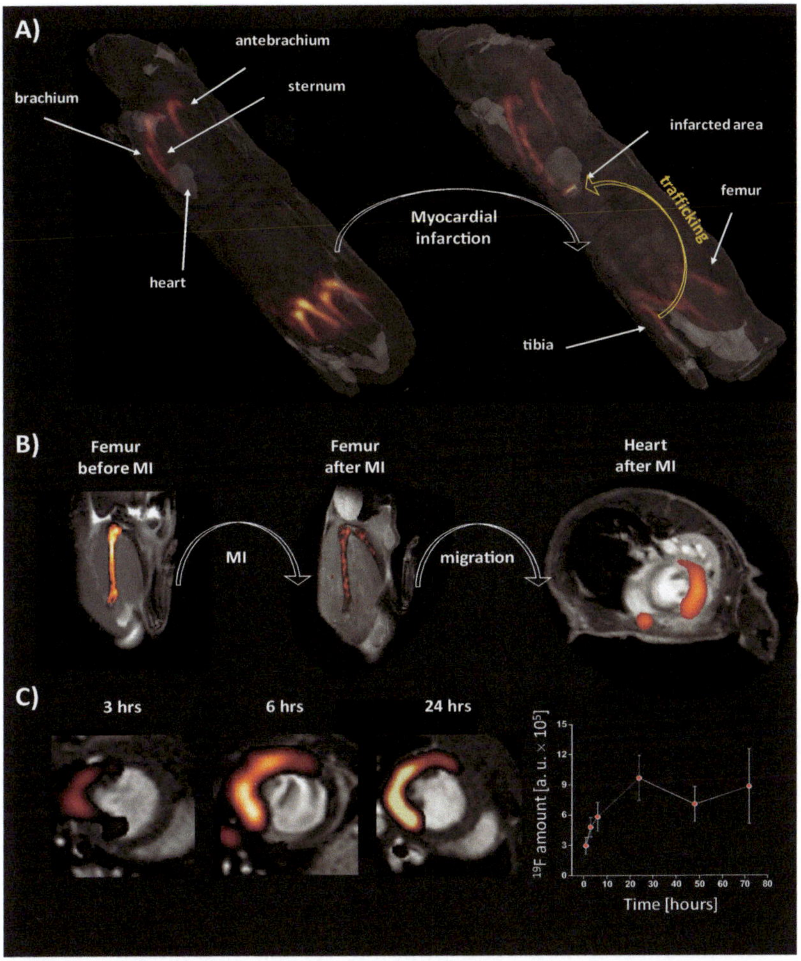

Figure 8.4 *In vivo* $^1H/^{19}F$ MRI of neutrophil trafficking. (A) 3D visualization of a mouse before (left) and after (right) induction of a myocardial infarction. Segmented ^1H datasets are merged with corresponding ^{19}F scans (hot iron). The ^{19}F signal from the liver and spleen is artificially attenuated to enhance the visibility of the bone marrow and the heart. Neutrophils in the bone marrow compartments are strongly labeled with ^{mNP}PFC, and egress from the bone marrow is accompanied by loss of fluorine signal in the bone marrow, especially the femur. Migration of those cells into the heart results in appearance of ^{19}F signal within the infarcted area. (B) Zoom-in shows the decrease in ^{19}F bone marrow signal after MI and the accumulation of the ^{19}F signal in the infarcted myocardium. (C) Time course of neutrophil infiltration into the infarcted heart. Source: Modified and reprinted with permission from Bouvain *et al.* [54].

8.3.2.4 ^{NP}PFC-targeting of CD177 does not alter cell function

A concern with cell tracking is the possibility of alterations in neutrophil functions due to targeting. Neutrophils are very sensitive to various stimuli, and functional modifications by nanoparticle uptake could either result in diminished or enhanced effector functions. This is evident in the targeting of Ly6G, for example, which modulates neutrophil migration into sites of inflammation [62]. In contrast, nanoparticle binding to CD177 does not cause any intracellular signaling and therefore does not induce major effects on cell activation and functionality [63]. Furthermore, we have observed that targeting of neutrophils by ^{NP}PFCs does not alter phagocytosis, degranulation, oxidative burst [54] or NETosis (unpublished). This imaging approach therefore provides an excellent method to follow pre-labeled neutrophils from the bone marrow into the blood and their immigration into inflammatory foci over several days without altering normal functionality.

8.4 Future Perspective

In 2008, the first use of ^{19}F MRI to detect cellular inflammation *in vivo* in the infarcted area of the heart and the brain using intravenously injected PFC. Nowadays, 15 years later, we are at a point where the detection of a specific immune cell subset, such as neutrophils, is possible. In the future, utilization of specific ligands against other cell types in combination with improved sensitivity may enable the imaging of T-cell subtypes (T_H17 or T_{reg}) or monocyte subtypes to further unravel their contribution to certain inflammatory diseases *in vivo*. With the just recently established technology for multi-color, multi-targeted ^{19}F MRI [64], it may be possible to investigate the interplay of several immune-cell subsets and thereby providing additional qualitative information about the local inflammatory status.

Moreover, targeted PFCs may also be combined with therapeutic agents as a theranostic platform for specific modulation of neutrophil functionality and parallel visualization of the consequences with ^{19}F MRI. As described herein, during sterile injury, such as in myocardial infarction or stroke, exaggerated effector functions of neutrophils are presumed to cause tissue damage. Thus, modulation of key neutrophil functions (phagocytosis, degranulation, oxidative burst, NETosis) may improve lesion resolution and the healing process. Cell-specific theranostics could therefore be a promising way to directly visualize the effect of a specific medical treatment on the desired target *in vivo*.

References

1. Phillipson, M., & Kubes, P. The healing power of neutrophils. *Trends Immunol.* **40**, 635–647 (2019).
2. Kay, A. B. Paul Ehrlich and the early history of granulocytes. *Microbiol. Spectr.* **4**, (2016).
3. Borregaard, N. Neutrophils, from marrow to microbes. *Immunity* **33**, 657–670 (2010).
4. Basu, S., Hodgson, G., Katz, M., & Dunn, A. R. Evaluation of role of G-CSF in the production, survival, and release of neutrophils from bone marrow into circulation. *Blood* **100**, 854–861 (2002).
5. Bleul, C. C., Fuhlbrigge, R. C., Casasnovas, J. M., Aiuti, A., & Springer, T. A. A highly efficacious lymphocyte chemoattractant, stromal cell-derived factor 1 (SDF-1). *J. Exp. Med.* **184**, 1101–1109 (1996).
6. Hendrix, C. W., *et al.* Safety, pharmacokinetics, and antiviral activity of AMD3100, a selective CXCR4 receptor inhibitor, in HIV-1 infection. *J. Acquir. Immune Defic. Syndr.* **37**, 1253–1262 (2004).
7. Martin, C., *et al.* Chemokines acting via CXCR2 and CXCR4 control the release of neutrophils from the bone marrow and their return following senescence. *Immunity* **19**, 583–593 (2003).
8. Kim, H. K., De La Luz Sierra, M., Williams, C. K., Gulino, A. V., & Tosato, G. G-CSF down-regulation of CXCR4 expression identified as a mechanism for mobilization of myeloid cells. *Blood* **108**, 812–820 (2006).
9. Ulich, T. R., del Castillo, J., & Souza, L. Kinetics and mechanisms of recombinant human granulocyte-colony stimulating factor-induced neutrophilia. *Am. J. Pathol.* **133**, 630–638 (1988).
10. Woodfin, A., *et al.* The junctional adhesion molecule JAM-C regulates polarized transendothelial migration of neutrophils *in vivo*. *Nat. Immunol.* **12**, 761–769 (2011).
11. Vols, S., Sionov, R. V., & Granot, Z. Always look on the bright side: anti-tumor functions of neutrophils. *Curr. Pharm. Des.* **23**, 4862–4892 (2017).

12. Häger, M., Cowland, J. B., & Borregaard, N. Neutrophil granules in health and disease. *J. Intern. Med.* **268**, 25–34 (2010).
13. Aratani, Y. Myeloperoxidase: its role for host defense, inflammation, and neutrophil function. *Arch. Biochem. Biophys.* **640**, 47–52 (2018).
14. Laforge, M., *et al.* Tissue damage from neutrophil-induced oxidative stress in COVID-19. *Nat. Rev. Immunol.* **20**, 515–516 (2020).
15. Brinkmann, V., *et al.* Neutrophil extracellular traps kill bacteria. *Science* **303**, 1532–1535 (2004).
16. Burgener, S. S., & Schroder, K. Neutrophil extracellular traps in host defense. *Cold Spring Harb. Perspect. Biol.* **12**, a037028 (2020).
17. Saffarzadeh, M., *et al.* Neutrophil extracellular traps directly induce epithelial and endothelial cell death: a predominant role of histones. *PloS One* **7**, e32366 (2012).
18. Xu, J., *et al.* Extracellular histones are major mediators of death in sepsis. *Nat. Med.* **15**, 1318–1321 (2009).
19. Silvestre-Roig, C., *et al.* Externalized histone H4 orchestrates chronic inflammation by inducing lytic cell death. *Nature* **569**, 236–240 (2019).
20. Wong, S. L., *et al.* Diabetes primes neutrophils to undergo NETosis, which impairs wound healing. *Nat. Med.* **21**, 815–819 (2015).
21. Peiseler, M., & Kubes, P. More friend than foe: the emerging role of neutrophils in tissue repair. https://www.jci.org/articles/view/124616/pdf (2019) doi:10.1172/JCI124616.
22. Szturmowicz, M., & Demkow, U. Neutrophil extracellular traps (NETs) in severe SARS-CoV-2 lung disease. *Int. J. Mol. Sci.* **22**, 8854 (2021).
23. Park, J.-S., *et al.* Importance of prognostic value of neutrophil to lymphocyte ratio in patients with ST-elevation myocardial infarction. *Medicine (Baltimore)* **97**, e13471 (2018).
24. Yellon, D. M., & Hausenloy, D. J. Myocardial reperfusion injury. *N. Engl. J. Med.* **357**, 1121–1135 (2007).
25. Westman, P. C., *et al.* Inflammation as a driver of adverse left ventricular remodeling after acute myocardial infarction. *J. Am. Coll. Cardiol.* **67**, 2050–2060 (2016).
26. Ong, S.-B., *et al.* Inflammation following acute myocardial infarction: multiple players, dynamic roles, and novel therapeutic opportunities. *Pharmacol. Ther.* **186**, 73–87 (2018).
27. Ge, L., *et al.* Neutrophil extracellular traps in ischemia-reperfusion injury-induced myocardial no-reflow: therapeutic potential of DNase-based reperfusion strategy. *APSselect* **2**, H500–H509 (2015).

28. Borissoff, J. I., *et al.* Elevated levels of circulating DNA and chromatin are independently associated with severe coronary atherosclerosis and a prothrombotic state. *Arterioscler. Thromb. Vasc. Biol.* **33**, 2032–2040 (2013).

29. Weiss, J. N., Goldhaber, J. I., & Ji, S. Oxygen free radicals in the pathophysiology of myocardial ischemia/reperfusion, in *Oxygen Free Radicals in Tissue Damage* (Tarr, M., & Samson, F., eds), pp. 250–266 (Birkhäuser, 1993). doi: 10.1007/978-1-4615-9840-4_13.

30. Chia, S., *et al.* Association of leukocyte and neutrophil counts with infarct size, left ventricular function and outcomes after percutaneous coronary intervention for ST-elevation myocardial infarction. *Am. J. Cardiol.* **103**, 333–337 (2009).

31. Guasti, L., *et al.* Neutrophils and clinical outcomes in patients with acute coronary syndromes and/or cardiac revascularisation. *Thromb. Haemost.* **106**, 591–599 (2011).

32. Horckmans, M., *et al.* Neutrophils orchestrate post-myocardial infarction healing by polarizing macrophages towards a reparative phenotype. *Eur. Heart J.* (2016) doi:10.1093/eurheartj/ehw002.

33. Yipp, B. G., & Kubes, P. Antibodies against neutrophil LY6G do not inhibit leukocyte recruitment in mice in vivo. *Blood* **121**, 241–242 (2013).

34. Gross, S., *et al.* Bioluminescence imaging of myeloperoxidase activity in vivo. *Nat. Med.* **15**, 455–461 (2009).

35. Hasenberg, A., *et al.* Catchup: a mouse model for imaging-based tracking and modulation of neutrophil granulocytes. *Nat. Methods* **12**, 445–452 (2015).

36. Signore, A., *et al.* Clinical indications, image acquisition and data interpretation for white blood cells and anti-granulocyte monoclonal antibody scintigraphy: an EANM procedural guideline. *Eur. J. Nucl. Med. Mol. Imaging* **45**, 1816–1831 (2018).

37. Eastlund, D. T., & Charbonneau, T. T. Superoxide generation and cytotactic response of irradiated neutrophils. *Transfusion (Paris)* **28**, 368–370 (1988).

38. Roca, M., de Vries, E. F. J., Jamar, F., Israel, O., & Signore, A. Guidelines for the labelling of leucocytes with (111)In-oxine. Inflammation/Infection Taskgroup of the European Association of Nuclear Medicine. *Eur. J. Nucl. Med. Mol. Imaging* **37**, 835–841 (2010).

39. Weiner, R., Hoffer, P. B., & Thakur, M. L. Lactoferrin: its role as a Ga-67-binding protein in polymorphonuclear leukocytes. *J. Nucl. Med. Off. Publ. Soc. Nucl. Med.* **22**, 32–37 (1981).

40. Tsan, M. F. Mechanism of gallium-67 accumulation in inflammatory lesions. *J. Nucl. Med. Off. Publ. Soc. Nucl. Med.* **26**, 88–92 (1985).
41. Segard, T., *et al.* Comparison between gallium-68 citrate positron emission tomography-computed tomography and gallium-67 citrate scintigraphy for infection imaging. *Intern. Med. J.* **49**, 1016–1022 (2019).
42. Front, D., & Israel, O. Present state and future role of gallium-67 scintigraphy in lymphoma. *J. Nucl. Med. Off. Publ. Soc. Nucl. Med.* **37**, 530–532 (1996).
43. Thackeray, J. T., *et al.* Molecular imaging of the chemokine receptor CXCR4 after acute myocardial infarction. *JACC Cardiovasc. Imaging* **8**, 1417–1426 (2015).
44. Hess, A., *et al.* Molecular imaging-guided repair after acute myocardial infarction by targeting the chemokine receptor CXCR4. *Eur. Heart J.* **41**, 3564–3575 (2020).
45. Wester, H. J. *et al.* Disclosing the CXCR4 expression in lymphoproliferative diseases by targeted molecular imaging. *Theranostics* **5**, 618–630 (2015).
46. Jung, K., *et al.* Targeting CXCR4-dependent immunosuppressive Ly6Clow monocytes improves antiangiogenic therapy in colorectal cancer. *Proc. Natl. Acad. Sci. U. S. A.* **114**, 10455–10460 (2017).
47. Bouvain, P., Temme, S., & Flögel, U. Hot spot 19 F magnetic resonance imaging of inflammation. *Wiley Interdiscip. Rev. Nanomed. Nanobiotechnol.* e1639 (2020) doi: 10.1002/wnan.1639.
48. Srinivas, M., Boehm-Sturm, P., Figdor, C. G., de Vries, I. J., & Hoehn, M. Labeling cells for in vivo tracking using 19F MRI. *Biomaterials* **33**, 8830–8840 (2012).
49. Flögel, U., *et al.* In vivo monitoring of inflammation after cardiac and cerebral ischemia by fluorine magnetic resonance imaging. *Circulation* **118**, 140–148 (2008).
50. Ebner Bernd, *et al.* Early assessment of pulmonary inflammation by ^{19}F MRI in vivo. *Circ. Cardiovasc. Imaging* **3**, 202–210 (2010).
51. Flögel, U., *et al.* Selective activation of adenosine A2A receptors on immune cells by a CD73-dependent prodrug suppresses joint inflammation in experimental rheumatoid arthritis. *Sci. Transl. Med.* **4**, 146ra108-146ra108 (2012).
52. Temme, S., *et al.* Technical advance: monitoring the trafficking of neutrophil granulocytes and monocytes during the course of tissue inflammation by noninvasive ^{19}F MRI. *J. Leukoc. Biol.* **95**, 689–697 (2014).

53. Mazzucchelli, L., *et al.* Cell-specific peptide binding by human neutrophils. *Blood* **93**, 1738–1748 (1999).

54. Bouvain, P., *et al.* Non-invasive mapping of systemic neutrophil dynamics upon cardiovascular injury. *Nat. Cardiovasc. Res.* 1–18 (2023) doi: 10.1038/s44161-022-00210-w.

55. Miettinen, H. M., Gripentrog, J. M., Lord, C. I., & Nagy, J. O. CD177-mediated nanoparticle targeting of human and mouse neutrophils. *PLOS ONE* **13**, e0200444 (2018).

56. Temme, S., *et al.* Noninvasive imaging of early venous thrombosis by ^{19}F magnetic resonance imaging with targeted perfluorocarbon nanoemulsions. *Circulation* **131**, 1405–1414 (2015).

57. Grapentin, C., *et al.* Optimization of perfluorocarbon nanoemulsions for molecular imaging by ^{19}F MRI, in *Nanomedicine* (Seifalin, A., de Mel, A., & Kalaskar, D. M., eds), One Central Press (2014), pp. 268–286.

58. Bouvain, P., *et al.* Dissociation of ^{19}F and fluorescence signal upon cellular uptake of dual-contrast perfluorocarbon nanoemulsions. *Magn. Reson. Mater. Phys. Biol. Med.* **32**, 133–145 (2019).

59. Bai, M., *et al.* CD177 modulates human neutrophil migration through activation-mediated integrin and chemoreceptor regulation. *Blood* **130**, 2092–2100 (2017).

60. Zhou, G., *et al.* CD177+ neutrophils as functionally activated neutrophils negatively regulate IBD. *Gut* **67**, 1052–1063 (2018).

61. Huang, Y.-H., Lo, M.-H., Cai, X.-Y., Liu, S.-F., & Kuo, H.-C. Increase expression of CD177 in Kawasaki disease. *Pediatr. Rheumatol. Online J.* **17**, 13 (2019).

62. Wang, J.-X., *et al.* Ly6G ligation blocks recruitment of neutrophils via a β2-integrin–dependent mechanism. *Blood* **120**, 1489–1498 (2012).

63. Goldschmeding, R., *et al.* Further characterization of the NB 1 antigen as a variably expressed 56–62 kD GPI-linked glycoprotein of plasma membranes and specific granules of neutrophils. *Br. J. Haematol.* **81**, 336–345 (1992).

64. Flögel, U., *et al.* Multi-targeted 1H/19F MRI unmasks specific danger patterns for emerging cardiovascular disorders. *Nat. Commun.* **12**, 5847 (2021).

Chapter 9

^{19}F MRI to Map the Spatio-Temporal Infiltration of Immune Cells in Stroke and Tissue Regeneration

Michel Modo[a,b,c] **and T. Kevin Hitchens**[d]
[a]*McGowan Institute for Regenerative Medicine, University of Pittsburgh, Pittsburgh, Pennsylvania, USA*
[b]*Department of Radiology, University of Pittsburgh, Pittsburgh, Pennsylvania, USA*
[c]*Department of Bioengineering, University of Pittsburgh, Pittsburgh, Pennsylvania, USA*
[d]*Department of Neurobiology, University of Pittsburgh, Pittsburgh, Pennsylvania, USA*
mmm154@pitt.edu

9.1 Stroke and Tissue Regeneration

Stroke remains a devastating neurological condition that affects almost 800,000 patients every year in the US. After stroke damage has occurred, neurorehabilitation is currently the only approved treatment for improving behavioral impairments. Although cell therapy continues to be investigated in clinical trials as a new

Fluorine Magnetic Resonance Imaging: Methods and Applications in Biomedicine
Edited by Eric T. Ahrens and Ulrich Flögel
Copyright © 2025 Jenny Stanford Publishing Pte. Ltd.
ISBN 978-981-5129-37-3 (Hardcover), 978-1-003-53004-6 (eBook)
www.jennystanford.com

treatment paradigm for stroke [1–3], it has yet to receive regulatory approval. In addition, none of these therapies replace the substantial amount of brain tissue that can be lost due to a stroke, with a tissue cavity found in 94% of stroke patients [4]. This dramatic loss of brain tissue is a major obstacle to achieve a more complete behavioral recovery and currently remains an unaddressed clinical issue.

Implantation of neural stem cells (NSCs) replaces lost cells in peri-infarct damaged tissue, but does not replace lost brain tissue [1, 5]. We hypothesize that the lack of structural support in the fluid-filled cavity, due to a loss of extracellular matrix (ECM), is the main biophysical determinant in the failure of the brain to regenerate lost tissue [6]. We have previously demonstrated that NSCs implanted on poly-lactic-glycolic acid (PLGA) microparticles [7, 8], as well as incorporated in polyethylene glycol (PEG) microspheres [9], can create a primitive *de novo* tissue in this cavity. However, these approaches are dependent on xenogenic or allogenic cells and it would be advantageous to engineer solutions that would support the brain to "heal" itself.

In other non-regenerative organs, such as the bladder and muscle, supplying ECM formulated as a bioscaffold exerts inductive properties that attract host cells into the tissue defect, while remodeling the bioscaffold and replacing it with new site-appropriate tissues [10–12]. By applying these principles to the brain [13, 14], we have been able to demonstrate that ECM-based hydrogel can be implanted into a tissue cavity caused by stroke to adapt to its unique topology, while inducing brain tissue regeneration [15–18].

9.2 The Role of Immune Cells in Tissue Regeneration

Any type of brain insult, such as a stroke or traumatic brain injury, causing cellular death will invoke a secondary inflammatory response driven by the immune system [19]. Microglia, brain-resident monocytes, as well as astrocytes activate and secrete cytokines that define a pro-inflammatory space to recruit bone marrow-derived, peripheral immune cells into the area of damage. A cascade of peripheral inflammatory cells, including neutrophils, monocytes/macrophages and subsets of T cells will invade the damaged region. Secretion of matrix metalloproteinases (MMPs) degrades the ECM that was supporting brain cells and blood vessels in the affected area. M1-like macrophages phagocytose the cleaved ECM proteins as well as cellular debris. However, the precise process and timescale of tissue cavitation remain poorly understood. Meanwhile, reactive astrocytes proliferate at the edge of viable and dying tissue to create a scar that seals off the area of cavitation. Glial scaring prevents cytotoxic fluids to reach surviving neurons and exacerbate the impact of the brain insult. By 2–3 weeks post-insult, this M1-like pro-inflammatory response gradually switches towards a M2-like pro-repair response [19]. Bioscaffold implantation needs to be considered in the context of this evolving backdrop of the damage-induced inflammatory response [6].

A bioscaffold implantation requires a tissue cavity that forms between 10–21 days post stroke and fills the voided area of damage to create an interface with host tissue [18]. However, the decrease in MMPs in the tissue cavity, as well as the maturation of the glial scar, reduce scaffold biodegradation and hinder the permeation of inductive signals into host tissue. Despite a pro-repair inflammatory response two weeks post insult, a delayed implantation of bioscaffolds can fail to induce an efficient tissue regeneration cascade due to an unfavorable inflammatory backdrop. Bioscaffold-induced tissue regeneration is highly dependent on an immune response that drives the biodegradation of the ECM hydrogel to release inductive signals, as well as structurally remodeling the scaffold to create space for host cells to invade and differentiate. As in other organs [12], *de novo* formation of tissue within the stroke cavity is dependent on hydrogel biodegradation [17]. A lack of biodegradation fails to induce a tissue replacement [20].

The infiltration of immune cells into a bioscaffold is therefore a critical determinant of the clinical outcome [21, 22]. In rodent models, implantation of ECM hydrogel within a stroke-induced tissue cavity in the brain invokes a substantial macrophage infiltration within 24 hours [15, 16]. The density (*i.e.*, concentration) of the bioscaffold hydrogel also affects the *in vivo* degradation properties. Stiffer (8 mg/mL) hydrogels tend to degrade more slowly with few macrophages retained long-term [20], whereas less stiff (4 mg/mL) hydrogels are consistently broken down by 90 days [17]. The infiltration of macrophages is the putative mode of action of ECM hydrogel to induce tissue restoration [14, 23]. Although macrophage depletion is known to prevent hydrogel biodegradation and the subsequent tissue replacement, other immune cells, such as neutrophils, have also been reported to acutely and transiently infiltrate bioscaffolds, potentially invoking and modulating the macrophage response [24]. Very little is known about the spatio-temporal dynamics of this immune cell trafficking during the regenerative process in the brain. To date, studies have relied on tedious histological time course studies [17, 18], with considerable inter-subject variability, which does not readily capture this highly dynamic process, nor does it allow the visualization of immune cell trafficking routes from peri-infarct tissues into the bioscaffold.

9.3 MR Imaging of Immune Cells

Longitudinal non-invasive imaging of immune cell infiltration to bioscaffolds would significantly improve our understanding of the roles these cells play in scaffold degradation. Translocator protein (TSPO) or PK11195 radioligands have a high specificity to image immune cells using positron emission tomography (PET) [25], but PET imaging lacks sufficient spatial resolution to resolve the tissue-bioscaffold interface [26, 27]. Optical imaging suffers from limited depth penetration in addition to scattering through the optically opaque skull. MRI is therefore advantageous, as it provides an adequate spatial resolution, as well as the ability to track immune cells *in vivo* [25, 28]. By exploiting the phagocytic activity of blood-circulating immune cells, such as peripheral macrophages, contrast agents or tracer materials can be incorporated into cells *in vivo* to non-invasively visualize their trafficking and distribution by MRI [29–31]. Cellular MRI has been used to visualize macrophage accumulation in graft rejection [32–34], as well as stroke in preclinical models [35–40] and translational settings [41–43]. Superparamagnetic iron oxide (SPIO) particles have been used for macrophage tracking due to the high sensitivity [32, 33, 44], which in some cases being able to detect single cells [45, 46]. However, the negative image contrast produced by SPIO can interfere with T_2-weighted analyses of lesioned and peri-infarct tissue volumes and therefore is not ideal to study immune cell infiltration in the context of bioscaffolds.

Alternatively, systemic perfluorocarbon (PFC) labeling of immune cells can visualize *in vivo* cell trafficking using ^{19}F MRI without interfering with anatomical (^{1}H) imaging [31, 34, 47, 48]. Although the sensitivity of directly detecting low concentrations of intracellular PFC by magnetic resonance can be significantly lower than visualization of SPIO-labeled macrophages by T2*-weighted MRI of tissue water, the ^{19}F signal is more readily quantifiable due to the unique magnetic resonance signal, with no physiological background [36]. Because ^{19}F has a different resonance frequency than ^{1}H, a dual resonance RF coil can be used to collect ^{19}F MRI for labeled cells and ^{1}H MRI for anatomy in the same imaging session, placing ^{19}F signals in anatomical context [43]. This "hot spot" imaging technique [49] was first reported over 30 years ago [50] and is increasingly used to investigate immune cell trafficking [27].

Phagocytic macrophages readily incorporate PFC-nanoemulsions *in vivo*, although cellular uptake is non-specific and some reports have also indicated some uptake by neutrophils and lymphocytes [51–54]. Incorporation into circulating immune cells is likely dependent on the immunological stimulus, inducing an infiltration cascade of different immune cell populations [55]. With direct intravenous injection, PFC nanoemulsion is found largely in the liver (Fig. 9.1A), indicating initial body clearance through the reticuloendothelial system (RES), with subsequent excretion *via* lung exhalation. Within the liver, PFC nanoemulsions are found in tissue-resident macrophage known as Kupffer cells (Fig. 9.1B). PFC nanoemulsion is also highly concentrated in the spleen which harbors monocytes that circulate through the blood prior to differentiation into macrophages as they invaded damaged tissues. This approach has been used to visualize the distribution and density of macrophages in a number of animal models [30, 31, 34, 36, 37, 56–59]. *In vivo* labeling of immune cells with PFC nanoemulsions therefore provides a unique opportunity to investigate immune cell trafficking into ECM.

In the context of regenerative medicine, a transient infiltration of neutrophils has been observed to precede and accompany the infiltration of macrophages [6, 21, 22, 60, 61]. A smaller population of lymphocytes has also been suggested to transiently infiltrate, but only macrophages are expected to persist beyond 7 days after implantation and correlate with the degree of hydrogel degradation

[17]. Still, a systematic spatio-temporal mapping of immune cell activity to implanted ECM hydrogel in the brain remains to be established. Although ^{19}F MRI has been used widely to visualize macrophages in peripheral tissues, only a few studies demonstrated the utility of this approach in the brain [36, 62–67]. The more limited invasion of circulating immune cells into the brain may pose challenges for the detection of PFC labeled cell clusters [66].

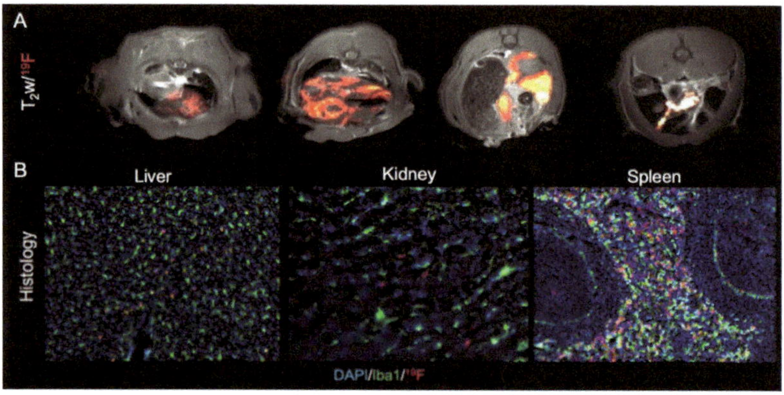

Figure 9.1 Intra-venous perfluorocarbon (PFC) labeling of immune cells. (A) An *ex vivo* whole-body scan revealed a tissue biodistribution of PFC in the liver, spleen, and colon, but not in the kidneys. The spleen is the source of blood-circulating monocytes which are a target of systemic PFC injections, whereas its localization to the liver and colon indicates its clearance path through the reticuloendothelial system (RES). (B) Immunohistochemistry of these organs further confirmed the presence of dual-mode (DM) PFC, using the red fluorescent moiety contained within the nanoemulsion.

9.4 Establishing ^{19}F MR Imaging to Visualize Perfluorocarbon (PFC) Nanoemulsion-Labeled Immune Cells

An individual perfluorocarbon (PFC) nanomulsion droplet (165 nm diameter = 2.35×10^{-12} mm^3 volume) is $\sim 10^8$–10^{10} smaller in volume than typical voxel volumes used in preclinical MRI ($\sim$0.01–1 mm^3). PFC nanoemulsions are not homogenously distributed throughout a tissue but are contained within cells. The PFC content and ^{19}F MRI signal of a voxel is hence determined by the number of labeled cells within a voxel (Fig. 9.2A). The signal is also dependent upon the PFC molecule, emulsion concentration and *in vivo* labeling efficiency of nanoemlusions. Molecules with a large number of chemically equivalent fluorine atoms (*i.e.*, having the same NMR chemical shift), such perfluor-15-crown-5-ether (PCE), with 20 chemically-equivalent fluorine atoms, or a perfluoropolyether (PFPE) with repeating $(-OCF_2FC_2-)_n$ subunits having 28–36 equivalent fluorines, help to amplify the ^{19}F NMR signal and detect labeled cells *in vivo* by ^{19}F MRI. Detailed reviews and discussions on different PFC nanoemulsions can be found here [68–70].

A variety of MRI sequences can be used to measure the ^{19}F signal. The most commonly used sequences consist of a gradient echo (Fast Low Angle SHot , or FLASH sequence), a spin-echo, (Multi-Slice Multi-Echo, or MSME), a Rapid Acquisition with Relaxation Enhancement (RARE), also called a Fast or Turbo Spin-Echo (FSE/TSE), a Zero Echo Time (ZTE), and a Fast Imaging with Steady-state Precession (FISP) sequence [65, 71–76]. To determine which sequence provides the most efficient acquisition strategy at a high magnetic field (*e.g.*, 9.4 T) used for animal imaging, a comparison of sequences can be performed with varying scan parameters, ideally using a Design-of-Experiment (DOE) approach [77]. A log2 series of PFPE nanoemulsion phantoms with fluorine atom concentrations of: 0, 0.02, 0.2, 2, 10, 20, 100, 200 mM, were used to calculate a limit-of-detection (LOD) for each sequence under identical conditions (Fig. 9.2B). At 9.4 T, using a 96 × 96 matrix size (0.42 mm in plane resolution) with a 2 mm slice thickness (0.35 mm^3 voxel) and a 60 min scanning time, the ^{19}F signal for each PFC concentration was measured to calculate an LOD for each sequence using a linear regression after a log2 transformation (Fig. 9.2C). An LOD of 5.46 mM was calculated for MSME, for 3.84 mM FLASH, 3.77 mM for ZTE, 2.85 mM for RARE, and 0.43 mM for FISP [77]. Others reported LODs for RARE (TSE) and FISP (Balance Steady-State Precession) at 3 T were previously estimated to be 40.8 mM and 23.5 mM ^{19}F atoms for PFPE [75]. The lower LOD compared to Ghuman *et al.* [77] can be explained by the lower 3T magnetic field and reduced signal averaging (10 min versus a 60 min acquisition time), but they are consistent in that FISP followed by RARE are the most efficient sequences to acquire a PFC signal. Still, a 2D FISP is only favorable

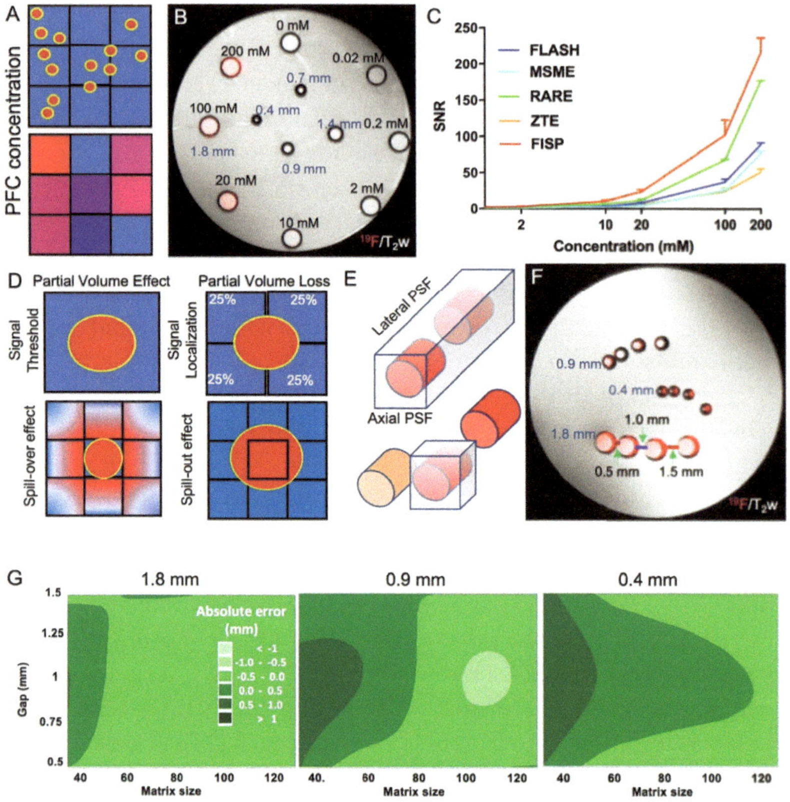

Figure 9.2 ^{19}F MRI limits of detection for perfluorocarbon nanoemulsions. (A) The ^{19}F MR signal of an individual voxel is a consequence of PFC nanoemulsion content. In the case of cell tracking, nanoemulsion is contained within cells. The number of cells within a voxel is hence related to its signal intensity. (B) ^{19}F/^{1}H MRI overlay of phantom used to measure the limit of detection (LOD) of PFC nanoemulsion. Here the PFC nanoemulsion was a commercially available (Celsense, Pittsburgh, PA) 20% v/v emulsion of perfluoropolyether (PFPE) with a major resonance at –91.5 ppm, corresponding to ~42 chemically equivalent ^{19}F atoms at a stock concentration of 10.92 M (^{19}F). (C) A comparison of MRI sequences using signal-to-noise revealed FISP to be the most efficient acquisition paradigm with a LOD of 0.65 mM of PFC in a voxel of 0.2 mm^3 [77]. (D) Localization of cells can be affected by partial volume effects (PVE). Although individual cells are much smaller than a voxel, PVEs can occur if only a small portion of a large voxel contains a PFC-positive cell cluster, or that this cluster is located at the intersection of multiple voxels and their signal is hence distributed across a larger area than its actual location. Spill-over and spill-out effects can also occur with very dense localize cell clusters. (E) Considerations of the point spread function (PSF) is hence required in 3 dimensions to ensure that cell clusters are appropriately located within their anatomical context. (F) PVEs can be thoroughly investigated in phantom experiments that control ^{19}F "hot spot" size

and distance from each other. (G) The separation of large objects/cell clusters (1.8 mm in diameter) can be reliably achieved using large voxel sizes (1.25 mm in plane), whereas smaller objects (0.4 mm diameter) required a smaller voxel sizes (<0.31 mm in plane) [77]. The image dimensions (*i.e.*, voxel/matrix size) hence need to be defined based on the size of 3D cell clusters to be localized within their anatomical context. (H) Within an image volume of bioscaffold-implanted stroke brain, only a small proportion (~2.5%) of voxels will contain PFC-labeled cells. (I) Although there is no ^{19}F signal in the brain, the low signal intensity of some voxels containing PFC-labeled cells requires the definition of a threshold at which a voxel is classified to be noise or containing a ^{19}F signal. (J) By balancing Type I (5%) and II (20%) error rates, a signal intensity at 35% higher than noise can be used to define ^{19}F positive voxels, with an SNR of 2 being a more conservative threshold that completely minimize the inclusion of voxels containing noise.

for a small number of slices, as the total acquisition time increases with the number of slices. FISP is also prone to banding and chemical shift artifacts, which can affect image quality, measurements and interpretations [78]. RARE will be favorable for a greater number of slices, as multiple slices are acquired during a single recycle time (TR).

One challenge with ^{19}F detection of PFC labeled cells is the relatively low ^{19}F concentration in tissue, $\sim 10^4$ lower than the water-based ^{1}H concentration (~80 M) [79]. Because of this, ^{19}F MRI is generally collected at a lower resolution than anatomical images to pool the signal from labeled cells in larger voxels, reflecting hot spots superimposed on brain anatomy [63]. Considering the larger voxel sizes, partial volume effects (PVE) need to be considered [80]. PVE can occur if insufficient PFC is present within a large voxel (*i.e.*, signal threshold effects), but a cluster of cells could also be distributed over many voxels, diluting out the ^{19}F signal (Fig. 9.2D). These PVE can occur in three dimensions; ideally one should employ isotropic voxels to accounting for the axial and lateral point spread function (PSF) of the ^{19}F (Fig. 9.2E). Phantom experiments that have an array of hot spot characteristics, such as size and separation, can be used to measure how the PVE affect impacts the identification of PFC hot spots in ^{19}F MR images (Fig. 9.2F). Voxel geometry and volume can hence be titrated to minimize PVE for a given imaging application. For instance, large cell clusters (1.8 mm) in size can be identified and separated using large voxel sizes, whereas smaller clusters (0.4 mm) require a higher voxel resolution (Fig. 9.2G). Cell clusters of a few hundred immune cells can be visualized using a 96 × 96 matrix size (0.42 mm resolution), which is double the in-plane voxel size (0.208 mm) of T_2-weighted images typically used to visualize the stroke cavity in rat models. With anatomical MRI, it is not uncommon for slice thickness to be at least 2–3× the in-plane resolution as this is sufficient to sample the coronal anatomy of the brain [20, 81, 82]. To ensure an appropriate contextualization and interpretation of ^{19}F hot spot images, PVE should be minimized in the context of the observed anatomy.

As there are no endogenous fluorocarbon molecules in the brain, the ^{19}F MR signal is specific for the injected PFC nanoemulsion; uptake by macrophages through phagocytosis offers a unique biomarker to visualize their participation in inflammatory events, such as the biodegradation of an ECM hydrogel. However, the relatively small number of ^{19}F atoms present in tissues limit the detection of labeled cell infiltrates (Fig. 9.3A). The detection threshold defines a value that affords signal to be accurately and reliably distinguished from background noise [83]. The noise component in ^{19}F MR images is primarily thermal (Johnson-Nyquist) noise produced by the subject, and in magnitude images, follows a Rician distribution [84]. Due to the lack of a normal distribution, the use of the standard deviation of noise as a divisor is not ideal for SNR calculations in the low signal regime (Fig. 9.3B). A scaling factor cannot appropriately correct for the nature of the noise distribution and the large disparity in voxels containing signal (<3%) versus noise (65%), which affects the Kurtosis of the distribution curves. Dividing mean signal by mean noise hence calculates an easily interpretable SNR. Setting of the detection threshold is governed by minimizing type I (*i.e.*, false positives, typically set at 5%) and II errors (*i.e.*, false negatives, typically set at 80%). The intersection point between both distributions was at 52% above the mean of the noise intensity, indicating that thresholds of 25%, or 1 standard deviation (37%) above mean noise, are inappropriate as detection thresholds (Fig. 9.3C). An SNR = 2 defined a conservative detection threshold with an acceptable type 1 error (<1%) and type II error (16.7%)

[77]. A comparison of MR pulse sequences of a brain phantom with PFC-labeled immune cells invading an ECM bioscaffold applying this detection threshold supported the conclusion that FISP provided the most robust detection of cell clusters, followed by RARE (Fig. 9.3D). Quantitation of these results indicates that FISP achieves an SNR almost 4× higher than the RARE sequence (Fig. 9.3E).

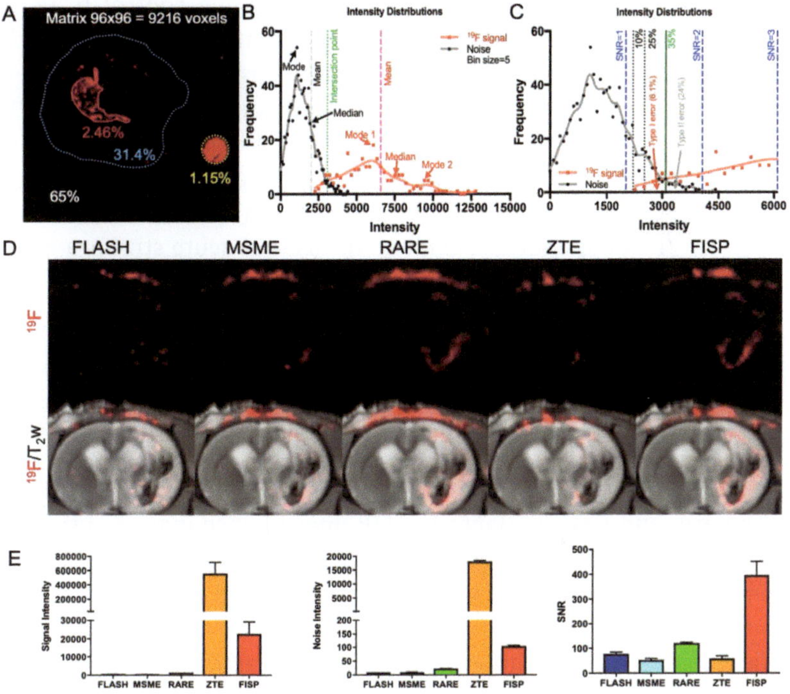

Figure 9.3 Selection of an MRI sequence to maximize ^{19}F signal detection. (A) Using optimized parameters for each sequence affords a comparison of these on an *ex vivo* head sample where systemically tagged PFC-immune cells invaded a bioscaffold implanted into a stroke cavity [77]. (B) The number of voxel for each ^{19}F signal intensity mapped according to their location (noise versus ECM hydrogel) with identification of key descriptive statistics. (C) Consideration of different detection threshold that are aimed at separating noise from the ^{19}F signal. (D) Comparison of ^{19}F MR sequences using an *ex vivo* head phantom. A rat with stroke damaged was injected with PFC nanoemlusions 48 hours prior to ECM hydrogel implantation. The animal was perfusion-fixed 24 hours after ECM implantation. (E) Measurement of signal and noise revealed that the FISP sequence achieved the highest signal-to-noise ratio and hence provides the most efficient method to visualize the distribution of PFC-labeled immune cells.

9.5 Visualizing the Spatio-Temporal Characteristics of Immune Cells in Bioscaffold-Induced Tissue Regeneration after a Stroke

^{19}F MRI can detect the presence of PFC nanoemulsions post-injection in circulating blood. This constitutes a positive control to verify PFC nanoemulsion delivery and can be quantified to determine the contrast material available for uptake by circulating immune cells over time. In rats with stroke damage, blood vessels containing PFC nanoemulsions produce a robust baseline ^{19}F signal (Fig. 9.4A) [54]. Brain tissue, 2 weeks post-stroke, only registered a weak ^{19}F signal due to PFC nanoemulsions circulating through the neurovasculature, but no ^{19}F hot spots were evident. In cases of acute stroke, a major time and location dependent invasion of immune cells to site of damage has been observed using ^{19}F MRI [62].

Implantation of an ECM hydrogel, based on anatomical MRI guidance, produced a unique ^{19}F MR signal increase in the peri-infarct region, as well as the bioscaffold (Fig. 9.4A). A major ^{19}F signal was also evident in the skin overlying the burr hole drilled for ECM injection, providing a positive control for immune cell accumulation. The pre- and post-implantation MR image hence indicated that the ^{19}F signal change in the skull wound and the lesion cavity are caused by the implantation of ECM hydrogel. A high-resolution 3D MR image reveals a more detailed anatomical distribution of infiltrating immune cells. In addition to a peri-infarct infiltration of immune cells, a strong ^{19}F signal was evident along stump of the middle cerebral artery, which was occluded to induce the stroke, as well as at the site of the burr hole where remnant ECM hydrogel created a bridge into the cavity-filling bioscaffold (Fig. 9.4B). Importantly, the route of immune cell invasion were previously not identified in histological studies [15–18, 20], as the brain is typically excised from the skull and perfusion-fixation clears blood vessels, such as the MCA. The isotropic and higher resolution of these *ex vivo* ^{19}F MR images, as well as increased SNR, also provided a more nuanced visualization of immune cell distribution within the ECM. Especially boundary regions interfacing with host tissue were more densely colonized than areas at the center of the bioscaffold. This outside-in migration of immune cells from host tissue into the bioscaffold is typical of inductive ECM hydrogels [16]. A verification of the ^{19}F MR-based distribution of immune cells was achieved histologically using the red fluorescent moiety contained within the PFC nanoemulsions (Fig. 9.4C). Macroscopically, the densities of hot spots on ^{19}F MRI corresponded to high density PFC nanoemulsions areas, which also reflected macrophage density.

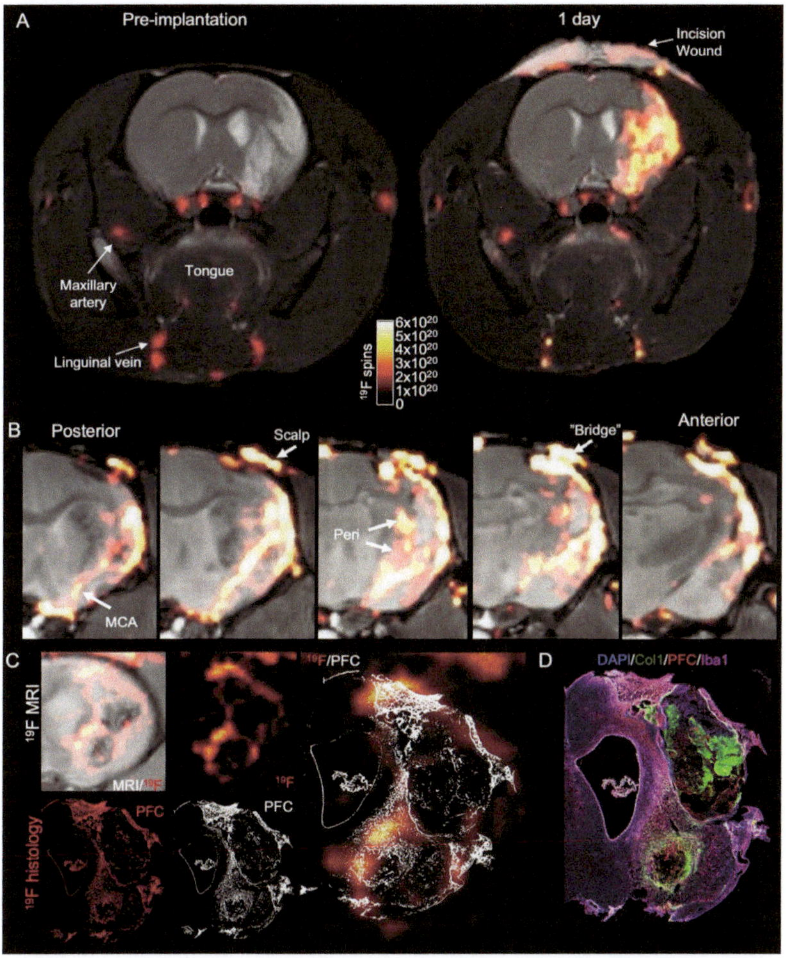

Figure 9.4 *In vivo* and *ex vivo* ^{19}F imaging of immune cell infiltration and migration after ECM implantation. (A) *In vivo* MRI afford the visualization of ^{19}F signal distribution prior to a bioscaffold implantation into the stroke cavity. The ^{19}F signal, 24 hours post-PFC administration, is exclusively localized to blood

(*Caption continued*)

vessels. In contrast 1-day post-bioscaffold implantation, a major ^{19}F signal is localized within the bioscaffold-filled stroke cavity. The incision wound created by the stereotactic injection potential can be used as a positive control to verify that immune cells were labeled in case of implantation of a bioscaffold that does not invoke an immune response. (B) Higher resolution *ex vivo* ^{19}F MR images provide greater detail about the localization of the PFC-labeled cells within their anatomical context. In addition to the peri-infarct tissue, the stump of the middle cerebral artery (MCA), as well as a bioscaffold-based bridge between the cavity and the scalp, facilitated the invasion of immune cells. A denser rim of PFC-labeled immune cells at the edge of the bioscaffold was also evident, whereas a lower cell density was evident within the core of the bioscaffold. (C) To verify that the ^{19}F MR image corresponded to the histological distribution of PFC-labeled immune cells, ^{19}F MR and PFC-histology images were co-registered and overlaid. MRI hot spots corresponded to denser PFC-labeled cell clusters. Some area with sparse distribution of PFC-labeled cells did not reveal a strong ^{19}F MR signal. (D) Immunohistochemistry for Iba-1 positive monocytes revealed further that ^{19}F hot spots corresponded to areas with a high monocyte density in the peri-infarct area, as well as the collagen I positive bioscaffold.

Post-mortem histological analysis further documented the infiltration of PFC nanoemulsion labeled cells into brain tissues (Fig. 9.5A) through extravasation from blood vessel (Fig. 9.5B) [54]. The PFC tag is only found in monocytes infiltrating from the blood (*i.e.*, macrophages), but is not present with microglia (Fig. 9.5C). Although there is leakage of the blood-brain barrier during the acute phase after stroke, it is intact at 2 weeks post-stroke when ECM hydrogel is implanted into the lesion cavity. PFC nanoemulsions in brain tissue were hence sequestered in infiltrating immune cells and not available for uptake by the brain's resident microglia. In the peri-infarct area, approximately 1% of peripheral immune cells did not contain a PFC tag (false negative Type I error), but in the ECM a greater proportion (12%) were not detectable by ^{19}F MRI. Nevertheless, this finding is considerably below an acceptable 20% type II error rate [85]. ^{19}F MRI therefore truthfully represented the distribution of PFC-labeled immune cells 24 hours post-ECM implantation [54]. Still, apoptosis of labeled cells, as well as exocytosis of the nanoemulsions, could potentially lead to transfer of the agent to microglia. Conversely, PFC nanoemulsions have a tendency to coalesce with time in a process known as Ostwald ripening [69]. This nanoemulsion instability increases droplet size which reduces the potential to become incorporated in non-phagocytic cells. Overall, the truthfulness and reliability of long-term tracking of PFC-labeled cells needs further validation.

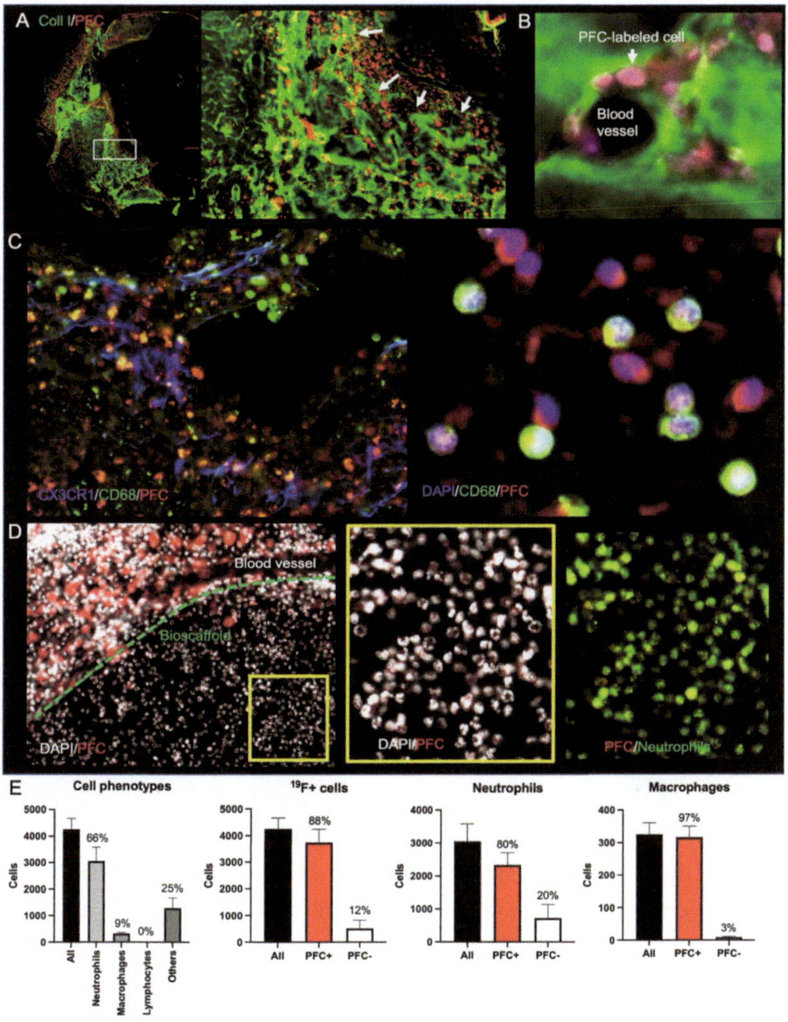

Figure 9.5 Immunohistochemistry of PFC-labeled immune cells. (A) PFC-labeled immune cells invade the collagen I bioscaffold from surrounding peri-infarct tissues. (B) PFC-labeled cells infiltrate tissue and vascularized bioscaffolds through the vascular wall. (C) PFC nanoemulsions are not present with microglia (CX3CR1) resident within brain tissue at 24 hours post-bioscaffold implantation. There is no leakage of the PFC into the brain as the blood-brain barrier damaged due to stroke has been re-established. Infiltration of peripheral immune cells, such as macrophages (CD68), contain PFC nanoemulsions, but red fluorescent PFC nanoemulsion are also present in other immune cells. (D) During the acute phase post-bioscaffold implantation, a major influx of neutrophils occurs. A

(Caption continued)

high proportion of these contain PFC. (E) Quantification of cell types within the bioscaffold indicated that 24 hours post-implantation, neutrophils are the dominant cell type with <10% of cells being macrophages. A high proportion (80%) of neutrophils and almost all (97%) of macrophages contained PFC nanoemulsions.

Although a major focus has been on visualizing macrophages using PFC nanoemulsions, other immune cells also incorporate this contrast material. In the blood, PFC nanoemulsions are present in 48% monocytes, which will differentiate into macrophages upon tissue infiltration, 4% neutrophils and 86% lymphocytes [54]. However, the amount (mass) of media taken up by lymphocytes is minimal compared to much larger, highly phagocytic monocytes/macrophages. The dynamics of immune cells and those labeled with PFC in the blood remain unknown. However, it is likely that different waves of neutrophils, macrophages and lymphocytes occur and that these are further reflected in the downstream infiltration into the brain. At 24 hours post-ECM implantation, most cells (66%) within the bioscaffold are actually neutrophils (Fig. 9.5D), which have been known to precede the infiltration of macrophages into ECM [86]. A high proportion (80%) of these neutrophils contained PFC tags (Fig. 9.5E). Only a small fraction of macrophages (9%) invaded the ECM at 24 hours and no lymphocytes were present within bioscaffold. Over 97% of infiltrating macrophages contained PFC nanoemlusions providing a very robust visualization of these cells. Considering the presence of PFC nanoemulsions in different immune cells and their trafficking in and out of the ECM hydrogel, it is essential to perform immunohistochemistry to determine the phenotypic identity of PFC-labeled cells in a target tissue. A greater specificity for individual immune cell types can potentially be achieved by further engineering PFC nanoemulsions or nanoparticles [87, 88]. Ideally, a combination of greater specificity with multi-spectral fluorine agents could result in multi-color ^{19}F imaging of the immune response [89]. The technological advancements would dramatically improve the *in vivo* characterization of the spatio-temporal dynamics of a host immune response to the implantation of ECM hydrogels.

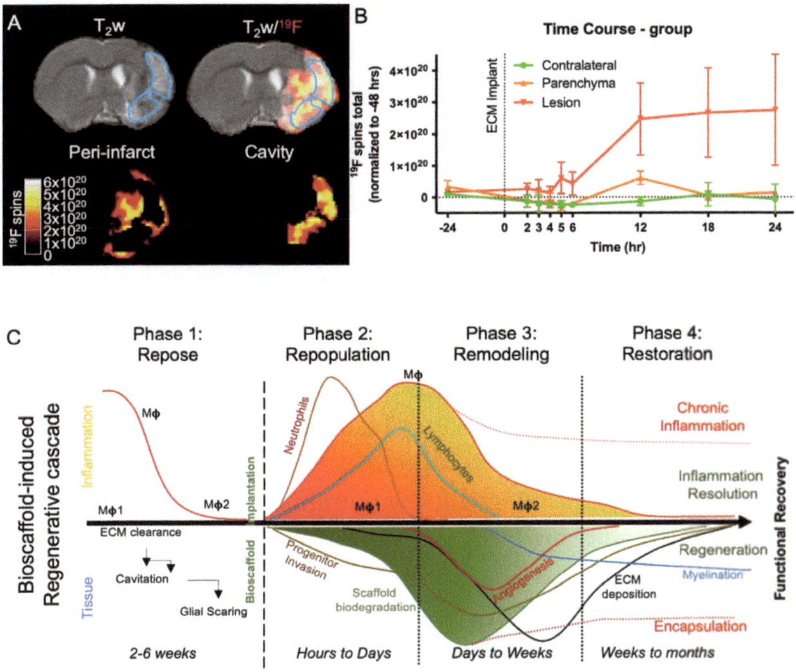

Figure 9.6 Longitudinal ^{19}F MR imaging of immune cells. (A) To quantitatively evaluate the spatio-temporal distribution of bioscaffold-infiltrating immune cells, ^{19}F signal intensity can be measured in the peri-infarct parenchyma and contrasted with the signal changes in the bioscaffold-filled cavity. (B) Invasion of immune cells into the bioscaffold occurs 5–6 hours post-implantation, with a major influx occurring between 6–12 hours and peaking at 18 hours post-implantation. However, the extent of invasion between 12–24 hours is variable and ^{19}F MRI can be used to further investigate how particular aspects of the bioscaffold implantation can affect immune cell infiltration. (C) Immune cell infiltration following a bioscaffold-implantation is key process to promote tissue regeneration in the stroke-damaged brain. Immune cells drive the biodegradation of the bioscaffold, with the pro-inflammatory (M1-like) phagocytic activity of macrophages being the main determinant. Neurotrophils and lymphocytes can further play crucial roles in modulating macrophage activity, including its polarization towards a pro-repair (M2-like) phenotype. The pro-repair phenotype in stroke has been associated with improved angiogenesis and remyelination, two key processes required in tissue regeneration. Failure of tissue regeneration due to bioscaffold implantation could lead to a chronic state of inflammation and bioscaffold encapsulation that can prevent functional recovery and potentially lead to a worse outcome.

Longitudinal ^{19}F MR imaging provides a unique view of the spatio-temporal dynamics of immune cells responding to ECM implantation that cannot easily be gained using other methods.

The ability to repeatedly visualize the immune cell distribution in the same individual allows us to monitor the gradual infiltration of immune cells into the brain and its accumulation in the ECM hydrogel. To analyze these "*in vivo* time-lapse" images, anatomical MRI scans are used to define regions-of-interest, such as the lesion cavity, peri-infarct tissue and contralateral controls tissues (Fig. 9.6A). Parcellation of the ^{19}F images into these ROIs further affords a measurement of ^{19}F spins at each time point. A time course of PFC-tagged cell invasion can hence be plotted. Immune cell infiltration into the peri-infarct area is only evident between 5–6 hours post-ECM implantation and peaks approximately 12 hours post-implantation (Fig. 9.6B). The response to the implantation injury in the skull was much faster than the response to the bioscaffold and occurred almost immediately (within 1 hour). The difference in infiltration timing between the skull wound and the ECM hydrogel further suggests that inductive properties of the bioscaffold invoke a time delay to recruit immune cells from the blood. It is however unclear if the initial immune response is generated due to the implantation injury or due to inductive signals released from the ECM hydrogel. The spatio-temporal characterization of the immune cell response to ECM implantation nevertheless clearly indicates a unique spatio-temporal response of the immune system to the bioscaffold in the stroke cavity as there are no ^{19}F hot spots in other brain regions. Longitudinal ^{19}F MR imaging hence provides a unique opportunity to characterize and investigate immune cell trafficking and its role in bioscaffold-induced tissue regeneration.

9.6 Conclusions

Currently, very little is known about the acute recruitment of immune cells into an ECM bioscaffold [60]. Macrophages have been determined to be key effectors of ECM biodegradation. A transition from M1-like pro-inflammatory towards a M2-like pro-repair phenotype is considered an essential step, determining if tissue regeneration occurs or if the biomaterial is encapsulated [90]. The involvement of other immune cells has been poorly documented, although there is some evidence of neutrophils also responding to ECM bioscaffolds. In the context of wound healing, neutrophils are known to infiltrate rapidly and to precede the infiltration of macrophages [91, 92]. A similar time course of infiltration is consistent with our observation of bioscaffold-induced brain tissue regeneration. Neutrophils represent the most common phenotype during the acute infiltration phase, whilst macrophages are a smaller and gradually increasing cell fraction that persist throughout the tissue regeneration process (Fig. 9.6C). The small number of lymphocytes present in the peri-infarct area at 24 hours post-implantation are hypothesized to constitute the next wave of immune cells invading the ECM bioscaffold and contributing to a shift in macrophage polarization [10]. PFC tagging of blood-borne immune cells and their visualization and quantification by ^{19}F MRI presents an excellent technical platform to investigate the spatio-temporal characteristic of immune cell trafficking. Longitudinal studies incorporating immune cell-selective contrast material with different ^{19}F chemical shifts (*i.e.*, multi-color imaging) offer new perspectives that would afford *in vivo* quantifications of mechanistic interventions aimed at controlling the immune response and the tissue regeneration process.

Disclosure

The authors declare no conflicts of interest.

Acknowledgments

The authors thank NINDS (R21NS118399; R01NS122768), the Pennsylvania Department of Health (4100061184) and the Department of Radiology, University of Pittsburgh for their generous support.

References

1. H. Ghuman, N. Perry, L. Grice, M. Gerwig, J. Moorhead, Jr., F. Nitzsche, A.J. Poplawsky, F. Ambrosio, M. Modo. Physical therapy exerts sub-additive and suppressive effects on intracerebral neural stem cell implantation in a rat model of stroke. *J Cereb Blood Flow Metab* **42**(5) (2022), 826–843, doi: 10.1177/0271678X211062955.

2. G.K. Steinberg, D. Kondziolka, L.R. Wechsler, L.D. Lunsford, A.S. Kim, J.N. Johnson, D. Bates, G. Poggio, C. Case, M. McGrogan, E.W. Yankee, N.E. Schwartz. Two-year safety and clinical outcomes in chronic ischemic stroke patients after implantation of modified bone marrow-derived mesenchymal stem cells (SB623): a phase 1/2a study. *J Neurosurg* (2018), 1–11.
3. K.W. Muir, D. Bulters, M. Willmot, N. Sprigg, A. Dixit, N. Ward, P. Tyrrell, A. Majid, L. Dunn, P. Bath, J. Howell, P. Stroemer, K. Pollock, J. Sinden. Intracerebral implantation of human neural stem cells and motor recovery after stroke: multicentre prospective single-arm study (PISCES-2). *J Neurol Neurosurg Psychiatry* **91**(4) (2020), 396–401.
4. F. Moreau, S. Patel, M.L. Lauzon, C.R. McCreary, M. Goyal, R. Frayne, A.M. Demchuk, S.B. Coutts, E.E. Smith. Cavitation after acute symptomatic lacunar stroke depends on time, location, and MRI sequence. *Stroke* **43**(7) (2012), 1837–1842.
5. E.J. Smith, R.P. Stroemer, N. Gorenkova, M. Nakajima, W.R. Crum, E. Tang, L. Stevanato, J.D. Sinden, M. Modo. Implantation site and lesion topology determine efficacy of a human neural stem cell line in a rat model of chronic stroke. *Stem Cells* **30**(4) (2012), 785–796.
6. M. Modo. Bioscaffold-induced brain tissue regeneration. *Front Neurosci* **13** (2019), 1156.
7. E. Bible, O. Qutachi, D.Y. Chau, M.R. Alexander, K.M. Shakesheff, M. Modo. Neo-vascularization of the stroke cavity by implantation of human neural stem cells on VEGF-releasing PLGA microparticles. *Biomaterials* **33**(30) (2012), 7435–7446.
8. E. Bible, D.Y. Chau, M.R. Alexander, J. Price, K.M. Shakesheff, M. Modo. The support of neural stem cells transplanted into stroke-induced brain cavities by PLGA particles. *Biomaterials* **30**(16) (2009), 2985–2994.
9. H. Ghuman, R. Matta, A. Tompkins, F. Nitzsche, S.F. Badylak, A.L. Gonzalez, M. Modo. ECM hydrogel improves the delivery of PEG microsphere-encapsulated neural stem cells and endothelial cells into tissue cavities caused by stroke. *Brain Res Bull* **168** (2021), 120–137.
10. S. Badylak, S. Arnoczky, P. Plouhar, R. Haut, V. Mendenhall, R. Clarke, C. Horvath. Naturally occurring extracellular matrix as a scaffold for musculoskeletal repair. *Clin Orthop Relat Res* (367 Suppl) (1999), S333–S343.
11. S.F. Badylak. Xenogeneic extracellular matrix as a scaffold for tissue reconstruction. *Transpl Immunol* **12**(3–4), (2004), 367–377.

12. L.T. Saldin, M.C. Cramer, S.S. Velankar, L.J. White, S.F. Badylak. Extracellular matrix hydrogels from decellularized tissues: structure and function. *Acta Biomater* **49** (2017), 1–15.
13. F. Meng, M. Modo, S.F. Badylak. Biologic scaffold for CNS repair. *Regen Med* **9**(3) (2014), 367–383.
14. M. Modo, S.F. Badylak. A roadmap for promoting endogenous in situ tissue restoration using inductive bioscaffolds after acute brain injury. *Brain Res Bull* **150** (2019), 136–149.
15. A.R. Massensini, H. Ghuman, L.T. Saldin, C.J. Medberry, T.J. Keane, F.J. Nicholls, S.S. Velankar, S.F. Badylak, M. Modo. Concentration-dependent rheological properties of ECM hydrogel for intracerebral delivery to a stroke cavity. *Acta Biomater* **27** (2015), 116–130.
16. H. Ghuman, A.R. Massensini, J. Donnelly, S.M. Kim, C.J. Medberry, S.F. Badylak, M. Modo. ECM hydrogel for the treatment of stroke: characterization of the host cell infiltrate. *Biomaterials* **91** (2016), 166–181.
17. H. Ghuman, C. Mauney, J. Donnelly, A.R. Massensini, S.F. Badylak, M. Modo. Biodegradation of ECM hydrogel promotes endogenous brain tissue restoration in a rat model of stroke. *Acta Biomater* **80** (2018), 66–84.
18. C. Damian, H. Ghuman, C. Mauney, R. Azar, J. Reinartz, S.F. Badylak, M. Modo. Post-stroke timing of ECM hydrogel implantation affects biodegradation and tissue restoration. *Int J Mol Sci* **22**(21) (2021).
19. R.L. Jayaraj, S. Azimullah, R. Beiram, F.Y. Jalal, G.A. Rosenberg. Neuroinflammation: friend and foe for ischemic stroke. *J Neuroinflammation* **16**(1) (2019), 142.
20. H. Ghuman, M. Gerwig, F.J. Nicholls, J.R. Liu, J. Donnelly, S.F. Badylak, M. Modo. Long-term retention of ECM hydrogel after implantation into a sub-acute stroke cavity reduces lesion volume. *Acta Biomater* **63** (2017), 50–63.
21. P. Cravedi, S. Farouk, A. Angeletti, L. Edgar, R. Tamburrini, J. Duisit, L. Perin, G. Orlando. Regenerative immunology: the immunological reaction to biomaterials. *Transpl Int* **30**(12) (2017), 1199–1208.
22. S.F. Badylak, T.W. Gilbert. Immune response to biologic scaffold materials. *Semin Immunol* **20**(2) (2008), 109–116.
23. M. Modo. Bioscaffold-induced brain tissue restoration. *Front. Neurosci* **13** (2019), 1156. PMID: 31787865.
24. J.E. Valentin, A.M. Stewart-Akers, T.W. Gilbert, S.F. Badylak. Macrophage participation in the degradation and remodeling of extracellular matrix scaffolds. *Tissue Eng Part A* **15**(7) (2009), 1687–1694.

25. S. Mukherjee, D. Sonanini, A. Maurer, H.E. Daldrup-Link. The yin and yang of imaging tumor associated macrophages with PET and MRI. *Theranostics* **9**(25) (2019), 7730–7748.

26. A.V. Naumova, M. Modo, A. Moore, C.E. Murry, J.A. Frank. Clinical imaging in regenerative medicine. *Nat Biotechnol* **32**(8) (2014), 804–818.

27. E.T. Ahrens, J.W. Bulte. Tracking immune cells in vivo using magnetic resonance imaging. *Nat Rev Immunol* **13**(10) (2013), 755–763.

28. M. Gauberti, A. Montagne, A. Quenault, D. Vivien. Molecular magnetic resonance imaging of brain-immune interactions. *Front Cell Neurosci* **8** (2014), 389.

29. E.T. Ahrens, J. Zhong. In vivo MRI cell tracking using perfluorocarbon probes and fluorine-19 detection. *NMR Biomed* **26**(7) (2013), 860–871.

30. J.M. Gaudet, A.M. Hamilton, Y. Chen, M.S. Fox, P.J. Foster. Application of dual (19) F and iron cellular MRI agents to track the infiltration of immune cells to the site of a rejected stem cell transplant. *Magn Reson Med* **78**(2) (2017), 713–720.

31. C. Jacoby, S. Temme, F. Mayenfels, N. Benoit, M.P. Krafft, R. Schubert, J. Schrader, U. Flogel. Probing different perfluorocarbons for in vivo inflammation imaging by 19F MRI: image reconstruction, biological half-lives and sensitivity. *NMR Biomed* **27**(3) (2014), 261–271.

32. Q. Ye, Y.L. Wu, L.M. Foley, T.K. Hitchens, D.F. Eytan, H. Shirwan, C. Ho. Longitudinal tracking of recipient macrophages in a rat chronic cardiac allograft rejection model with noninvasive magnetic resonance imaging using micrometer-sized paramagnetic iron oxide particles. *Circulation* **118**(2) (2008), 149–156.

33. Y.L. Wu, Q. Ye, D.F. Eytan, L. Liu, B.L. Rosario, T.K. Hitchens, F.C. Yeh, N. Rooijen van, C. Ho. Magnetic resonance imaging investigation of macrophages in acute cardiac allograft rejection after heart transplantation. *Circ Cardiovasc Imaging* **6**(6) (2013), 965–973.

34. T.K. Hitchens, Q. Ye, D.F. Eytan, J.M. Janjic, E.T. Ahrens, C. Ho. ^{19}F MRI detection of acute allograft rejection with in vivo perfluorocarbon labeling of immune cells. *Magn Reson Med* **65**(4) (2011), 1144–1153.

35. J.B. Williams, Q. Ye, T.K. Hitchens, C.L. Kaufman, C. Ho. MRI detection of macrophages labeled using micrometer-sized iron oxide particles. *J Magn Reson Imaging* **25**(6) (2007), 1210–1218.

36. S. Temme, F. Bonner, J. Schrader, U. Flogel. ^{19}F magnetic resonance imaging of endogenous macrophages in inflammation. *Wiley Interdiscip Rev Nanomed Nanobiotechnol* **4**(3) (2012), 329–343.

37. A. Khurana, F. Chapelin, H. Xu, J.R. Acevedo, A. Molinolo, Q. Nguyen, E.T. Ahrens. Visualization of macrophage recruitment in head and neck carcinoma model using fluorine-19 magnetic resonance imaging. *Magn Reson Med* **79** (2017), 1972–1980.

38. L.M. Foley, T.K. Hitchens, C. Ho, K.L. Janesko-Feldman, J.A. Melick, H. Bayir, P.M. Kochanek. Magnetic resonance imaging assessment of macrophage accumulation in mouse brain after experimental traumatic brain injury. *J Neurotrauma* **26**(9) (2009), 1509–1519.

39. N. Beckmann, C. Cannet, A.L. Babin, F.X. Ble, S. Zurbruegg, R. Kneuer, V. Dousset. In vivo visualization of macrophage infiltration and activity in inflammation using magnetic resonance imaging. *Wiley Interdiscip Rev Nanomed Nanobiotechnol* **1**(3) (2009), 272–298.

40. J.W. Bulte, J.A. Frank. Imaging macrophage activity in the brain by using ultrasmall particles of iron oxide. *AJNR Am J Neuroradiol* **21**(9) (2000), 1767–1768.

41. H.E. Daldrup-Link, D. Golovko, B. Ruffell, D.G. Denardo, R. Castaneda, C. Ansari, J. Rao, G.A. Tikhomirov, M.F. Wendland, C. Corot, L.M. Coussens. MRI of tumor-associated macrophages with clinically applicable iron oxide nanoparticles. *Clin Cancer Res* **17**(17) (2011), 5695–5704.

42. K.G. Petry, C. Boiziau, V. Dousset, B. Brochet. Magnetic resonance imaging of human brain macrophage infiltration. *Neurotherapeutics* **4**(3) (2007), 434–442.

43. A. Saleh, M. Schroeter, C. Jonkmanns, H.P. Hartung, U. Modder, S. Jander. In vivo MRI of brain inflammation in human ischaemic stroke. *Brain* **127**(Pt 7) (2004), 1670–1677.

44. J.W. Bulte, D.L. Kraitchman. Iron oxide MR contrast agents for molecular and cellular imaging *NMR Biomed* **17**(7) (2004), 484–499.

45. E.M. Shapiro, K. Sharer, S. Skrtic, A.P. Koretsky. In vivo detection of single cells by MRI. *Magn Reson Med* **55**(2) (2006), 242–249.

46. E.M. Shapiro, S. Skrtic, K. Sharer, J.M. Hill, C.E. Dunbar, A.P. Koretsky. MRI detection of single particles for cellular imaging. *Proc Natl Acad Sci U S A* **101**(30) (2004), 10901–10906.

47. E.T. Ahrens, R. Flores, H. Xu, P.A. Morel. In vivo imaging platform for tracking immunotherapeutic cells. *Nat Biotechnol* **23**(8) (2005), 983–987.

48. F. Bonner, M.W. Merx, K. Klingel, P. Begovatz, U. Flogel, M. Sager, S. Temme, C. Jacoby, M. Salehi Ravesh, C. Grapentin, R. Schubert, J. Bunke, M. Roden, M. Kelm, J. Schrader. Monocyte imaging after myocardial infarction with 19F MRI at 3 T: a pilot study in explanted porcine hearts. *Eur Heart J Cardiovasc Imaging* **16**(6) (2015), 612–620.

49. J.W. Bulte. Hot spot MRI emerges from the background. *Nat Biotechnol* **23**(8) (2005), 945–946.

50. A.V. Ratner, R. Hurd, H.H. Muller, B. Bradley-Simpson, W. Pitts, D. Shibata, C. Sotak, S.W. Young. 19F magnetic resonance imaging of the reticuloendothelial system. *Magn Reson Med* **5**(6) (1987), 548–554.

51. D.V. Hingorani, F. Chapelin, E. Stares, S.R. Adams, H. Okada, E.T. Ahrens. Cell penetrating peptide functionalized perfluorocarbon nanoemulsions for targeted cell labeling and enhanced fluorine-19 MRI detection. *Magn Reson Med* **83**(3) (2020), 974–987.

52. F. Chapelin, C.M. Capitini, E.T. Ahrens. Fluorine-19 MRI for detection and quantification of immune cell therapy for cancer. *J Immunother Cancer* **6**(1) (2018), 105.

53. M.N. Bouchlaka, K.D. Ludwig, J.W. Gordon, M.P. Kutz, B.P. Bednarz, S.B. Fain, C.M. Capitini. (19)F-MRI for monitoring human NK cells in vivo. *Oncoimmunology* **5**(5) (2016), e1143996.

54. M. Modo, H. Ghuman, R. Azar, R. Krafty, S.F. Badylak, T.K. Hitchens. Mapping the acute time course of immune cell infiltration into an ECM hydrogel in a rat model of stroke using (19)F MRI. *Biomaterials* **282** (2022), 121386.

55. M.V. Russo, D.B. McGavern. Immune surveillance of the CNS following infection and injury. *Trends Immunol* **36**(10) (2015), 637–650.

56. A.V. Makela, J.M. Gaudet, P.J. Foster. Quantifying tumor associated macrophages in breast cancer: a comparison of iron and fluorine-based MRI cell tracking. *Sci Rep* **7** (2017), 42109.

57. D.K. Kadayakkara, S. Ranganathan, W.B. Young, E.T. Ahrens. Assaying macrophage activity in a murine model of inflammatory bowel disease using fluorine-19 MRI. *Lab Invest* **92**(4) (2012), 636–645.

58. C. Fink, J.M. Gaudet, M.S. Fox, S. Bhatt, S. Viswanathan, M. Smith, J. Chin, P.J. Foster, G.A. Dekaban. (19)F-perfluorocarbon-labeled human peripheral blood mononuclear cells can be detected in vivo using clinical MRI parameters in a therapeutic cell setting. *Sci Rep* **8**(1) (2018), 590.

59. A.V. Makela, P.J. Foster. Imaging macrophage distribution and density in mammary tumors and lung metastases using fluorine-19 MRI cell tracking. *Magn Reson Med* **80**(3) (2018), 1138–1147.

60. K. Sadtler, M.T. Wolf, S. Ganguly, C.A. Moad, L. Chung, S. Majumdar, F. Housseau, D.M. Pardoll, J.H. Elisseeff. Divergent immune responses to synthetic and biological scaffolds. *Biomaterials* **192** (2019), 405–415.

61. G. Martino, S. Pluchino, L. Bonfanti, M. Schwartz. Brain regeneration in physiology and pathology: the immune signature driving therapeutic plasticity of neural stem cells. *Physiol Rev* **91**(4) (2011), 1281–1304.

62. U. Flogel, Z. Ding, H. Hardung, S. Jander, G. Reichmann, C. Jacoby, R. Schubert, J. Schrader. In vivo monitoring of inflammation after cardiac and cerebral ischemia by fluorine magnetic resonance imaging. *Circulation* **118**(2) (2008), 140–148.

63. J. Zhong, K. Narsinh, P.A. Morel, H. Xu, E.T. Ahrens. In vivo quantification of inflammation in experimental autoimmune encephalomyelitis rats using fluorine-19 magnetic resonance imaging reveals immune cell recruitment outside the nervous system. *PLoS One* **10**(10) (2015), e0140238.

64. J. Zhong, M. Sakaki, H. Okada, E.T. Ahrens. In vivo intracellular oxygen dynamics in murine brain glioma and immunotherapeutic response of cytotoxic T cells observed by fluorine-19 magnetic resonance imaging. *PLoS One* **8**(5) (2013), e59479.

65. A. Mastropietro, E. De Bernardi, G.L. Breschi, I. Zucca, M. Cametti, C.D. Soffientini, M. de Curtis, G. Terraneo, P. Metrangolo, R. Spreafico, G. Resnati, G. Baselli. Optimization of rapid acquisition with relaxation enhancement (RARE) pulse sequence parameters for (1)(9)F-MRI studies. *J Magn Reson Imaging* **40**(1) (2014), 162–170.

66. H. Waiczies, S. Lepore, S. Drechsler, F. Qadri, B. Purfurst, K. Sydow, M. Dathe, A. Kuhne, T. Lindel, W. Hoffmann, A. Pohlmann, T. Niendorf, S. Waiczies. Visualizing brain inflammation with a shingled-leg radio-frequency head probe for 19F/1H MRI. *Sci Rep* **3** (2013), 1280.

67. M.N. Hasan, L. Luo, D. Ding, S. Song, M.I.H. Bhuiyan, R. Liu, L.M. Foley, X. Guan, G. Kohanbash, T.K. Hitchens, M.G. Castro, Z. Zhang, D. Sun. Blocking NHE1 stimulates glioma tumor immunity by restoring OXPHOS function of myeloid cells. *Theranostics* **11**(3) (2021), 1295–1309.

68. A. Mali, E.L. Kaijzel, H.J. Lamb, L.J. Cruz. (19)F-nanoparticles: platform for in vivo delivery of fluorinated biomaterials for (19)F-MRI. *J Control Release* **338** (2021), 870–889.

69. J.M. Janjic, E.T. Ahrens. Fluorine-containing nanoemulsions for MRI cell tracking. *Wiley Interdiscip Rev Nanomed Nanobiotechnol* **1**(5) (2009), 492–501.

70. C. Grapentin, F. Mayenfels, S. Barnet, R. Suss, R. Schubert, S. Temme, C. Jacoby, J. Schrader, U. Flogel. Optimization of perfluorocarbon

nanoemlusions for molecular imaging by ^{19}F MRI, in *Nanomedicine* (A. Seifalian, A. de Mel, D. Kalaskar eds), One Central Press (2014), pp. 268–286.

71. M.J. Goette, G.M. Lanza, S.D. Caruthers, S.A. Wickline. Improved quantitative (19) F MR molecular imaging with flip angle calibration and B1-mapping compensation. *J Magn Reson Imaging* **42**(2) (2015), 488–494.

72. M.J. Goette, J. Keupp, J. Rahmer, G.M. Lanza, S.A. Wickline, S.D. Caruthers. Balanced UTE-SSFP for 19F MR imaging of complex spectra. *Magn Reson Med* **74**(2) (2015), 537–543.

73. A.J. Taylor, J. Granwehr, C. Lesbats, J.L. Krupa, J.S. Six, G.E. Pavlovskaya, N.R. Thomas, D.P. Auer, T. Meersmann, H.M. Faas. Probe-specific procedure to estimate sensitivity and detection limits for 19F magnetic resonance imaging. *PLoS One* **11**(10) (2016), e0163704.

74. N. Anisimov, M. Gulaev, O. Pavlova, D. Fomina, V. Glukhova, S. Batova, Y. Prirogov. Calculation of optimal parameters for ^{19}F MRI. *J Phys Conf Ser* **886** (2017), 012001.

75. R. Colotti, J.A.M. Bastiaansen, A. Wilson, U. Flogel, C. Gonzales, J. Schwitter, M. Stuber, R.B. van Heeswijk. Characterization of perfluorocarbon relaxation times and their influence on the optimization of fluorine-19 MRI at 3 tesla. *Magn Reson Med* **77**(6) (2017), 2263–2271.

76. B.Z. Gong, M. Gill, D.B. Washburn, W.C. Davenport, D. Adams, L. Kwock. Parameter optimization and calibration of 19F magnetic resonance imaging at 1.5 Tesla. *Magn Reson Imaging* **9**(1) (1991), 101–106.

77. H. Ghuman, T.K. Hitchens, M. Modo. A systematic optimization of (19) F MR image acquisition to detect macrophage invasion into an ECM hydrogel implanted in the stroke-damaged brain. *Neuroimage* **202** (2019), 116090.

78. W.R. Nitz. Fast and ultrafast non-echo-planar MR imaging techniques. *Eur Radiol* **12**(12) (2002), 2866–2882.

79. H. Zhu, P.B. Barker. MR spectroscopy and spectroscopic imaging of the brain. *Methods Mol Biol* **711** (2011), 203–226.

80. A. Alavi, T.J. Werner, P.F. Hoilund-Carlsen, H. Zaidi. Correction for partial volume effect is a must, not a luxury, to fully exploit the potential of quantitative PET imaging in clinical oncology. *Mol Imaging Biol* **20**(1) (2018), 1–3.

81. M. Ashioti, J.S. Beech, A.S. Lowe, M.B. Hesselink, M. Modo, S.C. Williams. Multi-modal characterisation of the neocortical clip model of focal

cerebral ischaemia by MRI, behaviour and immunohistochemistry. *Brain Res* **1145** (2007), 177–189.

82. M. Stille, E.J. Smith, W.R. Crum, M. Modo. 3D reconstruction of 2D fluorescence histology images and registration with in vivo MR images: application in a rodent stroke model. *J Neurosci Methods* **219**(1) (2013), 27–40.

83. S.W. Smith. *The Scientist and Engineer's Guide to Digital Signal Processing*. California Technical Publishing, San Diego, CA, 1997.

84. H. Gudbjartsson, S. Patz. The Rician distribution of noisy MRI data. *Magn Reson Med* **34**(6) (1995), 910–914.

85. M. Modo, J. Kolosnjaj-Tabi, F. Nicholls, W. Ling, C. Wilhelm, O. Debarge, F. Gazeau, O. Clement. Considerations for the clinical use of contrast agents for cellular MRI in regenerative medicine. *Contrast Media Mol Imaging* **8**(6) (2013), 439–455.

86. M.C. Cramer, S.F. Badylak. Extracellular matrix-based biomaterials and their influence upon cell behavior. *Ann Biomed Eng* **48**(7) (2020), 2132–2153.

87. E. Lambert, J.M. Janjic. Multiple linear regression applied to predicting droplet size of complex perfluorocarbon nanoemulsions for biomedical applications. *Pharm Dev Technol* **24**(6) (2019), 700–710.

88. S. Temme, P. Baran, P. Bouvain, C. Grapentin, W. Kramer, B. Knebel, H. Al-Hasani, J.M. Moll, D. Floss, J. Schrader, R. Schubert, U. Flogel, J. Scheller. Synthetic cargo internalization receptor system for nanoparticle tracking of individual cell populations by fluorine magnetic resonance imaging. *ACS Nano* **12**(11) (2018), 11178–11192.

89. U. Flogel, S. Temme, C. Jacoby, T. Oerther, P. Keul, V. Flocke, X. Wang, F. Bonner, F. Nienhaus, K. Peter, J. Schrader, M. Grandoch, M. Kelm, B. Levkau. Multi-targeted (1)H/(19)F MRI unmasks specific danger patterns for emerging cardiovascular disorders. *Nat Commun* **12**(1) (2021), 5847.

90. L. Huleihel, J.L. Dziki, J.G. Bartolacci, T. Rausch, M.E. Scarritt, M.C. Cramer, T. Vorobyov, S.T. LoPresti, I.T. Swineheart, L.J. White, B.N. Brown, S.F. Badylak. Macrophage phenotype in response to ECM bioscaffolds. *Semin Immunol* **29** (2017), 2–13.

91. M. Phillipson, P. Kubes. The healing power of neutrophils. *Trends Immunol* **40**(7) (2019), 635–647.

92. Z. Julier, A.J. Park, P.S. Briquez, M.M. Martino. Promoting tissue regeneration by modulating the immune system. *Acta Biomater* **53** (2017), 13–28.

Chapter 10

Sensitivity and ^{19}F MRI Cell Tracking

Eric T. Ahrens
Department of Radiology, University of California San Diego, La Jolla, California, USA
eta@ucsd.edu

10.1 Introduction

Although current generation materials and methods for ^{19}F MRI show promise for detecting therapeutic cell products and inflammation, next generation ^{19}F technologies will have substantially higher cell detection sensitivity and will unlock the technique's full potential. Cell tracking with ^{19}F tracers rely on *in situ* or *ex vivo* labeling of the cells of interest [1]. In *ex vivo* labeling, cell populations are co-cultured with an F-tracer to tag cells prior to patient infusion. This approach is of particular relevance to developers of live-cell immunotherapeutic products such as engineered T cells because clinically feasible imaging biomarkers for detecting the fate of transferred cells after they leave the bloodstream are of keen interest. In the case of *in situ* labeling, tracer agent is administered

Fluorine Magnetic Resonance Imaging: Methods and Applications in Biomedicine
Edited by Eric T. Ahrens and Ulrich Flögel
Copyright © 2025 Jenny Stanford Publishing Pte. Ltd.
ISBN 978-981-5129-37-3 (Hardcover), 978-1-003-53004-6 (eBook)
www.jennystanford.com

directly into circulation and endogenous cells take up the tagging agent. *In situ* labeling can, for example, provide diagnostic image readouts to elucidate inflammatory foci throughout the body.

An often-cited limitation of ^{19}F MRI cell tracers is the limit of detection (LOD) *in vivo*. In contrast to conventional ^{1}H MRI, where a high concentration water (>100 Molar ^{1}H) provides robust signal generation, ^{19}F MRI signal intensity is often restricted by the total amount and tissue distribution of the F-tracer after administration. In optimally designed experiments, the LOD often ranges from $\sim10^4$ to 10^5 cells per voxel [1, 2]. For a given experiment, achievable LOD is strongly dependent on specific details such as the nature of fluorous tracer molecule and its formulation, cell type of interest, cell activity, data acquisition methods, magnetic field strength, and MRI radio frequency (RF) detector design.

Over the past decade, advances in tracer design and image acquisition algorithms have collectively achieved dramatic leaps in ^{19}F MRI sensitivity, pushing the LOD boundaries by over an order of magnitude [3–5]. These achievements have enabled the detection of sparse cell populations administered for therapeutic purposes and early disease detection. In this chapter, we summarize the key factors that limit cell LOD and the chemical and algorithmic routes for achieving high cell sensitivity.

10.2 Molecular Substrates and Tracer Formulation

A critical factor in maximizing the sensitivity of ^{19}F-based cell tracking is the composition of the tracer materials. Fluorine-dense materials such as perfluorocarbon (PFC) polymers are most often employed, but other highly fluorinated molecules and nanoparticles show promise [6, 7]. The complexity of the fluoro-molecule's ^{19}F spectrum is also key to achieving sensitive detection. Ideally, the majority of fluorine sites should be magnetically equivalent and form a distinct resonance that is spectrally well separated from other molecular sites.

Because PFCs are insoluble in water, they are formulated as a colloidal suspension or emulsion for tracer use, often with ~100–200 nm diameter droplets (*i.e.*, 'nanoemulsion'). To improve the shelf life and impart desired biodistribution and cell uptake properties, emulsion droplets incorporate a stabilizing surfactant coating, commonly co-block polymers or phospholipids [8]. Optimal tracer compositions should maximize F-content and minimize the mass of MRI-invisible tracer excipients used in formulation, *i.e.*, the mass of emulsion surfactants used should be minimized.

To achieve adequate cell sensitivity, it is also essential that tracer agents are taken up intracellularly and aggregate inside the cell. Cell surface attachment of tracer without internalization (*e.g.*, *via* targeting ligand) often results in inadequate cell labeling due to the sparsity of attachment sites on cell surface. A key feature of effective PFC nanoemulsion (NE) tracers is their ability to become 'trapped' intracellularly for the lifetime of the cell. Non-targeted PFC droplets are endocytosed by cells and terminally trafficked to lysosomal vesicles, where droplets coalesce. Time-lapse microscopy of labeled cells indicated no evidence for tracer escape from lysosomal structures [9]. PFCs retain molecular integrity over time and remain impervious to agent degradation from low pH environments or enzymatic cleavage. Moreover, these molecules are strictly lipophilic and do not integrate into cellular components.

Published works use of a variety of emulsion droplet sizes for *ex vivo* cell labeling studies [10, 11]. Larger emulsion droplets (>200 nm) often yield higher intracellular cell labeling levels [11] and thus greater ^{19}F image sensitivity to sparse cells, but other factors are also important for practical implementation because emulsion droplet size can impact the processes for cell labeling *in vitro*. A product with larger emulsion droplets (>200 nm) is effective for flask-adherent cell types (*e.g.*, dendritic cells), where successful wash steps can be implemented to achieve high cell labeling levels [11]. For suspension-type cells such as lymphocytes, however, larger droplets spin-down with the cells under mild centrifugation, which stymies their removal. A smaller droplet size (<180 nm) allows excess agent not taken up by these cells to be removed with the supernatant during centrifugation and wash.

Small molecular weight fluorinated molecules are generally ineffective for *in vivo* cell tracking applications because the intracellular label does not tend to accumulate sufficiently within the cells, failing to achieve a concentration sufficient to enable *in vivo* detection. Small molecular weight molecules also have a tendency for cellular efflux because osmolality gradient forces drive the substance out of the cell, resulting in ^{19}F signal loss in regions of interest.

10.3 Cell Labeling

A crucial determinant of ^{19}F cell tracking sensitivity is the mass of fluorine taken up by cells populations of interest. Extensive *ex vivo* labeling studies using NEs [12] demonstrate that achievable labeling levels, often expressed as the mean ^{19}F/cell, measured by ^{19}F NMR spectroscopy of a cell pellet [13]. Labeling levels are predictable based on the intrinsic phagocytic tendencies of a cell type of interest in combination with a cell's cytoplasmic volume, its activation status, and culture viability during the labeling process. In macrophages, for example, one can readily achieve an uptake of order 10^{12} to 10^{13} mean ^{19}F/cell, while B or T lymphocytes take up ~100-fold lower mass by comparison [12] due to their intrinsically-low phagocytic activity and small cytoplasmic volume. Neutrophils, though highly phagocytic, are physically smaller than macrophages, with a ~4-fold lower volume, and thus have less intrinsic capacity to take up NE.

Due to their prominence in the field of cell therapy, fluorine-19 MRI cell tracking applications for T cells are of keen interest. Labeling T cells can be challenging, however, because low phagocytic activity and small cytoplasmic volume limits PFC NE uptake. Despite these challenges, several studies report labeling and imaging T cells [4, 14–19]. Importantly, optimal labeling efficiency is attained when cells are in a log phase of division. Because it is critical that the culture to be viable and actively undergoing mitosis, efficiency can be boosted by aggressive cytokine and co-stimulatory molecule treatments to culture media [17, 20].

To further enhance labeling levels in weakly phagocytic cell types, surfactants can be engineered to promote cell uptake [4, 10]. Early studies demonstrated that incorporation of cationic components into the surfactant is effective [10]. The addition of commercially available cationic transfection agents to the nanoemulsion prior to use also increases tracer uptake [14, 15]. Transfection agents may be cytotoxic, however, and can alter the phenotype and function of primary cells such as lymphocytes, thus they should generally be avoided for clinical use. As described in detail elsewhere, cell labeling must not alter viability, proliferation, phenotypic markers, or function [2, 21, 22].

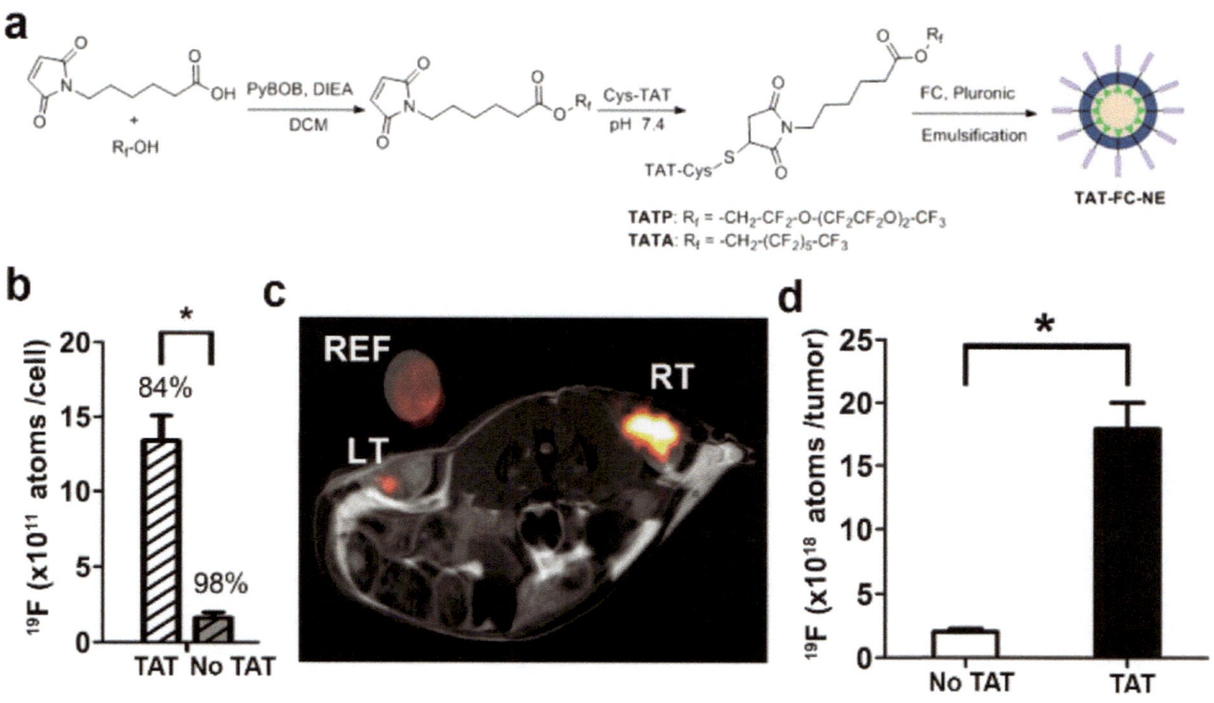

Figure 10.1 Fluorous TAT increases uptake of PFC NE in human T cells. Panel (a) displays synthesis and formulation of TAT-NE. The viability (%) and ^{19}F uptake (b) of NE with and without TAT is shown (*$p < 0.001$). Panel (c) shows composite $^{19}F/^{1}H$ MRI of mouse with bilateral gliomas in the flanks, where the left and right tumor (LT, RT) each received 1×10^7 T cells labeled with NE or TAT-NE, respectively. Panel (d) displays a comparison of apparent ^{19}F atoms measured in the LT and RT tumor measured *in vivo* ($N = 4$, *$p < 0.001$). Reproduced with permission from Ref. [4].

One highly effective approach to increase cellular uptake is described by Hingorani *et al.* [4]. A functionalized PFC NE was synthesized with cell penetrating peptides (Fig. 10.1a) displayed on the droplet surface [4]. Specifically, they used the positively charged transactivator of transcription (TAT) peptide, residues 49-58 (NH_2-CYGRKKRRQRRR-$CONH_2$), which carries a nuclear localization signal sequence facilitating endocytosis [23]. *Ex vivo* cell labeling with TAT-NE showed a labeling efficiency of $\sim 10^{12}$ ^{19}F atoms per cell, a >8-fold increase compared to NE without TAT, while retaining high cell viability (Fig. 10.1b). *In vitro* assays showed expression of T cell surface markers (CD3, CD4, CD8) were not altered by uptake of TAT-NE, and that anti-tumor T cell killing functions remained intact. *In vivo* ^{19}F MRI studies using TAT-NE-labeled T cells injected into flank tumors displayed an 8-fold higher signal compared to cells labeled using NE without TAT (Fig. 10.1c, d), consistent with *in vitro* results. This represents a major leap in sensitivity that would be difficult to achieve using prior materials and methods.

10.4 Paramagnetic Sensitivity Enhancement

Another avenue for boosting ^{19}F MRI sensitivity is by altering the ^{19}F longitudinal relaxation time (T_1) *via* the introduction of paramagnetic metal ions into the fluorous phase of the NE. Because ^{19}F images often exhibit a low signal-to-noise ratio (SNR < 10), signal averaging is frequently employed, and T_1 ultimately limits the number of averages attainable in a fixed scan time. Decreasing T_1 enables the collection of more signal averages, thereby increasing SNR per time and reducing the minimum number of detectable cells per voxel.

Although the same inertness of PFCs that leads to low toxicity also makes functionalization difficult. Our lab recently developed a family of novel fluorous metal chelates, displayed in Fig. 10.2, to functionalize PFC NEs [3, 24–26] with chelators that are soluble in PFCs and can bind a range of metal ions with high stability and biocompatibility. The chelate is first dissolved into PFC oil prior to NE formulation, after which NE can be rapidly metalated by the simple addition of metal salts to the aqueous phase at room temperature. The fluorous chelates in this family differ in metal binding affinity and solubilities in various PFCs, as described elsewhere [3, 24–26].

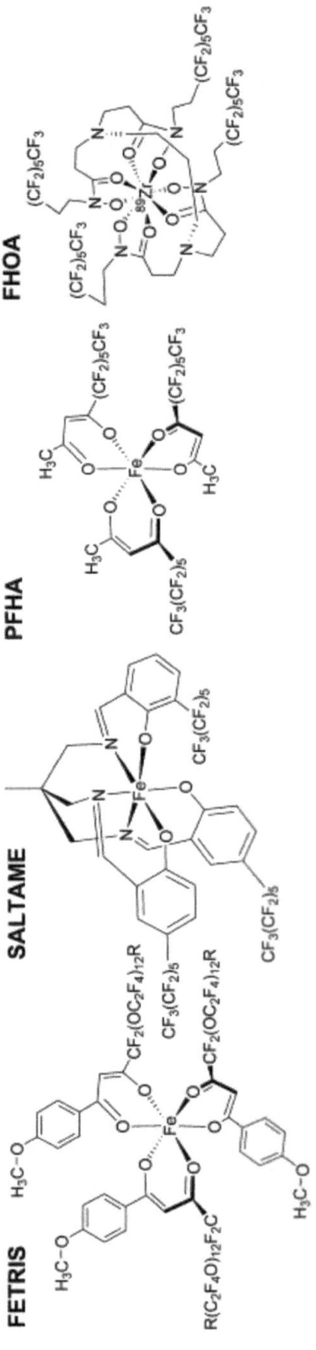

Figure 10.2 Survey of novel fluorous chelates used to functionalize PFC NEs. FETRIS [3], SALTAME [24], PFHA [25] and FHOA [26].

Initial work in this new area of chelate chemistry focused on doping PFC with paramagnetic Fe^{3+} to shorten ^{19}F T_1 *via* the paramagnetic relaxation enhancement (PRE) effect [3, 24, 25]. This strategy minimizes both the metal-fluorine distance, which is critical for PRE-based T_1 shortening, and the amount of metal additive needed. Moreover, metal encapsulation in the PFC phase imparts stability to the paramagnetic NE complex *in vivo*. The NE metalation and T_1 are stable in labeled cells, and T_1 is independent of concentration of NE in labeled cells.

Overall, the fluorous metal chelates (Fig. 10.2) are flexible with regard to coordinating different metals [3]. Relaxometry measurements with ^{19}F NMR (9.4 T) show that the formation of FETRIS (Fig. 10.2, 0.5 mM) dramatically increases the relaxation rate of ^{19}F from 2.37 s^{-1} to 30.6 s^{-1} with only moderate line-broadening (from 0.12 kHz to 0.25 kHz). In contrast, paramagnetic Gd^{3+} yields a relatively low T_1 relaxation rate and a very high spin-spin relaxation rate, resulting in severe line broadening (1.2 kHz for Gd^{3+} at 9.4 T) and signal loss undesirable for ^{19}F MRI.

FETRIS NE was formulated for intravenous injection and *in vivo* evaluation [27]. In Fig. 10.3a, we display $^{19}F/^{1}H$ MRI of intravenously delivered Fe^{3+} FETRIS NE in a granuloma inflammation model. The same study [27] also investigated the Fe^{3+} encapsulation stability in FETRIS NE *via in vivo* longitudinal ^{19}F T_1 of the liver signal. Intravenous PFC NE accumulate in cells of the reticuloendothelial system, including macrophages and Kupffer cells of the liver [28], making ^{19}F liver signal a reliable means to test the *in vivo* stability of the PRE effect in NE over time. Figure 10.3b shows the liver R_1 values (equivalent to $1/T_1$) of FETRIS NE, measured longitudinally in a mouse cohort using single-voxel ^{19}F spectroscopy. The values slowly decrease by ~30% over two weeks due to a slow loss of Fe^{3+} from the NE. This suggests adequate stability, considering the typical time interval between agent administration and imaging is ~24 h.

We have performed rudimentary SNR modeling for FETRIS and SALTAME (Fig. 10.3) using Fe^{3+} chelated NE relaxation rate data [3, 24]. Modeling results using the signal equations for a conventional gradient-echo (GRE) pulse sequence [29] show a multi-fold ^{19}F sensitivity enhancement compared to non-metalated PFC NE [3, 24, 27] at constant scan time. Modeling of SALTAME chelated Fe^{3+} in PFC (Fig. 10.4), for example, displays a sensitivity enhancement of nearly 4-fold using a conventional ^{19}F spin-density weighted GRE sequence at clinical magnetic field strengths [24], with the relative sensitivity enhancement of these materials greater at lower clinical magnetic field strengths.

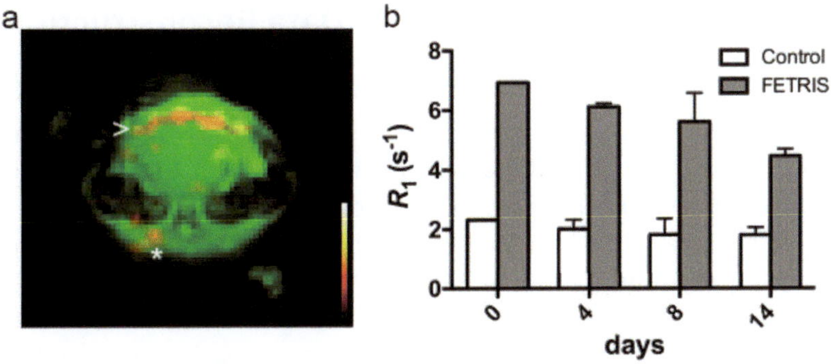

Figure 10.3 Study in mouse using FETRIS NE and ^{19}F MRI in granuloma inflammation model and tracer stability *in vivo* [27]. (a) Composite ^{19}F (hot-iron scale) and ^{1}H (green) MR image of *in situ* FETRIS NE labeled macrophage infiltration (arrow) in the scalp region induced by a subcutaneous lipopolysaccharide-Matrigel plug. Proximal lymph node involvement is also observed (asterisk). (b) FETRIS Fe^{3+} stability *in vivo* was measured in liver (Kupffer cells) using ^{19}F R_1 (=$1/T_1$) measurements over 14 days, where 'Control' is injected, chelate-free NE.

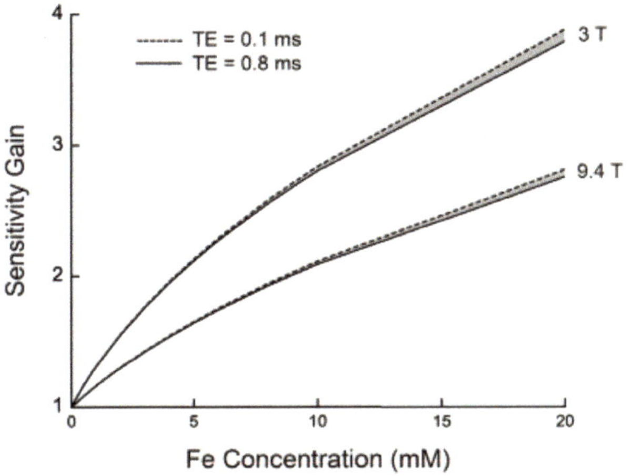

Figure 10.4 Fe-NE ^{19}F MRI signal intensity enhancement with gradient-echo sequence. Panel shows simulated sensitivity gain of Fe(+/–) NE at 3 T and 9.4 T as a function of NE-bound iron. Sensitivity gain is defined as SNR(Fe-NE)/SNR(NE) for constant scan time, acquired with optimum imaging parameters for each NE (*i.e.*, recycle time, tip-angle), and increased signal averages for Fe-NE due to short T_1. The model uses empirically measured relaxivities. The shaded regions represent a 0.1–0.8 ms echo-time (TE) parameter range. Figure adapted from Ref. [24] with permission.

10.5 Pulse Sequences and Data Reconstruction

The choice of ^{19}F MRI pulse sequence used is crucial for achieving satisfactory SNR and cell detection sensitivity. Ideally, ^{19}F acquisitions should provide three-dimensional (3D) data with isotropic voxels and whole-body images of tracer distribution, analogous to positron emission tomography methods. Conventional, isotropic, 3D spin-echo and gradient-echo methods have impractically long acquisition times, especially when used with signal averaging. More time-efficient methods such as echo-planar imaging can have severe artifacts for body imaging [30], particularly when working at high, preclinical magnetic field strengths. Consequently, image quantification becomes unreliable *in vivo*.

Novel 3D data acquisition and reconstruction methods have been developed to exploit the sparsity of tracer signals in ^{19}F MRI. Empirically, *in vivo* ^{19}F data often contains only ~0.5–5% voxels with significant signal, with a background comprised of pure noise. The nature of the ^{19}F data therefore lends itself to the application of compressed sensing (CS) theory [31, 32], which provides a framework for accurate recovery of sparse signals using undersampled k-space data. For ^{19}F Cartesian MRI, CS has been combined with increased signal averaging to improve image sensitivity for cell tracking [33] and chemical shift imaging [34], as well as to reduce motion artifacts [35]. In Zhong *et al.* [33], 3D ^{19}F CS-MRI was used to quantify *in vivo* macrophage burden in a wounding-inflammation rodent model following *in situ* labeling with PFC NE. The study demonstrated 8-fold acceleration in image acquisition time using undersampled Cartesian k-space lines and CS reconstruction, all while maintaining accurate ^{19}F signal distribution with no loss in image SNR.

Pulse sequence data reconstruction algorithms have also been developed to optimally image the PRE ^{19}F tracers that incorporate paramagnetic metal ions as described in Section 10.4. A limiting factor in sensitivity enhancement is T_2 line broadening, which constrains the maximum useful concentration of Fe^{3+} in the NE. Because this sensitivity barrier falls as TE → 0, we developed the 'zero' echo time (CS-ZTE) pulse sequence which relies on radial k-space sampling trajectories and undersampling of radial spokes, followed by CS reconstruction [5]. CS-ZTE can acquire 3D images in

~25 s with 64^3 image points and an 8-fold k-space undersampling factor [33]. The short acquisition time enables a commensurate increase in signal averages for a scan to boost sensitivity. The F-limits of detection (LOD) were evaluated using an approach that combined CS-ZTE with a paramagnetic (Fe^{3+}) NE tracer, 8-fold undersampling, and optimized CS reconstruction. Figure 10.5 demonstrates a ~30-

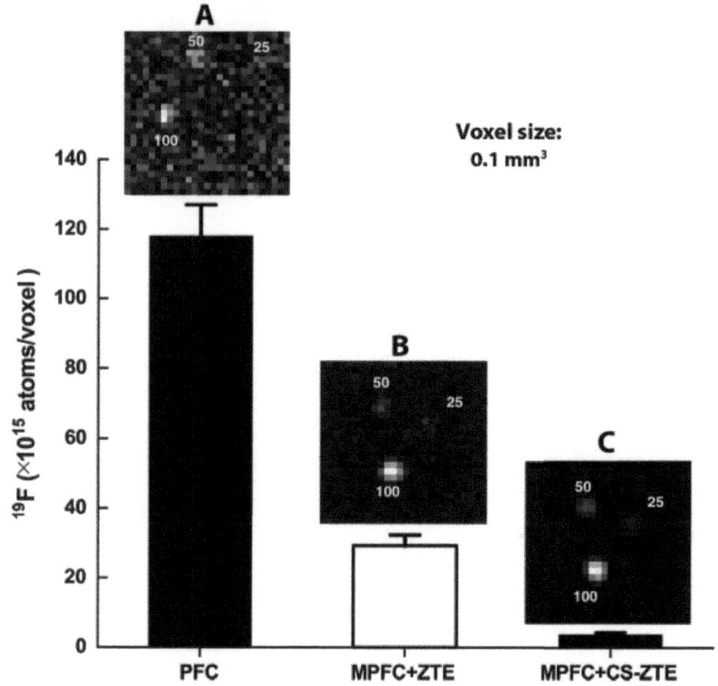

Figure 10.5 Limits of detection (LOD) of ^{19}F using the CS-ZTE method with fixed scan time of 3.3 min (11.7 T). LOD (y-axis) is ^{19}F concentration extrapolated to SNR = 2.5. Column (A) is PFC control (without Fe^{3+}) acquired with 3D GRE sequence using optimal parameters. Columns (B) and (C) are Fe-NE acquired with optimized and fully sampled ZTE and CS-ZTE (8-fold undersampling), respectively. Differing sample dilutions (100, 50 and 25, expressed as percentages) are shown. PFC control has an LOD $\sim 1.1 \times 10^{17}$ ^{19}F atoms/voxel (A). Overall, ^{19}F MRI detection sensitivity improves ~30-fold in Fe-NE compared to control. Importantly, ^{19}F image intensity analyses yield accurate estimates of absolute quantification of ^{19}F spins. Figure adapted from Ref. [5] with permission.

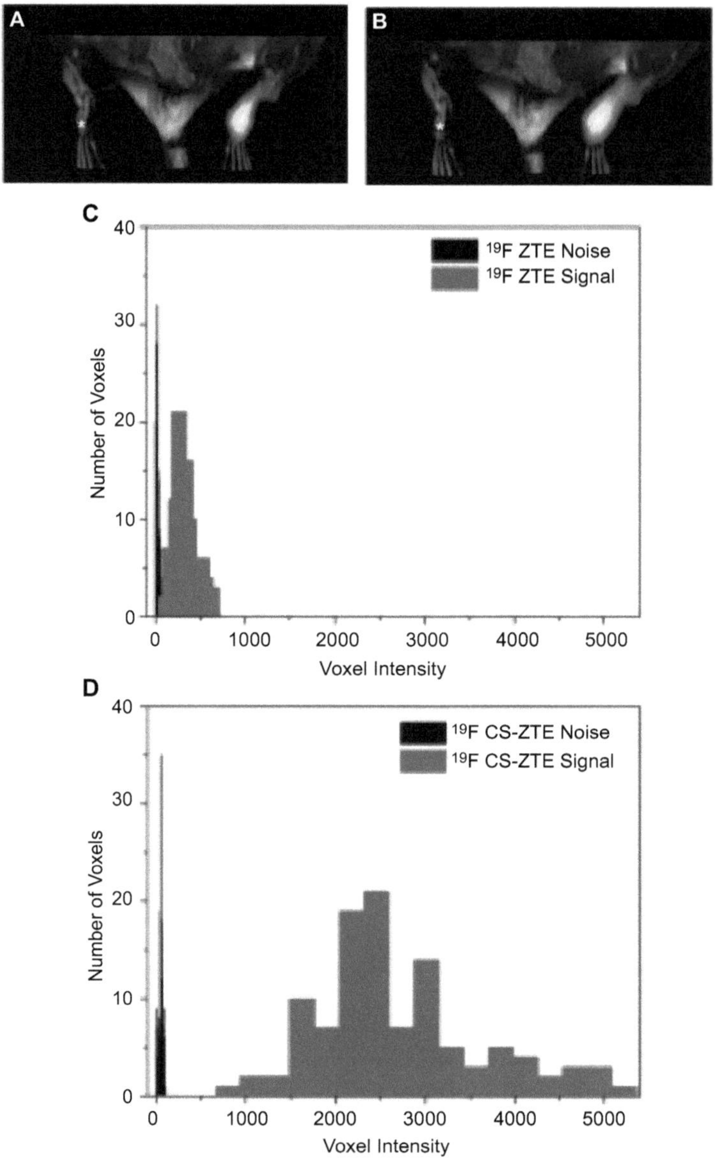

Figure 10.6 *In vivo* 3D ^{19}F MRI mouse data showing ^{19}F labeled DCs injected into the hind paw. DCs, labeled with either PFC (no Fe^{3+}) or Fe-NE, were injected into the left and right paw, respectively (5×10^6 cells each). The conventional ZTE pulse sequence (no undersampling) and CS-ZTE (8-fold undersampling) were used, with 30 and 240 averages, respectively, with total scan time of 10

min for each. Data were acquired using an isotropic 64^3 matrix. The ^{19}F images are displayed in pseudo-color and co-registered with grayscale ^1H anatomical images. In (A) and (B), DCs display hot-spots in the paw receiving Fe-NE-labeled DCs acquired by ZTE and CS-ZTE, respectively. Cells labeled with control PFC (left paw, asterisk) display no signal and falls below LOD. Graphs (C) and (D) show the raw histograms (no thresholding applied) of ^{19}F voxel intensities using the conventional ZTE and CS-ZTE methods, respectively. The histogram for CS-ZTE displays significantly higher signals compared to control ZTE and has a broader distribution with larger numbers of signal-positive voxels, consistent with enhanced sensitivity to sparse cells. Figure adapted from Ref. [5] with permissions.

fold relative decrease in the LOD for the same scan time when using a combined Fe-NE tracer and CS-ZTE as compared to NE (no metal) and standard GRE sequences. This suggests a dramatic leap in sensitivity may be feasible by pairing PRE NEs with optimal image acquisition and reconstruction methods. For example, given a myeloid cell population labeled with Fe-NE at a mean $\sim 10^{12}$ ^{19}F/cell, Fig. 10.5 suggests that $\sim 10^3$ cells per voxel could be detectable. Figure 10.6 shows preliminary *in vivo* ^{19}F data with Fe-NE and CS-ZTE using *ex vivo* labeled DCs injected into a rodent footpad.

We note that in Fig. 10.5, the LOD is calculated using a detection threshold of SNR = 2.5. This cutoff value was previously used in clinical ^{19}F cell studies, providing >99% confidence against false positive signals [2]. The Rose Model [36–38], with SNR = 4, could be employed as an alternative LOD criteria, although this model considers both statistical considerations and human perception. Because analysis of ^{19}F MRI data is most often performed computationally on a per-voxel basis, with no human interpretation as input except for the identification of ^1H-based anatomical context of ^{19}F signals, the Rose Model criteria is considered too conservative for ^{19}F MRI LOD analysis.

Overall, coupling CS-ZTE with PRE NE leads to sensitivity enhancement by three distinct mechanisms, namely (i) undersampling, and the commensurate increase in signal averaging for a fixed scan time, (ii) paramagnetic reduction of T_1, which accelerates data acquisition and enables further signal averaging in scan period, and (iii) the intrinsic 'denoising' properties of the CS algorithm [5].

10.6 Clinical Perspectives and Conclusions

Methods for implementation of ^{19}F MRI on a clinical scanner are established and involve installing appropriate pulse sequences and certain imaging hardware accessories, including an ^{19}F-capable coil. Numerous medical research institutions have implemented ^{19}F MRI on MRI platforms from major vendors like General Electric, Siemens, and Philips, indicating that any technical barriers are surmountable for clinical trial use. With increasing support from MRI vendors, as a function of market demand, clinical adoption of ^{19}F MRI will become more widespread.

Clinical systems have intrinsically lower sensitivity compared to high-field preclinical scanners due to decreased ^{19}F spin polarization at lower magnetic field strength and hardware-dependent differences in RF signal detection sensitivity. For clinical use, localized surface- or saddle-type coils, configurations commonly used in clinical scans, are preferred for higher ^{19}F MRI sensitivity compared to larger body volume coils.

Detection of ^{19}F in current generation PFC NEs using 1.5 and 3 Tesla clinical scanners has been demonstrated in numerous animal studies [2, 39–45]. Clinical pilot studies to detect immunotherapeutic DC [2] and T cell therapy [22] products in cancer patients have also been reported, where cells were labeled with PFC nanoemulsion *ex vivo* prior to delivery to patient. In the case of the DC trial [2], the LOD was benchmarked at $\sim 10^5$ DCs/voxel using a first-generation PFC nanoemulsion, 3 T MRI, localized body ^{19}F/^{1}H RF coil, and a conventional GRE sequence with scan time <10 min [2].

Importantly, the avenues for sensitivity enhancement described herein, including cell targeting, paramagnetic relaxation enhancement, and the use of advanced pulse sequences and reconstruction methods, are all clinically feasible. Moreover, simultaneous use of these approaches enables multi-pronged enhancement of cell detection sensitivity. Our view is that $>10\times$ improvement in the LOD is readily achievable and will enable unprecedented precision in cell tracking. These sensitivity gains are conserved when going from high-field animal scanners to clinical magnetic field strengths, suggesting that a combination of advanced tracer materials and image generation algorithms offers a superior, low-cost solution to LOD enhancement compared to alternatives like ultra-high magnetic field MRI [46] or cryogenic RF detection hardware [47], which are challenging to implement clinically. However, the complexity of manufacturing and characterizing next-generation ^{19}F tracers will increase and may require the development of innovative pharmaceutical approaches.

References

1. Ahrens ET, Bulte JWM. Tracking immune cells in vivo using magnetic resonance imaging. *Nat Rev Immunol.* 2013; **13**(10): 755–763. doi: 10.1038/nri3531.
2. Ahrens ET, Helfer BM, O'Hanlon CF, Schirda C. Clinical cell therapy imaging using a perfluorocarbon tracer and fluorine-19 MRI. *Magn Reson Med.* 2014; **72**(6): 1696–1701.
3. Kislukhin AA, Xu H, Adams SR, Narsinh KH, Tsien RY, Ahrens ET. Paramagnetic fluorinated nanoemulsions for sensitive cellular fluorine-19 magnetic resonance imaging. *Nat Mater.* 2016; **15**(6): 662.
4. Hingorani DV, Chapelin F, Stares E, Adams SR, Okada H, Ahrens ET. Cell penetrating peptide functionalized perfluorocarbon nanoemulsions for targeted cell labeling and enhanced fluorine-19 MRI detection. *Magn Reson Med.* 2020; **83**(3): 974–987. Epub 2019/10/22. doi: 10.1002/mrm.27988.
5. Chen JW, Pal P, Ahrens ET. Enhanced detection of paramagnetic fluorine-19 magnetic resonance imaging agents using zero echo time sequence and compressed sensing. *NMR Biomed.* 2022; **35**(8). doi: 10.1002/nbm.4725.
6. Ashur I, Allouche-Arnon H, Bar-Shir A. Calcium fluoride nanocrystals: tracers for in vivo (19) F magnetic resonance imaging. *Angewandte Chemie (International ed in English).* 2018; **57**(25): 7478–7482. Epub 2018/03/09. doi: 10.1002/anie.201800838.
7. Tirotta I, Mastropietro A, Cordiglieri C, Gazzera L, Baggi F, Baselli G, Bruzzone MG, Zucca I, Cavallo G, Terraneo G, Baldelli Bombelli F, Metrangolo P, Resnati G. A superfluorinated molecular probe for highly sensitive in vivo (19)F-MRI. *J Am Chem Soc.* 2014; **136**(24): 8524–8527. Epub 2014/06/03. doi: 10.1021/ja503270n.
8. Janjic JM, Patel SK. Chapter 14 Perfluorocarbon theranostic nanomedicines: pharmaceutical scientist's perspective, in *Fluorine Magnetic Resonance Imaging*, Jenny Stanford Publishing, 2016, pp. 407–432.

9. Patrick MJ, Janjic JM, Teng H, O'Hear MR, Brown CW, Stokum JA, Schmidt BF, Ahrens ET, Waggoner AS. Intracellular pH measurements using perfluorocarbon nanoemulsions. *J Am Chem Soc.* 2013; **135**(49): 18445–184457.

10. Janjic JM, Srinivas M, Kadayakkara DK, Ahrens ET. Self-delivering nanoemulsions for dual fluorine-19 MRI and fluorescence detection. *J Am Chem Soc.* 2008; **130**(9): 2832–2841.

11. Waiczies H, Lepore S, Janitzek N, Hagen U, Seifert F, Ittermann B, Purfurst B, Pezzutto A, Paul F, Niendorf T, Waiczies S. Perfluorocarbon particle size influences magnetic resonance signal and immunological properties of dendritic cells. *PLoS ONE.* 2011; **6**(7): e21981. doi: 10.1371/journal.pone.0021981.

12. Chapelin F, Capitini CM, Ahrens ET. Fluorine-19 MRI for detection and quantification of immune cell therapy for cancer. *J Immunother Cancer.* 2018; **6**(1): 105. doi: 10.1186/s40425-018-0416-9.

13. Ahrens ET, Flores R, Xu H, Morel PA. In vivo imaging platform for tracking immunotherapeutic cells. *Nat Biotechnol.* 2005; **23**(8): 983–987. doi: 10.1038/nbt1121.

14. Srinivas M, Morel PA, Ernst LA, Laidlaw DH, Ahrens ET. Fluorine-19 MRI for visualization and quantification of cell migration in a diabetes model. *Magn Reson Med.* 2007; **58**(4): 725–734. doi: 10.1002/mrm.21352.

15. Srinivas M, Turner MS, Janjic JM, Morel PA, Laidlaw DH, Ahrens ET. In vivo cytometry of antigen-specific T cells using 19F MRI. *Magn Reson Med.* 2009; **62**(3): 747–753.

16. Kadayakkara DK, Beatty PL, Turner MS, Janjic JM, Ahrens ET, Finn OJ. Inflammation driven by overexpression of the hypoglycosylated abnormal mucin 1 (MUC1) links inflammatory bowel disease and pancreatitis. *Pancreas.* 2010; **39**(4): 510–515. Epub 2010/01/20. doi: 10.1097/MPA.0b013e3181bd6501.

17. Chapelin F, Gao S, Okada H, Weber TG, Messer K, Ahrens ET. Fluorine-19 nuclear magnetic resonance of chimeric antigen receptor T cell biodistribution in murine cancer model. *Sci Rep-Uk.* 2017; **7**(1): 17748. doi: 10.1038/s41598-017-17669-4.

18. Chapelin F, Leach BI, Chen R, Lister D, Messer K, Okada H, Ahrens ET. Assessing oximetry response to chimeric antigen receptor t-cell therapy against glioma with 19F MRI in a murine model. *Radiol Imaging Cancer.* 2021; **3**(1): e200062. doi: 10.1148/rycan.2021200062.

19. Dubois VP, Sehl OC, Foster PJ, Ronald JA. Visualizing CAR-T cell immunotherapy using 3 tesla fluorine-19 MRI. *Mol Imaging Biol.* 2022; **24**(2): 298–308. doi: 10.1007/s11307-021-01672-3.

20. Bouchlaka MN, Ludwig KD, Gordon JW, Kutz MP, Bednarz BP, Fain SB, Capitini CM. (19)F-MRI for monitoring human NK cells in vivo. *Oncoimmunology.* 2016; **5**(5): e1143996. Epub 2016/07/29. doi: 10.1080/2162402x.2016.1143996.

21. Helfer BM, Balducci A, Nelson AD, Janjic JM, Gil RR, Kalinski P, de Vries IJ, Ahrens ET, Mailliard RB. Functional assessment of human dendritic cells labeled for in vivo (19)F magnetic resonance imaging cell tracking. *Cytotherapy.* 2010; **12**(2): 238–250. doi: 10.3109/14653240903446902.

22. Ahrens ET, Helfer BM, O'Hanlon CF, Lister DR, Bykowski J, Messer M, Leach BI, Chen J, Xu H, Daniels GA, Cohen EEW. Method for estimation of apoptotic cell fraction of cytotherapy using in vivo fluorine-19 magnetic resonance: pilot study in a patient with head and neck carcinoma receiving tumor-infiltrating lymphocytes labeled with perfluorocarbon nanoemulsion. *J Immunother Cancer.* 2023; **11**(6): e007015. doi: 10.1136/jitc-2023-007015.

23. Patel LN, Zaro JL, Shen WC. Cell penetrating peptides: Intracellular pathways and pharmaceutical perspectives. *Pharm Res.* 2007; **24**(11): 1977–1992.

24. Jahromi AH, Wang C, Adams SR, Zhu W, Narsinh K, Xu H, Gray DL, Tsien RY, Ahrens ET. Fluorous-soluble metal chelate for sensitive fluorine-19 magnetic resonance imaging nanoemulsion probes. *ACS Nano.* 2019; **13**(1): 143–151. doi: 10.1021/acsnano.8b04881.

25. Wang C, Adams SR, Xu H, Zhu W, Ahrens ET. β-diketonate-iron(III) complex: a versatile fluorine-19 MRI signal enhancement agent. *ACS Appl Bio Mater.* 2019; **2**(9): 3836–3842. doi: 10.1021/acsabm.9b00455.

26. Wang C, Leach BI, Lister D, Adams SR, Xu H, Hoh C, McConville P, Zhang J, Messer K, Ahrens ET. Metallofluorocarbon nanoemulsion for inflammatory macrophage detection via PET and MRI. *J Nucl Med.* 2021; **62**(8): 1146–1153. Epub 2020/12/06. doi: 10.2967/jnumed.120.255273.

27. Rho J, Stares E, Adams SR, Lister D, Leach B, Ahrens ET. Paramagnetic fluorinated nanoemulsions for in vivo F-19 MRI. *Mol Imaging Biol.* 2020; **22**(3): 665–674. Epub 2019/09/05. doi: 10.1007/s11307-019-01415-5.

28. Ahrens ET, Young W-B, Xu H, Pusateri LK. Rapid quantification of inflammation in tissue samples using perfluorocarbon emulsion and

fluorine-19 nuclear magnetic resonance. *Biotechniques*. 2011; **50**(4): 229–234. doi: 10.2144/000113652.

29. Haase A, Frahm J, Matthaei D. FLASH imaging: rapid NMR imaging using low flip-angle pulses. *J Magn Reson*. 1986; **67**: 258.
30. Stehling MK, Turner R, Mansfield P. Echo-planar imaging magnetic resonance imaging in a fraction of a second. *Science*. 1991; **254**(5028): 43–50.
31. Candes EJ, Romberg J, Tao T. Robust uncertainty principles: exact signal reconstruction from highly incomplete frequency information. *IEEE Trans Inf Theory*. 2006; **52**(2): 489–509. doi: 10.1109/TIT.2005.862083.
32. Donoho DL. Compressed sensing. *IEEE Trans Inf Theory*. 2006; **52**(4): 1289–1306. doi: 10.1109/TIT.2006.871582.
33. Zhong J, Mills PH, Hitchens TK, Ahrens ET. Accelerated fluorine-19 MRI cell tracking using compressed sensing. *Magn Reson Med*. 2013; **69**(6): 1683–1690. doi: 10.1002/mrm.24414.
34. Kampf T, Fischer A, Basse-Lusebrink TC, Ladewig G, Breuer F, Stoll G, Jakob PM, Bauer WR. Application of compressed sensing to in vivo 3D F-19 CSI. *J Magn Reson*. 2010; **207**(2): 262–273. doi: 10.1016/j.jmr.2010.09.006.
35. Darçot E, Yerly J, Hilbert T, Colotti R, Najdenovska E, Kober T, Stuber M, van Heeswijk RB. Compressed sensing with signal averaging for improved sensitivity and motion artifact reduction in fluorine-19 MRI *NMR Biomed*. 2021; **34**(1): e4418. doi: https://doi.org/10.1002/nbm.4418.
36. Shosa D, Kaufman L. Methods for evaluation of diagnostic imaging instrumentation. *Phys Med Biol*. 1981; **26**(1): 101–112. Epub 1981/01/01. doi: 10.1088/0031-9155/26/1/010.
37. Tanifum EA, Patel C, Liaw ME, Pautler RG, Annapragada AV. Hydrophilic fluorinated molecules for spectral (19)F MRI. *Sci Rep*. 2018; **8**(1): 2889. Epub 2018/02/15. doi: 10.1038/s41598-018-21178-3.
38. Bouvain P, Temme S, Flögel U. Hot spot 19F magnetic resonance imaging of inflammation. *Wiley Interdiscip Rev Nanomed Nanobiotechnol*. 2020; **12**(6): e1639. doi: https://doi.org/10.1002/wnan.1639.
39. Fink C, Smith M, Sehl OC, Gaudet JM, Meagher TC, Sheikh NA, Dikeakos JD, Rieder MJ, Foster PJ, Dekaban GA. Quantification and characterization of granulocyte macrophage colony-stimulating factor activated human peripheral blood mononuclear cells by fluorine-19 cellular MRI in an immunocompromised mouse model. *Diag Interv Imaging*. 2020; **101**(9): 577–588. doi: 10.1016/j.diii.2020.02.004.

40. Makela AV, Foster PJ. Imaging macrophage distribution and density in mammary tumors and lung metastases using fluorine-19 MRI cell tracking. *Magn Reson Med.* 2018; **80**(3): 1138–1147. doi: 10.1002/mrm.27081.

41. Fink C, Gaudet JM, Fox MS, Bhatt S, Viswanathan S, Smith M, Chin J, Foster PJ, Dekaban GA. (19)F-perfluorocarbon-labeled human peripheral blood mononuclear cells can be detected in vivo using clinical MRI parameters in a therapeutic cell setting. *Sci Rep.* 2018; **8**(1): 590. doi: 10.1038/s41598-017-19031-0.

42. Makela AV, Gaudet JM, Foster PJ. Quantifying tumor associated macrophages in breast cancer: a comparison of iron and fluorine-based MRI cell tracking. *Sci Rep.* 2017; **7**: 42109. doi: 10.1038/srep42109.

43. Gaudet JM, Ribot EJ, Chen Y, Gilbert KM, Foster PJ. Tracking the fate of stem cell implants with fluorine-19 MRI. *PLoS One.* 2015; **10**(3): e0118544. doi: 10.1371/journal.pone.0118544.

44. Ribot EJ, Gaudet JM, Chen Y, Gilbert KM, Foster PJ. In vivo MR detection of fluorine-labeled human MSC using the bSSFP sequence. *Int J Nanomedicine.* 2014; **9**: 1731–1739. Epub 2014/04/22. doi: 10.2147/ijn.s59127.

45. Partlow KC, Chen J, Brant JA, Neubauer AM, Meyerrose TE, Creer MH, Nolta JA, Caruthers SD, Lanza GM, Wickline SA. 19F magnetic resonance imaging for stem/progenitor cell tracking with multiple unique perfluorocarbon nanobeacons. *FASEB J.* 2007; **21**(8): 1647–1654.

46. Waiczies S, Rosenberg JT, Kuehne A, Starke L, Delgado PR, Millward JM, Prinz C, Dos Santos Periquito J, Pohlmann A, Waiczies H, Niendorf T. Fluorine-19 MRI at 21.1 T: enhanced spin-lattice relaxation of perfluoro-15-crown-5-ether and sensitivity as demonstrated in ex vivo murine neuroinflammation. *Magma.* 2019; **32**(1): 37–49. Epub 2018/11/14. doi: 10.1007/s10334-018-0710-z.

47. Waiczies S, Millward JM, Starke L, Delgado PR, Huelnhagen T, Prinz C, Marek D, Wecker D, Wissmann R, Koch SP, Boehm-Sturm P, Waiczies H, Niendorf T, Pohlmann A. Enhanced fluorine-19 MRI sensitivity using a cryogenic radiofrequency probe: technical developments and ex vivo demonstration in a mouse model of neuroinflammation. *Sci Rep.* 2017; **7**(1): 9808. Epub 2017/08/31. doi: 10.1038/s41598-017-09622-2.

Index